Na-ion Batteries

SCIENCES

Energy, Field Director – Alain Dollet

Electrical Energy Storage, Subject Heads – Laure Monconduit and Laurence Croguennec

Na-ion Batteries

Coordinated by

Laure Monconduit

Laurence Croguennec

iSTE

WILEY

First published 2020 in Great Britain and the United States by ISTE Ltd and John Wiley & Sons, Inc.

ISTE Ltd
27-37 St George's Road
London SW19 4EU
UK

www.iste.co.uk

John Wiley & Sons, Inc.
111 River Street
Hoboken, NJ 07030
USA

www.wiley.com

Library of Congress Control Number: 2020938693

British Library Cataloguing-in-Publication Data
A CIP record for this book is available from the British Library
ISBN 978-1-78945-013-2

ERC code:
PE4 Physical and Analytical Chemical Sciences
PE4_8 Electrochemistry, electrodialysis, microfluidics, sensors

MIX
Paper from responsible sources
FSC® C013604

Contents

Introduction

Laure MONCONDUIT[1,3,4] and Laurence CROGUENNEC[2,3,4]

[1] *Institut Charles Gerhardt Montpellier, University of Montpellier, CNRS, Montpellier, France*

[2] *CNRS, University of Bordeaux, Bordeaux INP, ICMCB UMR 5026, Pessac, France*

[3] *RS2E, Réseau Français sur le Stockage Electrochimique de l'Energie, FR CNRS 3459, Amiens, France*

[4] *ALISTORE-ERI European Research Institute, FR CNRS 3104, Amiens, France*

I.1. Why Na-ion batteries?

Climate changes and constraints of fossil resources result in the exploration of green and sustainable resources from renewable energy resources (solar, wind, hydro, tidal and geothermal energies), and their storage has become a critical issue in recent years. The development of large-scale energy storage systems has push the research on Li-ion batteries (LIBs), recognized as the electrochemical energy storage system showing the higher energy density since their first commercialization in 1991. Today, LIBs still dominate the market as the high-energy and power source for portable electronic devices. The limited abundance (Figure I.1) and especially the uneven distribution of lithium, added to the emerging huge demand for electrical energy storage, could increase the cost and sustainability of lithium. Almost 25% of Li reserves are expected to be depleted by electric vehicle (EV) applications by 2050 (Vikström *et al.* 2013). Moreover, the lithium supply is concentrated in a few areas, mainly in South America and Australia. Only a few companies dominate the lithium market such as Albemarle (ALB US), Rockwood Lithium, SQM (SQM US), FMC (FMC US) and Chengdu Tianqi (China). As a result, the price of Li_2CO_3 is continuously growing and is foreseen to dramatically rise with that of the EV

Na-ion Batteries,
coordinated by Laure MONCONDUIT and Laurence CROGUENNEC.

market. Li recycling could support the supply of Li with a recycling rate of 50–100% in 2050. Since 2006, Umicore is providing recycling services for lithium-ion (Li-ion) and nickel metal hydride batteries (NiMH) from all possible applications. They are now able to recycle lithium from LIB waste through both pyrometallurgical as hydrometallurgical flowsheets. Among other actions, Eramet, Suez and BASF have formed an alliance, which is supported by Europe for an innovative research program to recycle the LIBs of used electric vehicles, and recycle the active materials such as lithium, nickel and cobalt in a form that is sufficiently pure to be reusable in a new battery (https://www.basf.com/fr/fr/media/Communiques-de-presse-France).

This situation increased the necessity to explore alternative battery chemistries. Na is the sixth most abundant element in the Earth's crust (≈2.6%), and unlimited Na resources are available from sea water (Figure I.1).

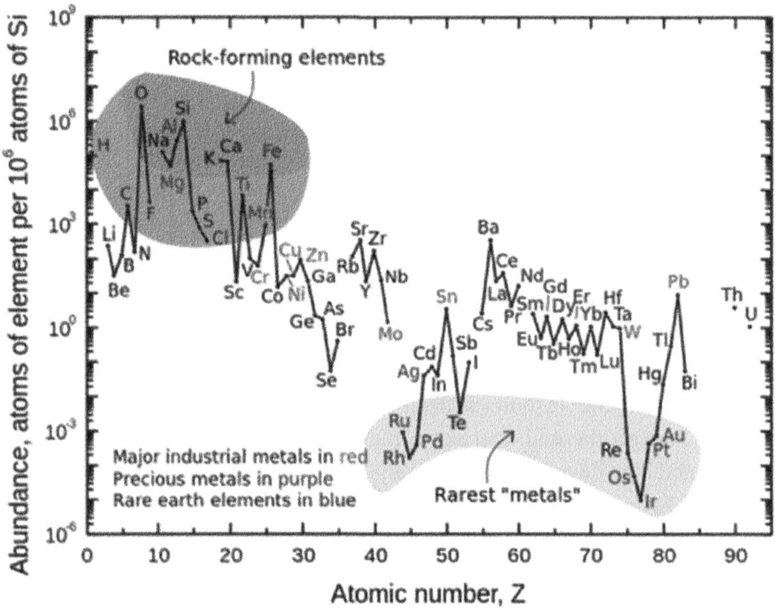

Figure I.1. *Abundance of elements in the Earth's crust (Pan* et al. *2013). For a color version of this figure, see www.iste.co.uk/monconduit/batteries.zip*

Sodium is the second lightest and smallest alkali metal next to lithium (1.02 Å for Na^+ vs. 0.76 Å for Li^+), however its larger ionic radius can result in sluggish reaction kinetics in electrode materials commonly used in LIBs such as layered oxides and graphite, usually producing lower capacity, lower rate capability and poor cycling stability. Moreover, the higher standard redox potential of Na (–2.71 V vs. SHE) than Li (–3.04 V vs. SHE) and its larger atomic weight represent other disadvantages when considering Na-ion batteries (NIBs), since they will trigger a lower energy density.

Na layered oxides intercalation compounds were first studied in the 1980s (Delmas *et al.* 1981) and have reemerged since 2012 at the positive electrode of NIBs for large-scale electrochemical storage systems at room temperature (Kim *et al.* 2012). Huge efforts were made to use graphite in NIBs at the beginning of the 1920s (Stevens and Dahn 2001), without success, leading then to the development of hard carbons at the negative electrode of NIBs. NIBs have received remarkable attention and are now recognized as the most promising alternative to LIBs, especially for large-scale energy storage applications. Much effort has been devoted to developing novel electrode materials and electrolytes for NIBs with high performance.

In this book, we summarize the most recent progress and the main challenges in the field: (1) for the positive and negative electrode materials, from the layered oxides to the polyanionic frameworks and from the carbons to the alloys, but also considering the development of new binders, additives and electrode formulation; (2) for the electrolytes, with a state of the art of currently used liquid and solid electrolytes in NIBs, and a special focus on those based on ionic liquids (IL); and (3) in the knowledge of solid electrolyte interphase (SEI) layer formation and evolution upon cycling of NIBs. Specific attention has been paid to understanding the sodiation/desodiation processes and ageing phenomena. Maturity and understanding of the LIB technology are obviously helping and accelerating the comprehension and the rapid development of NIBs. Last but not least, two chapters are devoted to demonstrating that NIB technology is not only a viable alternative to LIBs, but it is also competitive in terms of cost, performance and safety and is already developed to be introduced rapidly into the market.

I.2. From the electrodes to the electrolyte for NIBs

I.2.1. *Positive electrodes*

The first two chapters are dedicated to the positive and negative electrodes for a NIB. Among the materials intensively studied as prospective positive electrode for NIBs are polyanionics, layered oxides, Prussian blue analogues and organic compounds.

With their high theoretical capacity and ionic mobility, layered transition metal oxides are likely the most promising positive electrodes. It is interesting to mention that the alkali deintercalation/intercalation reactions in layered compounds were studied for the first time in Na_xCoO_2 (Braconnier *et al.* 1981; Delmas *et al.* 1981) before the demonstration of lithiation/delithiation reactions in the 2D lithium layered oxides. This structure presents a huge versatility of stackings and compositions by playing with the composition in 3*d* transition metals and in sodium that gives the potential to modulate their electrochemical properties. Their performance and the remaining challenges are detailed in Chapter 1.

Among the promising positive electrode materials for NIBs, although achieving lower specific capacity than the layered oxides, polyanionic compounds represent an important class of materials. Indeed, they are characterized by high operating voltage, excellent chemical stability and life span. A large panel of structures and compositions exist, with as for layered oxides the possibility to tailor significantly the operating voltage, and thus the energy density, by varying the nature of the involved redox couple. Today, the most attractive materials are vanadium-containing phosphates, characterized by two different structures, $Na_3V_2(PO_4)_3$ and $Na_3V_2(PO_4)_2F_3$. Chapter 2 shows the versatility of these families of compounds, highlighting their differences, their interesting properties and the challenges for their development in the next generations of NIBs.

I.2.2. *Negative electrodes*

Carbon materials present great interest as negative electrodes because of their low cost and environmental benignity, as well as their excellent electrochemical stability. In LIBs, graphite is the standard negative electrode. It exhibits a low operating voltage and 370 mAh/g specific capacity based on reversible Li deintercalation and reintercalation from/in between the graphene layers. Huge efforts were devoted to using it in NIBs (Stevens and Dahn 2001). Nevertheless, due to the thermodynamic instability of stage 1 Na-GIC formation, graphite was rapidly abandoned as the negative electrode. More recently, it was shown that when used in linear ether solvents, co-intercalation of ether molecules was observed and beneficial for Na^+ intercalation in between graphene layers even though it does not lead to acceptable cycles performance (Kim *et al.* 2015).

As the utilization of graphite as the negative electrode for NIBs failed, non-graphitic carbon materials such as hard carbon and soft carbon have been extensively investigated as alternatives (Dahbi *et al.* 2014). Chapter 3 discusses the main synthesis routes used to prepare hard carbons and the impact of both the nature of the precursor and synthesis conditions on the hard carbon features. Relationships between these synthesis conditions and the electrochemical properties are discussed in details, with a specific focus on the sodium storage mechanism, still a debate in literature.

Metal oxides have also been intensively studied as potential alternatives to carbon. Due to their low operation voltage as well as low cost and environmental benignity, Ti-based oxide compounds have been widely studied. The conversion reaction compounds based on metal oxides, and extended to sulfides, phosphides, fluorides, etc., have also recently attracted attention because of their high specific capacity involving the exchange of more than one electron per transition metal. These conversion type materials MX_n (M active (X) or inactive (Fe, Co, Ni, Cu, Mn, Mo) vs. Na and X = O, S, P, Ge, Sn, Sb, etc.) will be described in Chapter 4. High theoretical capacities, higher than those obtained for carbon-based materials, are also expected for Na-rich alloy phases from the phase diagram of most of p-block elements with Na. Silicon, which is the most studied alloy material for LIB due to the extremely high theoretical specific capacity, has been shown to be quite inactive versus Na. On the contrary, the 14 and 15 group elements with Ge, Sn, P and Sb can show excellent electrochemical behavior, sometimes better than in LIBs, even though they all present a huge volume expansion when alloyed with Na.

I.2.3. *Electrolytes and the solid electrolyte interphase*

The electrolytes are essential for the proper functioning of any battery technology, with an important focus to minimize interface electrolyte/electrode reactions and enhance both performance and safety. First studies surveyed electrolytes prepared using classical alkyl carbonates solvents and mixtures, in combination with different Na salts (Ponrouch *et al.* 2012). Their viscosity, ionic conductivity, thermal and electrochemical stability were evaluated to establish some intrinsic trends to identify the first electrolyte formulations with the widest range of applicability for NIBs.

More recently, IL, defined as room temperature molten salts and composed mainly of organic cations and (in)organic anions and presenting a huge versatility of structural variations, were proposed as major alternatives for the development of optimized electrolytes. Indeed, IL offer unique physical and chemical properties associated with low volatility that make them extremely interesting for the development of electrolytes with higher electrochemical and thermal stability. Chapter 5 will give the major trends for the IL-based electrolytes developed for NIBs.

The concept of solid electrolyte interphase (SEI), which is formed from the contact of the electrode with the electrolyte, was first introduced by Peled (1979) and then widely accepted in the LIB community, before being logically extended to NIBs as the nature of the electrodes and electrolyte are antecedent. Chapter 6 describes the SEI formation in NIBs, which has to be passivating for the electrode surface and stable upon cycling to get good cycling performance. Although the SEI layer and its formation have been intensely investigated during the last decades, even in LIBs it is still considered to be misunderstood being strongly dependant on a series of parameters such as the composition of the electrolyte, the nature of the electrodes and the cycling conditions. Moreover, regarding NIBs, to date there are only a few reports addressing the fundamental aspects of SEI chemistry. Recent works showed than even though the knowledge coming from LIBs is helpful for understanding the electrolyte/electrode interfaces modifications, a direct transfer of knowledge from LIBs to NIBs is not guaranteed. Indeed, many parameters have to be taken into account: the higher standard potential of Na^+/Na than Li^+/Li (~0.33 V), the higher solubility of carbonates from the SEI in NIBs, the lower desolvation energy of Na^+, the larger coordination shell of Na^+, the higher water reactivity of Na^+, etc. (Eshetua *et al.* 2019). These different features tailoring the nature and stability of the SEI in NIBs will be described in detail in Chapter 6.

I.3. Future commercialization of NIBs

Although LIBs are unquestionably the leader for electrochemical energy storage, their costs remain high and dependent on the demand that continuously and exponentially increases. In contrast, owing to the abundance of sodium, the price of NIBs could be preserved even with a large demand for energy storage. Furthermore, due to the proximity of their chemistry, the technology developed for LIBs should be promptly transferable to NIBs and easily upscaled. Widespread commercialization and mass production of NIBs will still depend to further developments, optimizations and market acceptance (Bauer *et al.* 2018). Chapters 7 and 8 are devoted to present the key achievements realized by NATRON and FARADION companies to develop and scale-up NIBs from the laboratory scale to the prototype level and commercialized battery stage.

NATRON develops Prussian blue-based NIBs. Chapter 7 provides a practical introduction to the use of Prussian blue and its analogues (PBA) as electrodes in NIBs. The relevant physical and electrochemical properties of PBA-based electrodes and their use in batteries are described, and their performance compared with the current commercial state-of-the-art. The open framework of PBAs allows them to get high-rate capability and a long cycle life. Capacities over 150 mAh/g at potentials above 3 V versus sodium are achieved, placing them among the highest energy density positive electrodes of NIBs. The first commercialization of products based on PBAs occurred in 2019.

FARADION has collaborated with commercial partners to produce large-scale quantities of optimized active materials, electrolytes and electrodes to be incorporated into prototype NIBs cells. The technology is based on non-aqueous electrolyte and layered nickelate positive electrode coupled to a hard carbon negative electrode. Due to the proximity of the electrodes with those used in LIBs, these NIB prototype cells can be fabricated on existing Li-ion manufacturing lines that should be cost effective. Note that 12 Ah prototype pouch NIB cells delivering a specific energy higher than 140 Wh/kg have been produced and higher specific energy are announced in the near future (Chapter 8).

The development of these very promising NIB technologies by emerging and proactive battery companies all over the world should accelerate the adoption of NIBs into many new markets and applications.

I.4. References

Bauer, A., Song, J., Vail, S., Pan, W., Barker, J., and Lu, Y. (2018). The scale-up and commercialization of nonaqueous Na-ion battery technologies. *Adv. Energy Mater.*, 8, 1702869.

Braconnier, J.J., Delmas, C., Fouassier, C., and Hagenmuller, P. (1981). Comportement electrochimique des phases Na_xCoO_2. *Materials Research Bulletin*, 15(12), 1797–1804.

Dahbi, M., Yabuuchi, N., Kubota, K., Tokiwac, K., and Komaba, S. (2014). Negative electrodes for Na-ion batteries. *Phys. Chem. Chem. Phys.*, 16, 15007–15028.

Delmas, C., Braconnier, J.-J., Fouassier, C., and Hagenmuller, P. (1981). Electrochemical intercalation of sodium in $NaxCoO_2$ bronzes. *Solid State Ionics*, 3–4, 165.

Delmas, C. (2018). Sodium and sodium-ion batteries: 50 years of research. *Adv. Energy Mater.*, 8(17), 1703137.

Eshetua, G.-G., Diemant T., Hekmatfar, M., Grugeon, S., Behma, R., Laruelle, J., Armand, S., and Passerini, M., (2019). The scale-up and commercialization of nonaqueous na-ion battery technologies. *Nano Energy*, 55, 327–340.

Kim, H., Jihyun Hong, J., Yoon, G., Kim, H., Park, K.-Y., Park, M.-S., Yoon W.-S., and Kang, K. (2015). Sodium intercalation chemistry in graphite. *Science Energy Environ. Sci.*, 8, 2963–2969.

Kim, S.-W., Seo, D.-H., Ma, X., Ceder, G., and Kang K. (2012). Electrode materials for rechargeable sodium-ion batteries: Potential alternatives to current lithium-ion batteries. *Adv. Energy Mater.*, 2, 710.

Pan, H., Hu, Y.S., and Chen, L. (2013). Room-temperature stationary sodium-ion batteries for large-scale electric energy storage. *Energy Environ. Sci.*, 6, 2338.

Peled, E., (1979). The electrochemical behavior of alkali and alkaline earth metals in nonaqueous battery systems—The solid electrolyte interphase model. *Journal of The Electrochemical Society*, 126(12), 2047.

Ponrouch, A., Marchante, E., Courty, M., Tarascon, J.-M., and Palacin, M.R. (2012). In search of an optimized electrolyte for Na-ion batteries. *Energy Environ. Sci.*, 5, 8572.

Stevens, D.A. and Dahn, J.R. (2001). The mechanisms of lithium and sodium insertion in carbon materials. *J. Electrochem. Soc.*, 148, A803.

Vikström, H., Davidsson, S., and Höök, M. (2013). Lithium availability and future production outlooks, *Appl. Energy* 110, 252.

1

Layered $NaMO_2$ for the Positive Electrode

Shinichi KOMABA and Kei KUBOTA

Department of Applied Chemistry,
Tokyo University of Science, Japan

1.1. Research history of layered transition metal oxides as electrode materials for Na-ion batteries until 2009

Studies on room temperature Na-ion batteries started in the 1970s and electrochemical properties of Na//Na_xCoO_2 and Na//TiS_2 cells were reported in 1980 (Braconnier *et al.* 1980; Newman Klemann 1980) when properties of Li//$LiCoO_2$ were also first reported (Mizushima *et al.* 1980). The Li-ion batteries commercialized by Sony in 1991 consisted of $LiCoO_2$ as a positive electrode material and carbon as a negative one, and have attracted much attention as high-voltage rechargeable batteries to date. On the other hand, Na-ion batteries are essentially based on the same chemistry and technology as Li-ion batteries, except for the charge carriers as Li^+ and Na^+ ions. Na-containing (sodiated) and Na-free (desodiated) materials are used for positive and negative electrodes, respectively, which correspond to the discharge state of the battery. The working voltage of the Na//Na_xCoO_2 cell was, however, ca. 1 V lower than that of the Li//$LiCoO_2$ cell, resulting in lower energy density (Goodenough *et al.* 1980). Allied Corp. in the United States and Showa Denko K. K. and Hitachi, Ltd. in Japan undertook collaborative work on Na-ion batteries and filed patents of Na-Pb alloy//γ-Na_xCoO_2 cells (Shacklette *et al.* 1985, 1988; Shishikura *et al.* 1989) exhibiting good cycle stability but commercialization of the Na-ion batteries has not been achieved so far. The Na-Pb alloy//γ-Na_xCoO_2 cells needed a pre-sodiation

Na-ion Batteries,
coordinated by Laure MONCONDUIT and Laurence CROGUENNEC.

process for the Pb negative electrode. A primary drawback of Na-ion batteries was the fact that the pre-sodiation process was required for practical use of any negative electrode materials at that time. In the 2010s, there was a general consensus that non-graphitizable carbon, so-called hard carbon, is known to deliver large reversible capacities and good capacity retention without any pre-sodiation (Stevens and Dahn 2000; Komaba *et al.* 2011).

Thus, another issue of the practical use of Na-ion is the low working potential of the positive electrode materials. The voltage of $NaCoO_2$, isostructural to α-$NaFeO_2$, in a Na cell at the voltage plateau region close to the end of discharge is ca. 2.5 V and much lower than ca. 3.9 V for α-$NaFeO_2$ type $LiCoO_2$ in a Li cell as shown in Figure 1.1. The large difference cannot be explained only by the standard redox potential of Na metal, which is lower than that of Li metal by ca. 0.3 V (Marcus 1985; Komaba *et al.* 2015). The voltage difference is much smaller than that between Na_xCoO_2 and $LiCoO_2$ at the end of discharge (ΔV = ca. 1.5 V) (Kubota *et al.* 2014). The large voltage difference is probably due to larger ionic size and lower Lewis acidity of Na^+ in comparison to Li^+ as discussed by Goodenough *et al.* (1980).

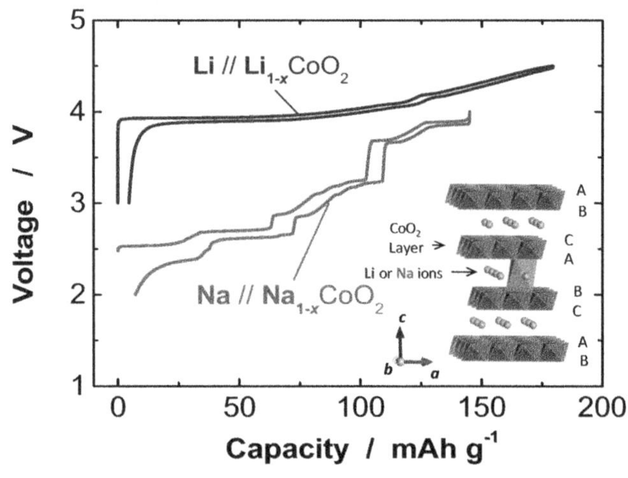

Figure 1.1. *Comparison of charge/discharge curves for Li//LiCoO₂ and Na//NaCoO₂ cells. A schematic illustration of the crystal structure of isostructural LiCoO₂ and NaCoO₂ is also shown. Reproduced with permission from Yabuuchi* et al. *(2014). Copyright 2014, American Chemical Society. For a color version of this figure, see www.iste.co.uk/monconduit/batteries.zip*

Since our group first demonstrated hard carbon//$NaNi_{1/2}Mn_{1/2}O_2$ full cells exhibiting acceptable cycle stability in 2009 (Komaba *et al.* 2009) and the results were published in *Advanced Functional Materials* in 2011 (Komaba *et al.* 2011), yearly numbers of scientific papers on sodium batteries have rapidly increased (Kubota *et al.* 2018a) and significant efforts have been devoted to attaining high-voltage and high-capacity positive electrode materials for Na-ion batteries (Delmas *et al.* 1982; Pan *et al.* 2013; Slater *et al.* 2013; Xu *et al.* 2013; Yabuuchi *et al.* 2014; Clement *et al.* 2015; Kundu *et al.* 2015; Fang *et al.* 2016; Kim *et al.* 2016; Deng *et al.* 2017; Hwang *et al.* 2017b; Ortiz-Vitoriano *et al.* 2017).

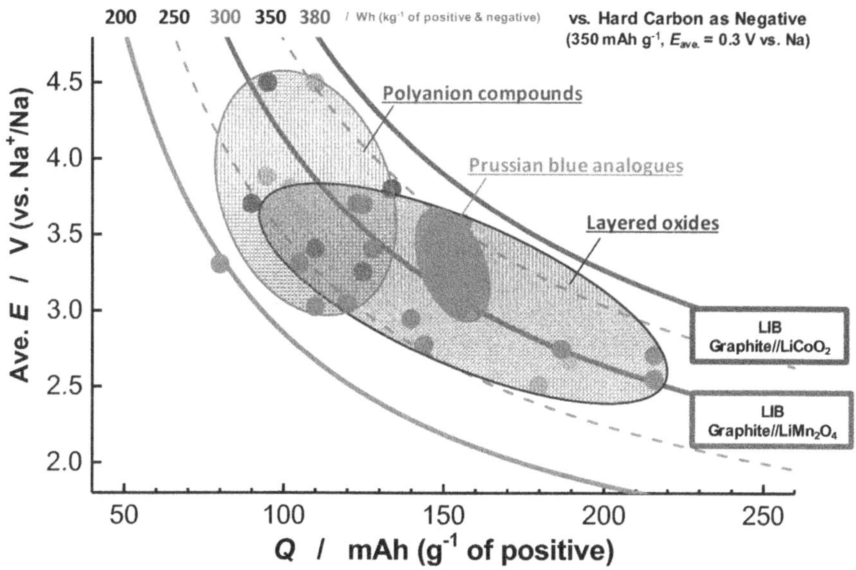

Figure 1.2. *Average voltage (V) and energy density (Wh kg^{-1}) versus gravimetric capacity (mAh g^{-1}) for selected positive electrode materials for Na-ion batteries. Energy density was calculated with the hard carbon (reversible capacity of 350 mAh g^{-1} with E_{ave} = 0.3 V vs. Na metal) as negative electrode materials. Reproduced with permission from Kubota* et al. *(2018b). Copyright 2018, Wiley-VCH. For a color version of this figure, see www.iste.co.uk/monconduit/batteries.zip*

Figure 1.2 shows a comparison of reversible capacity and working voltage among the positive electrode materials. The positive electrode materials for Na-ion batteries can be basically categorized into layered oxides, polyanionic compounds, and Prussian blue analogues. Polyanion materials seem to exhibit relatively higher working voltage and smaller capacities compared to the others. However, the high-voltage polyanion materials typically contain costly cobalt

(Nose *et al.* 2013) or vanadium (Jian *et al.* 2012; Kang *et al.* 2012; Lim *et al.* 2014; Zhang *et al.* 2016) except for $Na_2Fe_2(SO_4)_3$ (Barpanda *et al.* 2014). On the other hand, Na-containing layered transition metal oxides exhibit relatively lower working voltage but larger capacities on the basis of redox activity of Fe^{3+}/Fe^{4+} or manganese Mn^{3+}/Mn^{4+}. Because of the high-performance redox of iron and manganese, abundant mineral resources of iron and manganese are available, and the resultant cost-reduction will be thus advantageous for stationary applications of Na-ion batteries. We emphasize that the layered transition metal oxides meet this important demand for future application. Of course, higher energy density (i.e. large capacities and high working potential), long-term cycle life, high coulombic and energetic efficiencies, thermal-stability, and ease in handling and production, toxic-free chemistry and so on are desired for the practical use as positive electrode materials for Na-ion batteries.

In this chapter, developments of Na-containing layered 3*d* transition metal oxides are reviewed for the application as active materials of Na-ion batteries based upon the authors' experience since 2003 (Komaba 2019). The electrochemical performances, phase transitions during the charge/discharge, surface chemistry in the batteries, key factors influencing the battery performances and future prospective are discussed mainly based on our leading studies on the layered oxides since 2005.

1.2. Crystal structures of layered materials

1.2.1. *Crystal structures of synthesizable Na_xMO_2*

An iron based oxide, α-$NaFeO_2$, is one of the most well-known Na-containing layered transition metal oxides in inorganic chemistry and in the battery research community. This is because $LiCoO_2$, $LiNi_{0.8}Co_{0.15}Al_{0.05}O_2$ and $LiNi_{1/3}Mn_{1/3}Co_{1/3}O_2$ used as positive electrode in commercialized Li-ion batteries are isostructural to α-$NaFeO_2$. That is, the layered rock salt–type structure with space group of *R*-3*m* is called α-$NaFeO_2$ type. α-$NaFeO_2$-type materials are found in $NaMO_2$ (M = Co, Cr, Fe, Ti, Sc, etc.), as shown in Figure 1.3. In contrast, α-$NaFeO_2$-type $LiM'O_2$ ones are crystallized only for M' = Co, Ni, Cr and V, which can be explained by the difference in ionic radii between Li^+ and Na^+ ions (Shirane *et al.* 1995; Kanno *et al.* 1997). The ionic radius of a Li^+ ion at an octahedral site of sixfold coordination is 0.76 Å and is quite close to those of transition metal ions (Shannon 1976). Because of the similar ion size, lithium ion is often mixed with transition metal ions during a solid-state synthesis reaction of $LiMO_2$, resulting in the formation of a cation-ordered rock salt–type (γ-$LiFeO_2$ type) or a cation-disordered rock salt–type (NaCl type) phase as seen in Figure 1.3 (Hoffmann Hoppe 1977; Shirane *et al.* 1995; Kanno *et al.* 1997). An ionic radius of Na^+ (1.02 Å) is larger than those of Li^+ and transition metal ions, leading to the obvious separation of the Na^+ and transition metal layers. A variety of transition metals can be accommodated in α-$NaFeO_2$ type. This fact implies that cation mixing between Na^+ and transition metal ions is suppressed and avoided in the synthesis of $NaMO_2$ and various transition metals are simultaneously adopted to form α-$NaFeO_2$-type solid solutions, which is advantageous for optimizing composition of Na_xMO_2 representing good electrochemical properties for Na-ion batteries.

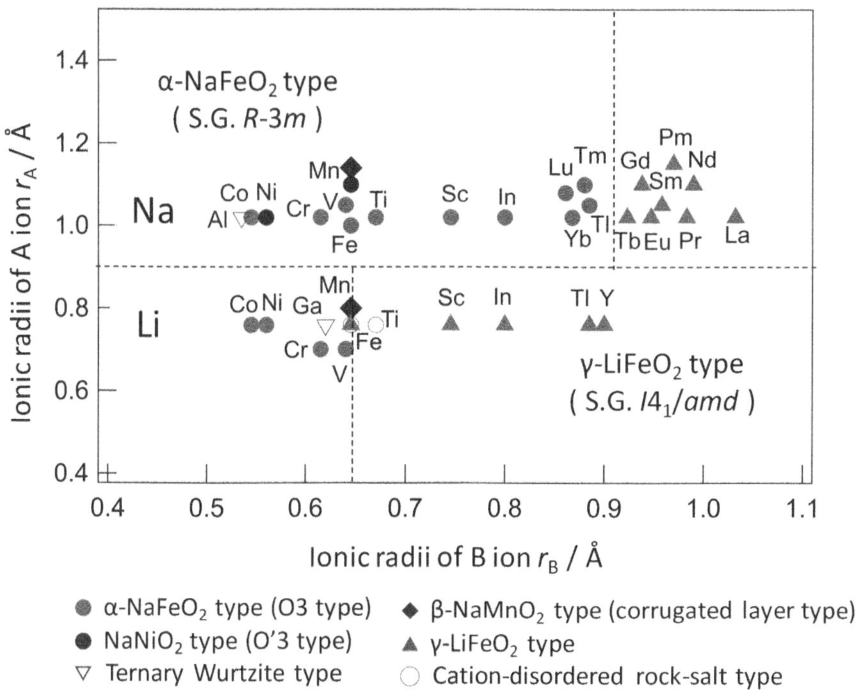

Figure 1.3. *Structure field map of ABO_2 compounds. Modified with permission from Kanno* et al. *(1997). Copyright 1997 Elsevier. For a color version of this figure, see www.iste.co.uk/monconduit/batteries.zip*

A systematic notation system for layered transition metal oxides containing alkali metal was proposed by Delmas *et al.* (1977, 1980). Layered oxides of α-$NaFeO_2$ and α-$NaCoO_2$ are categorized into O3-type materials, and β- and γ-Na_xCoO_2 are P'3- and P2-type materials, respectively. Schematic illustrations of typical layered structures for sodium transition metal oxides were drawn using the program VESTA (Momma and Izumi 2011) and are shown in Figure 1.4 (Kubota *et al.* 2014).

In O3-type (α-$NaFeO_2$-type) structure, MO_2 slabs consisting of edge-shared MO_6 octahedra stack along *c*-axis with cubic close-packed oxygen as AB CA BC array and alkali metal ions are accommodated at *octahedral* sites in the interslab space. The number of MO_2 slabs is *three* in the hexagonal unit cell. Namely, O in O3-type means the octahedral site accommodating alkali metal ions and following 3 is the number of MO_2 slabs included in a hexagonal unit cell. When the hexagonal lattice is distorted into monoclinic or orthorhombic lattice, a prime symbol is added between the alphabet and number, but the number of MO_2 layers is counted in pseudo-hexagonal unit cells, such as O'3-type $NaMnO_2$ with a monoclinic lattice (space group [S.G.], *C2/m*) (Parant *et al.* 1971), P'3-type Na_xCoO_2 (S.G., *C2/m*) (Fouassier *et al.* 1973) and P'2-type Na_xMnO_2 with an orthorhombic lattice (S.G., *Cmcm*) (Parant *et al.* 1971). O3-type $NaCoO_2$ reversibly transforms into P3-type Na_xCoO_2 by electrochemical Na extraction during the charging process (Braconnier *et al.* 1980).

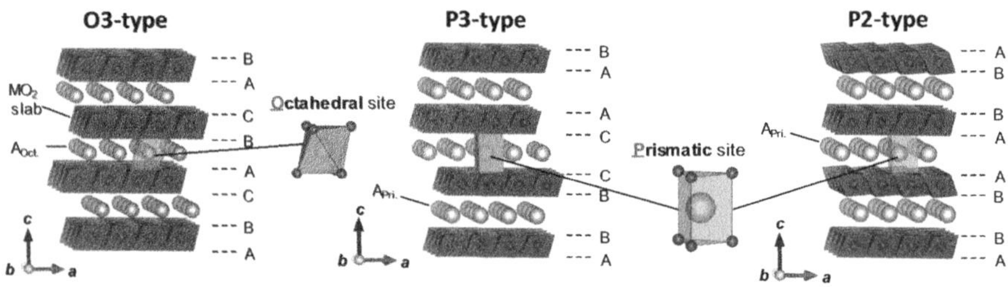

Figure 1.4. *Schematic illustrations of the crystal structures of O3-, P3-, and P2-type A_xMO_2. For a color version of this figure, see www.iste.co.uk/monconduit/batteries.zip*

P3-type Na_xCoO_2 can also be obtained as a Na deficient ($0.5 \leq x < 1$) and low-temperature phase by a solid state reaction (Fouassier *et al.* 1973). In the P3-type layered structure, alkali metal ions occupy *prismatic* sites in the interslab space between MO_2 slabs stacking with AB BC CA array of oxygen packing along the *c*-axis and the number of MO_2 slabs is *three* in the hexagonal unit cell with space group of *R*3*m*. The P3 type is generally formed by electrochemical Na extraction from O3-type accompanied by the gliding of MO_2 slabs without breaking the M-O bonds. P3-type materials are often observed as Na-deficient intermediate phases transform from O3-type ones by Na extraction. Compared to the P3-type phase, the P2-type (β-$RbScO_2$-type) phase is recognized as a high-temperature one, and P2-type (γ-)Na_xCoO_2 is actually obtained by heating at a higher temperature than that for P'3-Na_xCoO_2 (Fouassier *et al.* 1973). The phase transition from P3- to P2-type phase is accompanied by breaking Co-O bonds by heating at high temperature and does not occur in a Na (de)intercalation reaction at room temperature. In the P2-type structure, alkali metal ions occupy *prismatic* sites in interslab space between MO_2

slabs stacking with an AB BA array of oxygen packing along the *c*-axis and the number of MO_2 slabs is *two* in the hexagonal unit cell with space group of $P6_3/mmc$. The P2-type phase can electrochemically transform into an O2-type phase by the gliding of MO_2 slabs without breaking the M-O bonds (Lu and Dahn 2001a) as explained in detail in section 1.2.3.

The average valence of transition metals in O3-$NaMO_2$ (M = Ti, V, Cr, Mn, Fe, Co, Ni) and P2-$Na_{2/3}MO_2$ (M = V, Mn, Co) is +3 and +3.33, respectively. Tetravalent or higher valent metals are necessary for the crystallization of P2-type materials, and only V, Mn and Co oxides have been reported as single 3*d* transition metal P2 systems so far due to difficulty in formation of trivalent Ti and tetravalent Cr, Fe and Ni under ambient air or oxygen pressure. In contrast to the single 3*d* transition metal P2 system, multiple transition metal P2 systems can be stabilized by combining divalent Ni, trivalent Cr and Fe, tetravalent Ti and Mn. Moreover, monovalent alkali metal ions, pentavalent Bi and Sb, and hexavalent Te are also includable for stabilization of O3- and P2-type layered oxides. These transition metal elements dominate the charge/discharge capacity and redox potential of O3-$NaMO_2$ and P2-$Na_{2/3}MO_2$ in Na cells as well as the structural changes accompanied by Na-extraction/insertion with the formation of Na/vacancy and charge orderings (or formation of dimers or trimers of transition metals).

1.2.2. *Structural changes of O3-$NaMO_2$ by Na extraction*

Most O3-type materials generally transform in the O3 → P3 → O3 (→ O1) sequence by gliding of the MO_2 slab during Na extraction during the charging process (Kaufman and Van der Ven 2019) as shown in Figures 1.5 and 1.6(a). Note that O1-type CoO_2 was reported to be obtained via electrochemical Li extraction from $LiCoO_2$ in a Li cell (Amatucci *et al.* 1996). However, defect-free O1-type MO_2 including O1-CoO_2 has not been reported so far, although some literature reports defective O1-type MO_2 having migrated transition metal ions into the interslab spacing after Na extraction from $NaMO_2$ (Mariyappan *et al.* 2018b; Wang *et al.* 2019b). Non-hexagonal structures of O'3- and P'3-type phases are also formed in the compositional regions surrounding those of non-distorted O3- and P3-type phases. Furthermore, Na^+/vacancy orderings in the interslab spacing are often observed for x = 2/3, 1/2 and 1/3 in Na_xMO_2 (Figure 1.6(b)), leading to voltage jumps in the charge-discharge profiles (Zandbergen *et al.* 2004; Toumar *et al.* 2015; Kaufman and Van der Ven 2019).

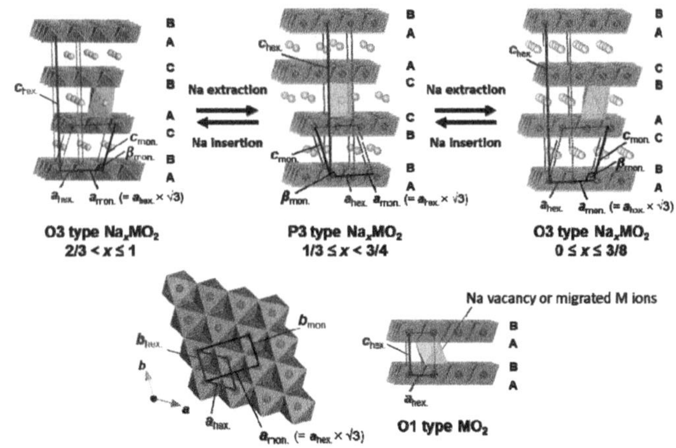

Figure 1.5. *Schematic illustrations of crystal structures reported for partly desodiated O3- and O'3-type Na_xMO_2. For a color version of this figure, see www.iste.co.uk/monconduit/batteries.zip*

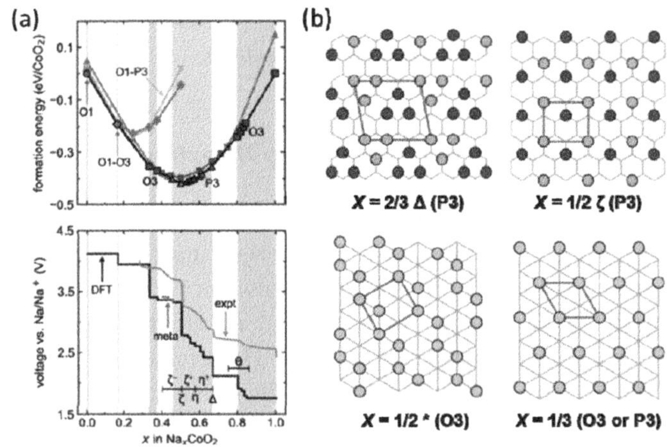

Figure 1.6. *(a) Calculated formation energies versus composition for Na_xCoO_2 configurations on the local convex hull of each host structure (top). Calculated zero-temperature equilibrium voltage curve (black) compared to experiment from Kubota* et al. *(2016) (gray) (bottom). (b) P3 ground-state orderings ζ and Δ, with Na shown in blue (top). O3 orderings with Na in yellow on the local convex hull of O3 for x = 1/3 and 1/2 in Na_xCoO_2 (bottom). Asterisks indicate that the ordering is above the global hull. Reprinted with permission from Kaufman and Van der Ven (2019). Copyright 2019, American Physical Society. For a color version of this figure, see www.iste.co.uk/monconduit/batteries.zip*

1.2.3. *Structural changes of P2-Na_xMO_2 by Na extraction*

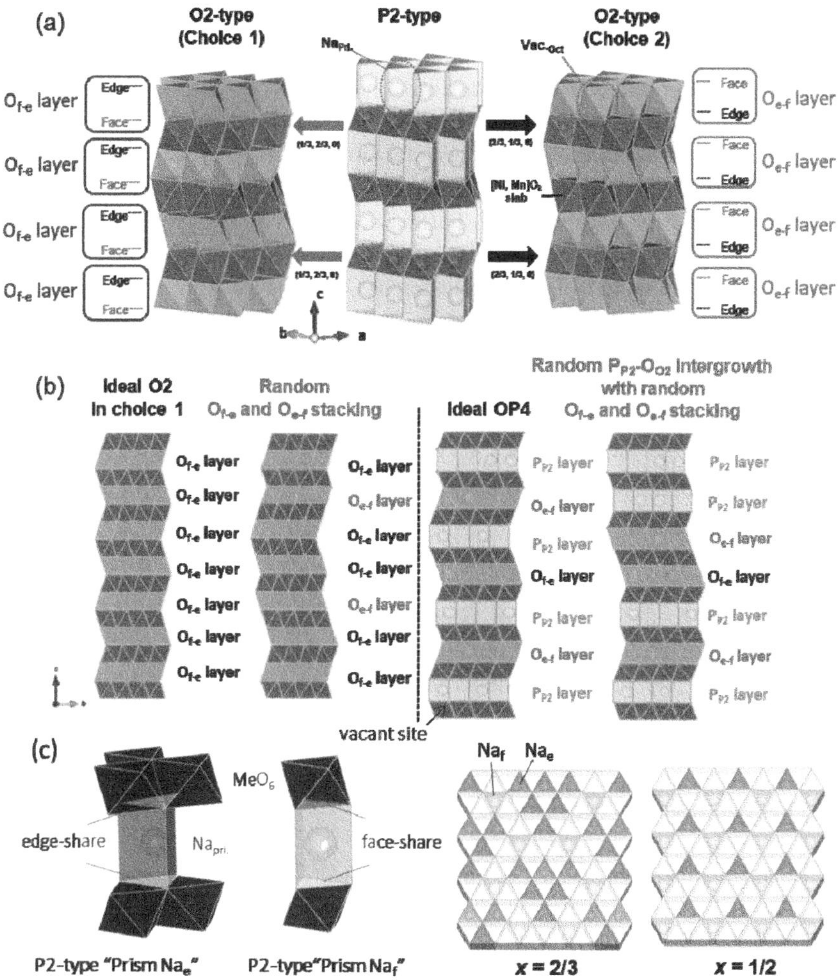

Figure 1.7. *Schematic illustrations of (a) P2-type and two choices of O2-type structures and (b) ideal O2 in choice 1 and a structural model with random stacking of $O_{f\text{-}e}$ and $O_{e\text{-}f}$ layers as completely desodiated [Ni,Mn]O_2 (left) and ideal OP4-type and a structural model with random stacking of P_{P2} and O_{O2} layers with simultaneous random stacking of $O_{f\text{-}e}$ and $O_{e\text{-}f}$ layers (right). Reprinted with permission from Kubota et al. (2018b) Copyright 2018, Wiley-VCH. (c) Na^+/vacancy ordering observed for P2-$Na_{2/3}CoO_2$ and P'2-$Na_{1/2}CoO_2$. For a color version of this figure, see www.iste.co.uk/monconduit/batteries.zip*

P2-type materials generally transform in the P2 → O2 sequence by gliding MO_2 slabs during Na extraction during the charging process (Lu and Dahn 2001a; Berthelot *et al.* 2011). However, the phase transition is not strictly accurate and is more complicated in the actual reaction. According to the literature (Lu and Dahn 2001a), O2-type phase contains stacking faults between two different O2-type stackings (Figure 1.7(a)). The two choices of slab-gliding, (1/3, 2/3, 0) and (2/3, 1/3, 0) vectors for the slab with BA oxygen-stacking in the P2-type AB BA stacking, form two types of O2-type structures possessing different oxygen stacking manners of AB CB and AB AC, respectively. Furthermore, the two O2-type stacking phases coexists with the P2-type one in the specific voltage plateau region at ~4.2 V. Our group proposed formation of the two different O2 types stacking and staking faults as a P2-O2 intergrowth structure for $Na_{2/3-x}Ni_{1/4}Mn_{2/3}Cu_{1/12}O_2$ instead of the ideal OP4-type one (Figure 1.7(b)) with simulated synchrotron X-ray diffraction (XRD) patterns using DIFFaX software (Treacy *et al.* 1991; Kubota *et al.* 2017). Similar to O3 type, P2-type materials also have Na^+/vacancy orderings in the interslab spacing typically for x = 2/3 and 1/2 in Na_xMO_2, as shown in Figure 1.7(c) (Hinuma *et al.* 2008; Berthelot *et al.* 2011; Lee *et al.* 2013).

These phase transitions were usually confirmed by *ex situ* and *in situ*/*operando* XRD and transmission electron microscopy (TEM) measurements. The phase transitions highly depend on the transition metals. In order to understand the characteristics of each transition metal in O3-$NaMO_2$ and P2-$Na_{2/3}MO_2$, single 3*d* transition metal O3 and P2 systems are reviewed before each of the multiple transition metal systems.

1.3. O3-type layered materials

1.3.1. *$NaMO_2$ (M = Sc, Ti, V, Cr, Mn, Fe, Co, Ni)*

O3-type layered sodium 3*d* transition metal oxides (including O'3-type ones) were discovered in 1930s–1975s (Goldsztaub 1935; Scholder and Kyri 1952; Dyer *et al.* 1954; Rüdorff and Becker 1954; Andersson and Wadsley 1962; Hoppe *et al.* 1965; Jansen and Hoppe 1972) and electrode performances of Ti, V, Cr, Mn, Co and Ni systems in Na cells were first reported by Delmas *et al.* in 1980–1985 (Braconnier *et al.* 1980; Braconnier *et al.* 1982; Delmas *et al.* 1982; Maazaz *et al.* 1983; Mendiboure *et al.* 1985). These oxides are electrochemically active and often evaluated with Na-metal half-cells consisting of the oxide as a positive electrode and metallic Na as a negative electrode for the primary tests. Since $NaMO_2$ contains Na in the structure, Na-free electrode materials such as carbon are applied to the negative electrode for the practical Na-ion full cells. Charge/discharge tests of $NaMO_2$ electrodes generally start in the Na-extraction way with simultaneous oxidation of transition metals in both the half-cell and full cell. In this section, electrochemical characteristics of the single 3*d* transition metal O3 and O'3 systems in Na cells are reviewed in order of increasing atomic number from Sc ($3d^0$) to Ni ($3d^9$).

1.3.1.1. *O3-$NaScO_2$, O3-$NaTiO_2$ and O3-$NaVO_2$*

As expected from the empty 3*d* orbital of Sc(III), O3-$NaScO_2$ (Hoppe *et al.* 1965) is electrochemically inactive and almost no capacity is observed in the voltage range of 1.5–4.5 V (Kubota *et al.* 2018b). In contrast, O3-$NaTiO_2$ is electrochemically active and delivers reversible capacity of ca. 150 mAh g^{-1} based on $Ti^{3+/4+}$ redox in the voltage range of 0.6–1.6 V in a Na cell (Maazaz and Delmas 1982; Maazaz *et al.* 1983; Wu *et al.* 2015). In the voltage range, the working voltage is low (ca. 1.0 V) and charging (oxidation) to >1.6 V leads to the irreversible change in the voltage profile and the capacity decay during charge-discharge cycles (Maazaz *et al.* 1983). O3-$NaTiO_2$ is, thus, possibly available as a negative electrode material for Na-ion batteries. In contrast to O3-$NaTiO_2$ (Maazaz and Delmas 1982; Maazaz *et al.* 1983), Na (de)intercalation properties of O3-$NaVO_2$ were first reported in 2011 by Didier *et al.* (2011) and Hamani *et al.* (2011). O3-$NaVO_2$ delivers a reversible capacity of 126 mAh g^{-1} corresponding to 0.5 mol Na extraction/insertion from/into $NaVO_2$ in the voltage range of 1.4–2.5 V in a Na cell. O3-$NaVO_2$ represents a very flat voltage plateau at 1.8 V and a large voltage jump from 1.8 to 2.1 V on the charging (Na-extraction) process. In general, the very flat voltage plateau originates from a large difference of formation energies between different Na contents of the terminal phases, e.g. O3-$NaVO_2$ and O'3-$Na_{0.5}VO_2$, and the two phases coexist on the voltage plateau region, while a small difference of formation energies between the different Na-content phases leads to monophasic solid solution–like reactions with sloping voltage curves. In addition, the voltage steps correspond to phase transitions and/or Na^+/vacancy orderings as described above. Similar to O3-$NaTiO_2$, charging to >2.5 V suffers capacity decay probably due to irreversible migration of vanadium ions into the interslab space (Didier *et al.* 2011). Thus, the low working voltage of O3-$NaVO_2$ is also unavailable as a positive electrode material.

1.3.1.2. *O3-$NaCrO_2$*

In contrast to O3-$NaScO_2$, O3-$NaTiO_2$ and O3-$NaVO_2$, O3-$NaMO_2$ (M = Cr, Mn, Fe, Co, Ni) exhibits relatively high working voltages suitable for the positive electrodes of Na-ion batteries. O3-$NaCrO_2$ delivers a reversible capacity of 110 mAh g^{-1} with an average working voltage of 3.02 V in a voltage range of 2.5–3.6 V (Figure 1.8) (Komaba *et al.* 2009, 2010).

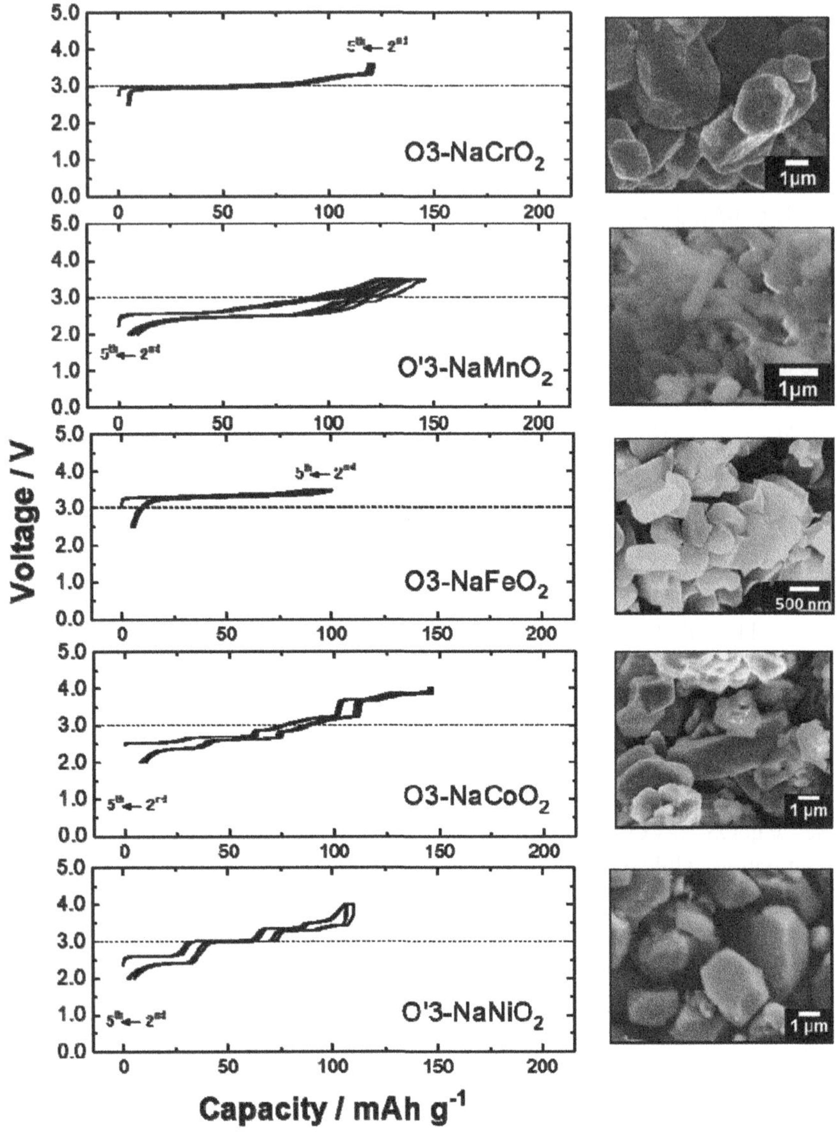

Figure 1.8. *Comparison of galvanostatic charge/discharge curves of layered O3 and O'3 type single 3d transition metal oxides (left). Morphology of particles for each sample is also compared (right)*

It should be noted that O3-$LiCrO_2$ is electrochemically inactive in a Li cell (Komaba *et al.* 2010) and electrochemical behaviors of layered CrO_2 are quite different in between Li and Na cells. O3-$NaCrO_2$ exhibits reversible structural changes in the O3 → O'3 → P'3 sequence by Na extraction to $x = 0.5$ in Na_xCrO_2 upon charging to 3.6 V versus Na (Komaba *et al.* 2009; Chen *et al.* 2013; Zhou *et al.* 2013; Kubota *et al.* 2015a). Further, Na extraction to $x < 0.5$ in Na_xCrO_2 causes an irreversible structural change, resulting in almost no reversible capacity (Figure 1.9(a)) (Kubota *et al.* 2015a; Bo *et al.* 2016).

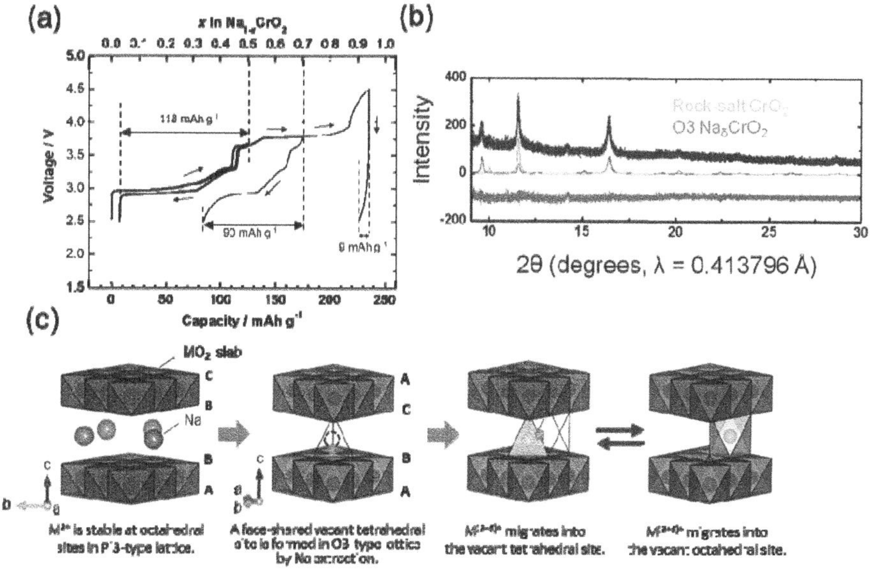

Figure 1.9. *(a) Initial charge and discharge curves of Na//$NaCrO_2$ cells at a rate of 12.5 mA g^{-1} in the ranges of $0.0 \leq x \leq 0.5$ and $0.0 \leq x \leq 0.7$ in $Na_{1-x}CrO_2$ and in a voltage range of 2.5–4.5 V. Reprinted with permission from Kubota* et al. *(2015a). Copyright 2015, American Chemical Society. (b) Rietveld refinements for the sample that was charged to the end of the 3.8 V plateau. Reprinted with permission from Bo* et al. *(2016). Copyright 2016, American Chemical Society. (c) Mechanism of transition metal migration on the sodium extraction process. Reprinted with permission from Kubota* et al. *(2015a). Copyright 2015, American Chemical Society. For a color version of this figure, see www.iste.co.uk/monconduit/batteries.zip*

The migrated chromium ions into the interslab space were detected with XRD (Figure 1.9(b)), electron diffraction (ED), and X-ray absorption spectroscopy (XAS) (Kubota *et al.* 2015a; Bo *et al.* 2016). The presumed migration mechanism is shown as schematic illustrations in Figure 1.9(c). Transition metal ions irreversibly migrate from the slab into interslab, resulting in disturbing Na insertion on discharge (Kubota *et al.* 2015a; Bo *et al.* 2016). Even in 2.5–3.6 V, capacity decay is observed in Na cells filled with carbonate ester-based electrolyte, which is probably due to electrolyte decomposition on the surface of O3-$NaCrO_2$. To suppress the side reaction on the surface, Du *et al.* synthesized large-grained O3-$NaCrO_2$ from $Na_2Cr_2O_7 \cdot 2H_2O$ and demonstrated a large reversible capacity of 123 mAh g^{-1} and a high tap density of 2.55 g cm^{-3} (Wang *et al.* 2019c). Yu *et al.* carried out surface coating on the O3-$NaCrO_2$ particles with carbon. Carbon-coated O3-$NaCrO_2$ delivers a larger reversible capacity of 120 mAh g^{-1} and exhibits superior capacity retention and rate performance in the Na cells and a hard carbon//O3-$NaCrO_2$ full cell (Yu *et al.* 2015). Among the single 3*d* transition metal O3 type systems, charge/discharge behaviors of O3-$NaCrO_2$ such as large reversible capacity, excellent capacity retention and rate performance, non-stepwise and slightly inclined voltage curves are suitable as a positive electrode material for practical use in Na-ion batteries. Actually, Sumitomo Electric Industries, Ltd. made prototype Zn-Na alloy//O3-$NaCrO_2$ cells with the molten salt of sodium bis(fluorosulfonyl)amide (NaFSA) and potassium bis(fluorosulfonyl)amide (KFSA) as electrolyte, and battery and safety performances were reported (Nitta *et al.* 2013). Despite the toxicity issue of chromium, O3-$NaCrO_2$ has been used as a standard positive electrode material to evaluate performance of Na-ion batteries. Furthermore, $Na_{0.5}CrO_2$ shows high thermal stability in $NaPF_6$-based electrolyte, higher than those of $Li_{0.5}CoO_2$ and $Li_xNi_{1/3}Mn_{1/3}Co_{1/3}O_2$ in a LiPF6-based electrolyte, which is evidenced by accelerating rate calorimetry (Xia and Dahn 2012) and explained with phase transitions without oxygen generation below 477°C (Yabuuchi *et al.* 2016).

1.3.1.3. *O'3-$NaMnO_2$*

O'3-$NaMnO_2$, generally known as α-$NaMnO_2$, has an advantage in the abundant resources of manganese. Its electrochemical performance in a Na cell was first reported in 1971 by Hagenmuller's group (Parant *et al.* 1971) and was recently re-investigated in 2011 by Ceder's group (Ma *et al.* 2011; Chen *et al.* 2018). O'3-$NaMnO_2$ delivers a reversible capacity of 185 mAh g^{-1} corresponding to ca. 0.8 Na extraction per the formula unit at an initial cycle in the voltage range of 2.0–3.8 V in a Na cell as shown in Figure 1.8. Although phase transitions during charge/discharge are still not fully clear, O3- and O'3-type structures seem to be retained without formation of P3 type (Kubota *et al.* 2015b; Chen *et al.* 2018). Despite there being no significant structural changes, the capacity inevitably decays during cycles. Even if the upper cut-off voltage is set to <3.5 V, the capacity fade is not completely avoided (Ma *et al.* 2011). Further detailed studies are necessary to elucidate the deterioration mechanism, which will be reported in the near future by our group. In addition to O'3-$NaMnO_2$, $NaMnO_2$ has another layered oxide polymorph of β-$NaMnO_2$, of which structure is known to be a zig-zag layered (or corrugated) type (Parant *et al.*

1971; Fouassier *et al.* 1975). Zig-zag layered $NaMnO_2$ has orthorhombic symmetry with space group of *Pnmm* (Abakumov *et al.* 2014). The zig-zag layered structure consists of corrugated MnO_2 slabs alternately stacked with Na along the <100> direction in the orthorhombic lattice and the MnO_6 octahedra is cooperatively distorted and elongated along <001> as shown in Figure 1.10(a) (Abakumov *et al.* 2014). A single phase of zig-zag layered $NaMnO_2$ is hardly obtained due to contamination of O'3-like stacking faults, which is caused by a small difference in the formation energies between O'3 and zig-zag layered ones (Mishra and Ceder 1999; Shishkin *et al.* 2018), resulting in the presence of local intergrowths of O'3-$NaMnO_2$ and stacking faults (twin planes) within the lattice of zig-zag layered $NaMnO_2$ (Abakumov *et al.* 2014; Billaud *et al.* 2014; Orlandi *et al.* 2018) (Figure 1.10(b)) as evidenced with the XRD, neutron diffraction (ND) and ED patterns (Abakumov *et al.* 2014; Orlandi *et al.* 2018) and ^{23}Na NMR spectra (Billaud *et al.* 2014). It should be noted that zig-zag layered like stacking locally exists even in the O'3-type phase and would influence the electrochemical properties of O'3-$NaMnO_2$ (Figure 1.10(c)).

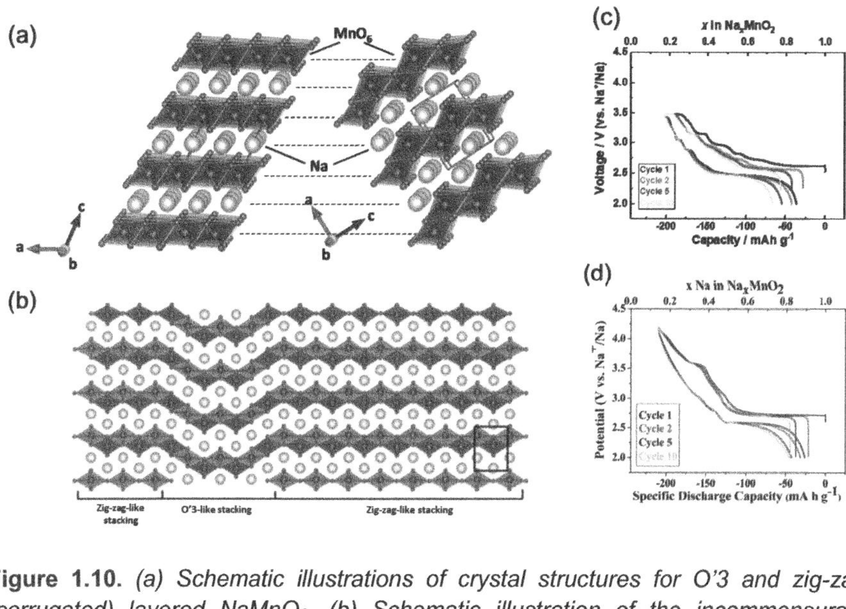

Figure 1.10. *(a) Schematic illustrations of crystal structures for O'3 and zig-zag (corrugated) layered $NaMnO_2$. (b) Schematic illustration of the incommensurate compositional modulated structure of the $NaMnO_2$ polymorphs. Modified with permission from Orlandi* et al. *(2018). Copyright 2018, American Physical Society. Galvanostatic charge/discharge curves of (c) O'3 and (d) zig-zag layered $NaMnO_2$. The first, second, fifth and tenth Na extraction/reinsertion cycles are represented in black, red, blue and green, respectively. Reprinted with permission from Billaud* et al. *(2014). Copyright 2014, American Chemical Society. For a color version of this figure, see www.iste.co.uk/monconduit/batteries.zip*

Zig-zag layered $NaMnO_2$ with the stacking faults delivers a reversible capacity of 190 mAh g^{-1} in a voltage range of 2.0–4.2 V at a current density of 10 mA g^{-1} as shown in Figure 1.10(d) (Billaud *et al.* 2014; Kubota *et al.* 2018a). Recently, our group successfully synthesized zig-zag layered $NaMnO_2$ using Cu-substitution (Shishkin *et al.* 2018). Differences in structures and electrochemical behaviors between O'3 and zig-zag layered $NaMnO_2$ will be reported in detail elsewhere in the near future.

In addition to the $NaMnO_2$ polymorphs, $Na_2Mn_3O_7$, which can be described as $Na_{4/7}[Mn_{6/7}\square_{1/7}]O_2$ ($\square$: Mn vacancy), is also a layered material (Chang and Jansen 1985). The structure is composed of $[Mn_{6/7}\square_{1/7}]O_2$ slabs and has a vacancy at the honeycomb center in the slab. Na^+ ions are accommodated at distorted-octahedral and prismatic sites in the interslab space and the structure is regarded as intermediate between O'3- and P'3-type ones with space group of *P*-1. Due to the Na-deficient composition, $Na_2Mn_3O_7$ is first reduced down to 1.5 V versus Na to accommodate Na^+ ions and delivers a reversible capacity of 160 mAh g^{-1} with a voltage plateau at 2.2 V based on $Mn^{3+/4+}$ redox in 1.5–3.0 V (Adamczyk and Pralong 2017) as shown in Figure 1.11(a). Recently, the study on oxygen-redox of layered oxide materials in the high-voltage region above 4.0 V is a hot topic in Li-ion and Na-ion battery fields. The transition metal vacancies in the slab were reported to enhance the redox activity of oxide ions for $Na_{0.78}[Ni_{0.23}Mn_{0.69}\square_{0.08}]O_2$ (Ma *et al.* 2017). Mortemard de Boisse *et al.* (2014) demonstrated highly reversible charge-discharge reactions delivering a capacity of 75 mAh g^{-1} with a distinct voltage plateau at ~4.1 V in $Na_2Mn_3O_7$ as shown in Figures 1.11(a) and (b) (Mortemard de Boisse *et al.* 2018). Surprisingly, voltage hysteresis between charge and discharge is negligible unlike Li_2MnO_3 and Li-rich materials having the $[Mn_{2/3}Li_{1/3}]O_2$ honeycomb structure in the slab (Singer *et al.* 2018). Oxygen redox was also reported for a Na-rich layered oxide of Na_2RuO_3, which can be described as O3-$Na[Ru_{2/3}Na_{1/3}]O_2$ and has a $[Ru_{2/3}Na_{1/3}]O_2$ honeycomb structure in the slab (Rozier *et al.* 2015; Mortemard de Boisse *et al.* 2016, 2019). Mortemard de Boisse *et al.* (2014) revealed from the *ex situ* XRD patterns that coulombic attractive interactions between the slabs are enhanced by the existence of ordered alkali vacancies, i.e. the $[Ru_{2/3}\square_{1/3}]O_2$ honeycomb structure formed by Na extraction during charge, leading to reduction of stacking faults and progressive ordered stacking of the $[Ru_{2/3}\square_{1/3}]O_2$ slabs upon charging. The cooperatively ordered vacancies generate the electro-active nonbonding 2p orbitals of neighboring oxygen anions and stabilize the phase transformation for highly reversible oxygen-redox reactions (Mortemard de Boisse *et al.* 2019). Although the operation voltage of O3-Na_2RuO_3 is relatively low as a positive electrode material and further studies are required to enhance the oxygen-redox capacity for $Na_2Mn_3O_7$, the findings provide a compelling future research direction toward reversible oxygen-redox positive electrode materials for high energy density batteries (Mortemard de Boisse *et al.* 2018).

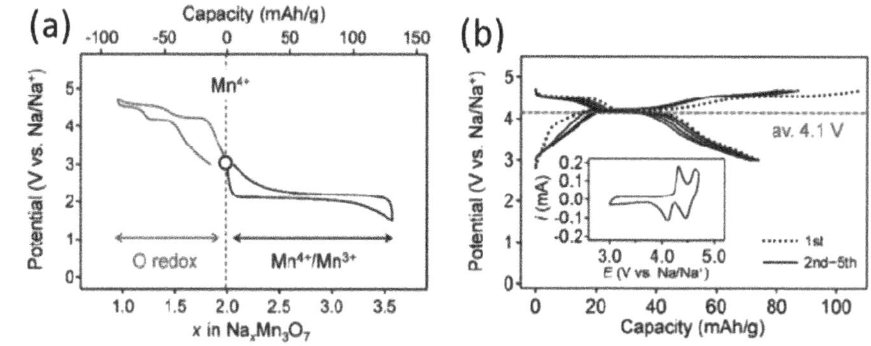

Figure 1.11. *(a) Potential profile (second cycle) of $Na_2Mn_3O_7$ upon (de)intercalation in 1.5–4.7 V versus Na/Na^+. The dashed gray box highlights the high voltage region, where O is the active redox process. (b) Galvanostatic charge/discharge curves in 3.0–4.7 V versus Na/Na^+ at a rate of C/20. The inset shows a cyclic voltammetry curve at the second cycle at a scan rate of 0.1 mV s^{-1}. Reprinted with permission from Mortemard de Boisse* et al. *(2018). Copyright 2018, Wiley-VCH. For a color version of this figure, see www.iste.co.uk/monconduit/batteries.zip*

1.3.1.4. *O3-$NaFeO_2$*

Another elementally attractive material in terms of abundant resources is O3-$NaFeO_2$. Although O3-$NaFeO_2$ is well known as a typical structural type of α-$NaFeO_2$, the electrochemical properties were first reported by Takeda *et al.* in 1994 (Takeda *et al.* 1994) and O3-$NaFeO_2$ delivers a discharge capacity of ca. 120 mAh g^{-1} based on $Fe^{3+/4+}$ redox in the Li cell. The electrochemical $Fe^{3+/4+}$ redox activity was a surprising fact because O3-$LiFeO_2$ delivers a reversible capacity of <10 mAh g^{-1} and is electrochemically inactive unlike O3-$NaFeO_2$. Moreover, O3-$LiFeO_2$ cannot be directly synthesized by a solid-state reaction and is prepared by Li^+/Na^+ ion-exchange from O3-$NaFeO_2$ (Ado *et al.* 1997) due to the relatively large ionic radius of Fe^{3+} as shown in Figure 1.3. As is the case in the Li cell, O3-$NaFeO_2$ is electrochemically active in a Na cell and delivers a reversible capacity of ca. 80 mAh g^{-1} based on $Fe^{3+/4+}$ redox with a flat potential plateau at ca. 3.3 V versus Na^+/Na in the voltage range of 2.5–3.4 V (Figure 1.8) (Takahashi *et al.* 2004; Okada *et al.* 2006; Yabuuchi *et al.* 2012b; Zhao *et al.* 2013; Li *et al.* 2018). $Fe^{3+}/^{4+}$ redox was confirmed with ^{57}Fe Mössbauer spectroscopy (Takeda *et al.* 1994; Zhao *et al.* 2013; Lee *et al.* 2015) and oxidation of oxide ions was recently revealed with soft XAS at O K-edge (Li *et al.* 2018). The mean working voltage of 3.3 V is the highest among single 3*d* transition metal O3 and O'3 systems. When Na^+ ions are extracted by >0.5 mole from Na_xFeO_2 upon a charging process, the discharge capacity significantly deteriorates due to an irreversible structural change accompanied by Fe migration into the interslab space (Yabuuchi *et al.* 2012b) as seen in $NaTiO_2$, $NaVO_2$ and $NaCrO_2$. Detailed phase transitions during charge/discharge were studied with *ex situ* ^{57}Fe Mössbauer spectroscopy and *operando* synchrotron XRD by Lee *et al.* and the results revealed non-equilibrium phase-transition behavior among original O3, secondary O3 and O'3 phases in which iron ions simultaneously migrate into interslab space even in $x \leq 0.5$ in Na_xFeO_2 (Lee *et al.* 2015). Furthermore, reduction of Fe^{4+} accompanied by electrolyte decomposition was also observed during storage for 2 days after charging to 3.6 V. Suppression of the electrolyte decomposition and of the migration of iron ions is required to utilize $Fe^{3+/4+}$ redox in layered oxides for Na-ion batteries.

1.3.1.5. *O3-NaCoO₂*

Electrochemical properties of O3-$NaCoO_2$ in Na cells were first reported by Braconnier *et al.* in 1980 (Braconnier *et al.* 1980) at almost the same time when those of O3-$LiCoO_2$ in Li cells were first reported by (Mizushima *et al.* 1980). O3-$NaCoO_2$ delivers a reversible capacity of ca. 140 mAh g^{-1} in the voltage range of 2.5–4.0 V (Figure 1.8) (Yoshida *et al.* 2013; Lei *et al.* 2014). As discussed above on O'3-$NaMnO_2$, O3-$NaCoO_2$ represents a stepwise voltage profile that is attributed to multiple phase transitions associated with CoO_2-slab gliding and in-plane Na^+/vacancy ordering in the interslab spacing as described in section 1.2.2.

1.3.1.6. *O'3-NaNiO₂*

Electrochemical properties and phase transitions of O'3-$NaNiO_2$ in Na cells were first reported by Braconnier *et al.* in 1982 (Braconnier *et al.* 1982) and re-investigated by Vassilaras *et al.* in 2013 (Vassilaras *et al.* 2013). O'3-$NaNiO_2$ delivers a reversible capacity of ca. 100 mAh g^{-1} with stepwise voltage profile in the voltage range of 1.25–3.75 V (Figure 1.8). Detailed phase transitions during charge-discharge were reported by Han *et al.* (2014) and Wang *et al.* (2017a) with operando XRD measurements. The results revealed that O'3-$NaNiO_2$ transforms into at least six phases of O'3 and P'3 during a charging process. Both the authors found that no original O'3 phase was observed on the discharging process and the irreversible phase transition was thought to cause capacity degradation during charge–discharge cycles. Unlike Ti, V, Cr, Mn and Fe, but like Co system, O'3-$NaNiO_2$ delivers a large discharge capacity of ca. 130 mAh g^{-1} even after charging to the high voltage of 4.5 V versus Na (Vassilaras *et al.* 2013; Wang *et al.* 2017a). Wang *et al.* revealed from *in situ* synchrotron XRD that the Na-extracted phase of $Na_{0.17}NiO_2$ is irreversibly formed upon charging to 4.5 V and is partly remained in the core of the particles even after discharging to 2.0 V versus Na. Migration of nickel ions into interslab space was often observed for O3-$LiNiO_2$ at the end of charging to 4.45 V versus Li (Croguennec *et al.* 2001). However, Li *et al.* mentioned no migration of nickel ions for O'3-$NaNiO_2$ after charging to 4.5 V versus

Na from the unpublished HRTEM image in the literature (Li *et al.* 2016). The reaction mechanism of O'3-$NaNiO_2$ might be different to that of O3-$LiNiO_2$.

These fundamental studies are very helpful to understand the electrochemical properties of the complicated multiple transition metal systems. Transition metals usually dominate redox potential and phase transitions of O3-type layered materials. However, those of multiple 3*d* transition metal systems are not simple and are influenced by difference in oxygen orbital contribution to the redox reaction (Nanba *et al.* 2016). Not only nominal valence of transition metals but also hybridization between oxygen 2p and transition metal 3*d* orbitals plays a critical role in determining the redox potential of layered transition metal oxides in Na batteries. Redox potential and electrochemical activity of nickel ions vary on the other transition metal elements in $NaMO_2$ (Nanba *et al.* 2016). Selection of the transition metals is, therefore, important to synthesize ternary and quaternary 3*d* transition metal systems exhibiting excellent electrochemical performance.

1.3.2. *O3-Na[M,M']O_2 (M, M' = transition metals)*

Cobalt-containing sodium layered oxides often exhibit good rate performance in Na cells. However, the less abundant elemental resources do not meet the demand for Na-ion batteries, and iron and/or manganese-based oxides have been extensively studied as positive electrode materials for Na-ion batteries in the past decade (Yabuuchi and Komaba 2014).

1.3.2.1. *O3-Na[Fe,M]O_2*

Electrochemical properties of iron-based multiple transition metal oxides in a Na cell were first reported by Okada and colleagues (2006). They compared electrochemical properties of O3-$NaFeO_2$, $NaFe_{1/2}Ni_{1/2}O_2$ and $NaNi_{1/2}Ti_{1/2}O_2$ in Na cells (Okada *et al.* 2006). O3-$NaFe_{1/2}Ni_{1/2}O_2$ delivers a relatively larger reversible capacity of ca. 110 mAh g^{-1} corresponding to 0.4 Na extraction from $NaFe_{1/2}Ni_{1/2}O_2$ in the voltage range of 2.0–3.8 V (Wang *et al.* 2014) compared to ca. 80 mAh g^{-1} for O3-$NaFeO_2$ (Yabuuchi *et al.* 2012b). Electronic states of Fe and Ni in O3-$NaFe_{1/2}Ni_{1/2}O_2$ are high spin (HS) Fe^{3+} ($t_{2g}^3e_g^2$) and LS Ni^{3+} ($t_{2g}^6e_g^1$) configurations respectively. Vassilaras *et al.* revealed by density functional theory (DFT) calculations and Electron Energy Loss Spectroscopy (EELS) that Fe^{3+} oxidation is responsible for most of the electrochemical charging reaction in 2.0–3.9 V versus Na with also a partial oxidation of Ni^{3+} (Vassilaras *et al.* 2017). Overcharge from 3.9 to 4.2 V results in an irreversible structural change and severe capacity fade. It indicates irreversible Fe-migration from the slab into interslab space for the desodiated $Na_xFe_{1/2}Ni_{1/2}O_2$ ($x < 0.5$). More severe capacity fade is observed for O3-Na[Fe,Ti]O_2;

actually, O3-$Na_{0.80}Fe_{0.80}Ti_{0.20}O_2$ delivers a reversible capacity of ca. 20 mAh g^{-1} with two distinct plateaus at 4.0 V on charge and at 2.0 V on discharge in the voltage range of 1.5–4.1 V (Thorne *et al.* 2014a). The obvious voltage hysteresis is also observed for O3-$NaFeO_2$ and indicates irreversible migration of Fe into the Na layer. Coexistence with Ti^{4+} or Ni^{3+} is worse for Fe^{3+} in O3-$Na_x[Fe,M]O_2$ and facilitates the irreversible structural change and Fe migration during the charging process.

$Na[Fe,Mn]O_2$ is one of the most attractive materials because of the elemental abundance of Fe and Mn in the Earth's crust (Rudnick and Gao 2014). Also, our group reported electrode performance of O3-$NaFe_{1/2}Mn_{1/2}O_2$ in a Na cell in 2012 (Yabuuchi *et al.* 2012a). Electronic states of Fe and Mn in O3-$NaFe_{1/2}Mn_{1/2}O_2$ are HS Fe^{3+} ($t_{2g}^{\ 3}e_g^{\ 2}$) and HS Mn^{3+} ($t_{2g}^{\ 3}e_g^{\ 1}$) configurations respectively. O3-$NaFe_{1/2}Mn_{1/2}O_2$ delivers a reversible capacity of 170 mAh g^{-1} based on $Mn^{3+/4+}$ redox in a low-voltage region and $Fe^{3+/4+}$ in a high-voltage region in the voltage range of 1.5–4.0 V (Figure 1.12).

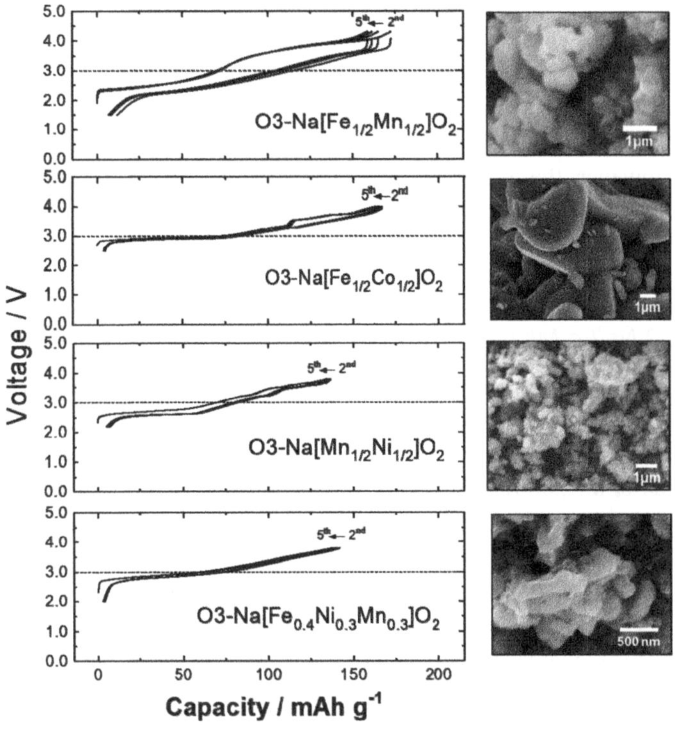

Figure 1.12. *Comparison of galvanostatic charge/discharge curves of layered O3 and O'3 type binary and ternary 3d transition metal oxides (left). Morphology of particles for each sample is also compared (right). Reprinted with permission from Kubota* et al. *(2018b). Copyright 2018, Wiley-VCH*

The O3-type phase transitions into P3 and then OP2-like phases upon charge (Yabuuchi *et al.* 2012a), which is different from those of the end-members of O3-$NaFeO_2$ and O'3-$NaMnO_2$ that show only O3 and O'3 phases during charge–discharge (Kubota *et al.* 2018b). Partial Mn substitution for Fe in O3-$NaFeO_2$ induces formation of a P3-type phase during charge and Mn-substitution would stabilize a P-type phase as estimated by calculations (Kim *et al.* 2012b). However, capacity degradation during cycles is not completely suppressed, which would be caused by migration of transition metal ions since the iron content of 50% in O3-$NaFe_{1/2}Mn_{1/2}O_2$ is higher than the critical values of 33%.

1.3.2.2. *O3-Na[Fe,Co,M]O$_2$*

Good effect of 3*d* metal coexistence is obtained for a cobalt containing material of $NaFe_{1/2}Co_{1/2}O_2$. O3-$NaFe_{1/2}Co_{1/2}O_2$ delivers a large reversible capacity of 160 mAh g^{-1} based on $Co^{3+/4+}$ redox in a low-voltage region and $Fe^{3+/4+}$ in a high-voltage region in the voltage range of 2.5–4.0 V (Figure 1.12) (Yoshida *et al.* 2013). Furthermore, relatively good cycling stability of the reversible capacities and excellent rate-capability are obtained. O3-$NaFe_{1/2}Co_{1/2}O_2$ undergoes reversible 0.7 Na extraction from the structure, implying that iron migration is suppressed by coexistence with cobalt compared to O3-type Fe, Fe-Ti, and Fe-Ni systems, but further Na extraction leads to large irreversible capacity and capacity degradation (Kubota *et al.* 2016). Further improvement and compositional modification have been conducted for O3-$NaFe_{1/2}Co_{1/2}O_2$. For examples, O3-$Na[Fe_{1/3}Co_{1/3}Ni_{1/3}]O_2$ (Vassilaras *et al.* 2014) and O3-$Na[Fe_{1/4}Co_{1/4}Ni_{1/4}Mn_{1/4}]O_2$ (Li *et al.* 2014) deliver reversible capacities of ca. 165 and 180 mAh g^{-1}, respectively, which are larger than 160 mAh g^{-1} for O3-$NaFe_{1/2}Co_{1/2}O_2$. Although Ti-substitution reduces the reversible capacity, O3-$Na[Fe_{1/4}Co_{1/4}Ni_{1/4}Mn_{1/8}Ti_{1/8}]O_2$ delivers a reversible capacity of ca. 130 mAh g^{-1} in the voltage range of 2.0–4.1 V and exhibits excellent cycle stability and rate-capability (Yue *et al.* 2015). Optimization of the composition including the iron and cobalt contents and particle morphology is expected to further enhance the electrode performances of $Na[Fe,Co,M]O_2$.

1.3.2.3. *O3-Na[Ni,Mn,M]O$_2$*

In $Na[Fe,M]O_2$, co-existence with Ni^{3+} or Mn^{3+} induces relatively negative effects for Na extraction/insertion. In contrast, Ni and Mn co-existence leads to enhanced electrode performance in Na cells. Our group first studied electrochemical properties of O3-$NaNi_{1/2}Mn_{1/2}O_2$ in 2009 (Komaba *et al.* 2009). Although strong influence of the synthesis methods, Ni and Mn are not randomly distributed in the slab in which LS Ni^{2+} ($t_{2g}^6e_g^2$) is surrounded by HS Mn^{4+} ($t_{2g}^3e_g^0$) in the first coordination shell and theoretically in a zig-zag arrangement (Breger *et al.* 2007). O3-$NaNi_{1/2}Mn_{1/2}O_2$ delivers a reversible capacity of ca. 130 mAh g^{-1} in a Na cell in the voltage range of 2.2–3.8 V, as shown in Figure 1.12, and ca. 200 mAh g^{-1} in the voltage range of 2.2–4.5 V (Komaba *et al.* 2012). Interestingly, relatively large discharge capacity is obtainable even after charging to the high voltage of 4.5 V corresponding to $x < 0.5$ in $Na_xNi_{1/2}Mn_{1/2}O_2$, which is different from the behaviors of O3-Na[Fe,M]O_2 and single 3*d* transition metal systems of O3-$NaTiO_2$, O3-$NaVO_2$, O3-$NaCrO_2$ and O3-$NaFeO_2$. O3-$NaMn_{1/2}Ni_{1/2}O_2$ transforms in the O3 → O'3 → P3 → P'3 → O3 sequence as Na is extracted during charge up to 4.5 V (Komaba *et al.* 2012; Mariyappan *et al.* 2018a). The P'3 → O3 transition in the charge voltage plateau region at 4.2 V versus Na involves significant shrinkage in the interslab distance ($\Delta d_{00l}/d_{00l}$ ~18% from pristine) and is believed to cause gradual capacity fading during cycles. Furthermore, potential jumps are observed at $x = 1/2$ and 1/3 in $Na_xNi_{1/2}Mn_{1/2}O_2$ during charge and discharge, which correspond to in-plane Na/vacancy orderings in interslab space (Vinckeviciute *et al.* 2016). Much effort has been devoted to suppression of the interslab shrinkage and Na/vacancy ordering.

Kim *et al.* first reported O3-$NaNi_{1/3}Fe_{1/3}Mn_{1/3}O_2$ in 2012 (Kim *et al.* 2012a) and our group systematically investigated and reported Na battery performance in O3 $NaNi_{1/2}Mn_{1/2}O_2$-$NaFeO_2$ solid solution in 2013 (Yabuuchi *et al.* 2013). O3-$Na(Ni_{1/2}Mn_{1/2})_{1-x}Fe_xO_2$ (x = 0.4 and 0.6) deliver reversible capacities of ca. 140 mAh g^{-1} with small polarization and smooth voltage curves in the relatively narrow voltage range of 2.0–3.8 V, as shown in Figure 1.12. Wang *et al.* fabricated an 1 Ah soft-packed Na-ion full cell and demonstrated good cycling stability with capacity retention over 73% after 500 cycles at the 1 C rate (Wang *et al.* 2016). Operando XRD patterns and *ex situ* XAS spectra of O3-$NaNi_{1/3}Fe_{1/3}Mn_{1/3}O_2$ revealed reversible phase transitions in the O3 → P3 sequence and only the $Ni^{2+/3+}$ oxidation in the range of 3.5–4.0 V and $Ni^{3+/4+}$ followed by $Fe^{3+/4+}$ in the range of 4.0–4.3 V during charge (Xie *et al.* 2016). Our group further studied effect of partial substitution of Ti for Mn in O3-$NaNi_{0.3}Fe_{0.4}Mn_{0.3}O_2$ (Komaba *et al.* 2014). O3-$NaNi_{0.3}Fe_{0.4}Mn_{0.3-x}Ti_xO_2$ delivers almost the same reversible capacity of ca. 130 mAh g^{-1} in 2.0–3.8 V in Na cells. The average operating voltage, however, raises up with increasing Ti content, x in O3-$NaNi_{0.3}Fe_{0.4}Mn_{0.3-x}Ti_xO_2$. Sun *et al.* (2014) also synthesized O3-$NaNi_{0.4}Fe_{0.2}Mn_{0.4-x}Ti_xO_2$ and O3-$NaNi_{0.4}Fe_{0.2}Mn_{0.2}Ti_{0.2}O_2$ exhibits superior cycling stability to that of O3-$NaNi_{0.4}Fe_{0.2}Mn_{0.4}O_2$. In addition to the Fe-substituted O3-Na[Mn,Ni]O_2, Qi *et al.* (2016), Zheng and Obrovac (2017), and Wang *et al.* (2017b) reported Ti-substituted one and the electrochemical performance of the Na cells. O3-Na_x[Mn,Ni,Ti]O_2 delivers almost the same discharge capacity of 120–135 mAh.g^{-1}: in general we are using a as O3-$NaMn_{1/2}Ni_{1/2}O_2$ (Wang *et al.* 2017b). The Ti-substitution changes the voltage profiles from stepwise into smooth ones and enhances capacity retention. Qi *et al.* (2016) and Yao *et al.* (2017) anticipated that Ti-substitution disturbs both the Na/vacancy ordering in the interslab space and the Ni^{2+}/Mn^{4+} ordering in the slab due to the substantial difference in Fermi level between Ti^{4+} and Ni^{2+}/Mn^{4+} and decrease in the electronic localization as Wang *et al.* (2015) proposed in P2-$Na_{0.6}Cr_{0.6}Ti_{0.4}O_2$. Mariyappan *et al.* (2018a) proposed that substitution of HS Mn^{4+} ($3d^3$; $t_{2g}{}^3e_g{}^0$) in O3-$NaMn_{1/2}Ni_{1/2}O_2$ by Ti^{4+} ($3d^0$; $t_{2g}{}^0e_g{}^0$) increases the electronic density on oxygen and enlarges the energy difference between Ni 3*d* and O 2p orbitals, leading to an increase in the M-O bond ionicity and redox potential. Note that higher Ti content leads to the smoother voltage curves and the higher average charge and discharge voltage, however voltage hysteresis increases (Zheng and Obrovac 2017). Furthermore, metal substitution was conducted to O3-Na_x[Mn, Ni, Ti]O_2 by Guo's (Yao *et al.* 2017), Deng's (Wang *et al.* 2019a) and Tarascon's (Wang *et al.* 2019b; Mariyappan *et al.* 2020) groups. Substitution by Ti^{4+} ($3d^0$; $t_{2g}{}^0e_g{}^0$) for HS Mn^{4+} ($3d^3$; $t_{2g}{}^3e_g{}^0$) and divalent metals such as Cu^{2+} ($3d^9$; $t_{2g}{}^6e_g{}^3$) or Zn^{2+} ($3d^{10}$; $t_{2g}{}^6e_g{}^4$) for LS Ni^{2+} ($3d^8$; $t_{2g}{}^6e_g{}^2$) in O3-$NaMn_{1/2}Ni_{1/2}O_2$ achieve excellent capacity retention and rate capability in the range of 2.0–4.0 V (Yao *et al.* 2017) and even in the range of 2.0–4.5 V (Wang *et al.* 2019b). O3-$NaNi_{1/2-y}Cu_yMn_{1/2-z}Ti_zO_2$ (y = 0, 0.05, 0.1; z = 0.1, 0.2) solid solution phases deliver reversible capacities of ca. 125 and 200 mAh g^{-1} in 2.0–4.0 V and 2.0–4.5 V, respectively, with smooth voltage profiles. Structures of O3-$NaNi_{0.4}Cu_{0.1}Mn_{0.4}Ti_{0.1}O_2$ are evolving in the O3 → O'3 → P3 → P'3 sequence by Na extraction during charging to 4.0 V (Yao *et al.* 2017; Wang *et al.* 2019b) and then from P'3 → P3-O3-O1 intergrowth by charging from 4.0 to 4.5 V, which is confirmed by operando XRD patterns and ED patterns as well as high-angle annular dark field (HAADF) STEM images (Wang *et al.* 2019b; Mariyappan *et al.* 2020). Wang *et al.* revealed with the HAADF-STEM images that the O1-type phase (Figure 1.5), having migrated transition metal cations in the interlayer octahedral interstices of ~5.1 Å in distance, randomly co-exist with P3 domains having a wider interlayer space with ~7.1 Å. The nucleation of the O1 stacking domains in the P3 ones and the significantly different interlayer distances reduce the lattice contraction and overall lattice changes, thus improving the cycle life (Wang *et al.* 2019b). Furthermore, DFT calculations estimate that the Cu and Zn substitution in O3-Na_x[Mn,Ni,Ti]O_2 decreases in the energy difference between P and O stackings and leads to a continuous transition between P- and O-stacking phases, explaining the sloping solid solution–like charging profile above 4.0 V (Wang *et al.* 2019b) unlike the flat charging voltage profiles corresponding to the biphasic P3 → O3 transition in O3-$NaMn_{1/2}Ni_{1/2}O_2$ (Figure 1.12). Tarascon's group fabricated the prototype 18 650 cells of the O3-phase positive electrode with a hard carbon negative electrode and demonstrated a reversible specific capacity of ca. 154 mAh g^{-1} with an operating voltage of ca. 3.1 V with gravimetric and volumetric energy densities of ca. 115 Wh kg^{-1} and ca. 250 Wh L^{-1} for the total cell weight and volume, respectively, comparable or slightly superior to the polyanionic $Na_3V_2(PO_4)_2F_3$||hard carbon cells (100 Wh kg^{-1}, 175 Wh L^{-1}) because of the higher tapped density of the layered oxide material (1.9 g cm^{-3}) than 1.1 g cm^{-3} of $Na_3V_2(PO_4)_2F_3$ (Mariyappan *et al.* 2020).

O3-type binary, ternary and quaternary transition metal systems exhibit a larger reversible capacity, better rate performance and capacity retention with smooth voltage profiles in Na cells compared to those of the single transition metal system. Most of the O3-type multiple transition metal systems reported so far overcome the severe issues for the practical use faced by the single transition metal systems: (1) an irreversible structural change related to the migration of transition metals into interlayers during the charging process; (2) stepwise voltage profiles originating from Na/vacancy orderings and phase transitions; (3) slow kinetics of Na (de)insertion; and (2) less moist air stability (Sathiya *et al.* 2012). Lowering the content of migrating cations such as Fe^{3+} reduces probability of the migration and enhances structural stability during charge and discharge (Li *et al.* 2016). The substitution by both Ti^{4+} and divalent cations such as Cu^{2+} and Zn^{2+} leads to sloping voltage profiles and reduces the number of phase transitions (Mariyappan *et al.* 2020). Furthermore, Ti^{4+}-substitution seems to stabilize a P3-type phase having a large interslab space (Mariyappan *et al.* 2018a), in which Na^+ ions at prismatic sites directly migrate to a neighbor face-shared prismatic site through the wide rectangle bottleneck in P3 and P'3 structures, suggesting high Na^+ diffusion in the P3-type stacking structure. Thus, the P3-type phase exhibits high Na^+ diffusion, and a wider compositional range for P3-type $Na_xNi_{0.4}Cu_{0.1}Mn_{0.4}Ti_{0.1}O_2$ stabilized by the metal-substitution leads to excellent rate capability (Yao *et al.* 2017). Consequently, O3-$NaNi_{0.4}Cu_{0.1}Mn_{0.4}Ti_{0.1}O_2$ also has higher moist air stability than that of the non-substituted O3-$NaMn_{1/2}Ni_{1/2}O_2$ (Yao *et al.* 2017).

1.3.3. *Moist air stability of O3-NaMO$_2$ and surface coating*

Most O3-type $NaMO_2$ are moisture sensitive and the stability of O3-$NaMO_2$ against a moist environment and the easiness of handling are important for practical applications (Kubota and Komaba 2015; Zhang *et al.* 2019). During exposure to moist air, the water molecules react with air-instable Na_xMO_2 by (1) H_2O insertion into the interlayer spacing between MO_2 slab and the Na layer (Park *et al.* 2007), (2) oxidation of transition metal ions with the simultaneous removal of Na^+ ions (Huang *et al.* 2005; Monyoncho and Bissessur 2013) and (3) H^+/Na^+ exchange (Blesa *et al.* 1993). (1) H_2O insertion is often observed for Na-deficient phases such as O3-related $Na_{0.3}NiO_2$ and P2-type Na_xMO_2 (Legoff *et al.* 1993; Paulsen and Dahn 1999) and leads to the expansion of the interlayer spacing and then to exfoliation and crack generation at the particle surface after a drying process for cell assembly. Na removal in the (2) and (3) processes results in the further reaction with CO_2 and H_2O in air to form NaOH, $NaHCO_3 \cdot nH_2O$, $Na_2CO_3 \cdot nH_2O$, etc. (Huang *et al.* 2005; Monyoncho and Bissessur 2013) and the sodium salts containing water are formed on the particle surface, which was confirmed by You *et al.* with subnanometer surface-sensitive time-of-flight secondary ion mass spectroscopy for O3-$NaNi_{0.7}Mn_{0.15}Co_{0.15}O_2$ (You *et al.* 2019). They further found that nickel ions gradually dissolve from the bulk to form NiO and accumulate on the particle surface. The accumulated species on the surface are insulating and lead to an increase in interfacial resistance and deteriorated electrochemical performance (You *et al.* 2019). Thus, surface coating is very effective and some researchers applied to O3-type materials. Carbon-coated O3-$NaCrO_2$ (Yu *et al.* 2015), Al_2O_3-coated O3-$Na[Ni_{0.6}Co_{0.2}Mn_{0.2}]O_2$ (Hwang *et al.* 2017a) and ZrO_2-coated O3-$NaNi_{0.7}Mn_{0.15}Co_{0.15}O_2$ (You *et al.* 2019) have demonstrated superior rate performance and suppression of the surface degradation owing to the inert carbon, Al_2O_3 and ZrO_2 layers. Hwang *et al.* (2017a) revealed that ca. 20-nm-thick Al_2O_3 coating on the periphery of the primary O3-$Na[Ni_{0.6}Co_{0.2}Mn_{0.2}]O_2$ particles remained intact and served as a protective layer against HF attack in the cell. NaF is deposited on bared O3-$Na[Ni_{0.6}Co_{0.2}Mn_{0.2}]O_2$ particles during charge–discharge cycles, resulting in thick layer on the surface evidenced with X-ray photoelectron spectroscopy (XPS). Al_2O_3 coating successfully suppresses formation of a thick NaF deposition layer on the surface of the O3 materials due to HF scavenging in the electrolyte to form AlF_3 by the proposed following reaction:

$$Al_2O_3 + 6HF \rightarrow 2AlF_3 + 3H_2O \qquad [1.1]$$

Formation of an AlF_3 layer on the outermost surface of the Al_2O_3 coating protects the O3 materials and improves rate performance, coulombic efficiency and capacity retention of the O3-type materials. In order to obtain a large reversible capacity by charging to high voltage, surface protection of O3-type materials is necessary to capture HF generated from the electrolyte as well as to supress the reaction with moist air, leading to superior electrode performance in Na cells.

Instead of the surface coating process, air-stable O3-type materials have been extensively studied recently. According to the discovery of air-stable P2-$Na_{7/9}[Cu_{2/9}Fe_{1/9}Mn_{2/3}]O_2$, Li *et al.* (2015) also synthesized air-stable O3-$Na_{0.9}[Fe_{0.30}Mn_{0.48}Cu_{0.22}]O_2$ (Mu *et al.* 2015) and then O3-$Na[Li_{0.05}Mn_{0.50}Ni_{0.30}Cu_{0.10}Mg_{0.05}]O_2$ (Deng *et al.* 2018), O3-$Na[Ni_{0.45}Cu_{0.05}Mn_{0.4}Ti_{0.1}]O_2$ (Yao *et al.* 2017) and O3-$NaNi_{0.45}Zn_{0.05}Mn_{0.4}Ti_{0.1}O_2$ (Mariyappan *et al.* 2020). Although the HF scavenging property of the air-stable materials is unknown, the air-stable surface leads to the suppressed surface degradation in contact with moist air and results in the stable surface even during charge–discharge (You *et al.* 2019). All these air-stable materials contain divalent metal ions such as Mg^{2+}, Cu^{2+} and Zn^{2+}, indicating that substitution of ionic divalent-metal ions is effective to enhance the moist air–resistant property. Selection of doping metals as well as transition metals is, therefore, important in designing ternary and quaternary 3*d* transition metal systems to enhance the air stability, suppressing the surface degradation in contact with moist air and to achieve excellent electrochemical performances with smooth voltage profiles without irreversible phase transitions during charge and discharge processes.

1.4. P2-type layered materials

1.4.1. *Practical issues of P2-type materials for Na-ion batteries*

Similar to the O3-type layered sodium 3*d* transition metal oxides (including O'3-type ones), P2-type sodium 3*d* transition metal oxides have been developed for achieving high-capacity, high-power, long-life and safe Na-ion batteries: (1) increase in a reversible capacity and operation voltage, (2) suppression of significant interslab shrinkage and irreversible phase transitions, (3) smoothing voltage profiles with suppressed Na^+/vacancy ordering, (4) suppressing insertion of water molecules to interslab spacing and (5) surface protection against moist air and HF attack. However, there is a significant difference between P2- and O3-type layered oxides with single 3*d* transition metal systems. In contrast to O3-type layered oxides in which most of 3*d* transition metals are able to be accommodated in the slab, P2 type phases crystallize with only V, Mn and Co as single 3*d* transition metal systems. P2-type materials are generally Na deficient and crystallize in the Na content of $0.5 \leq x \leq 0.8$ in Na_xMO_2. Many researchers synthesize P2-type materials in the typical composition of $Na_{2/3}MO_2$ ($Na_{2/3}M^{3+}{}_{2/3}M^{4+}{}_{1/3}O_2$). The average valence of transition metals in $Na_{2/3}MO_2$ is +3.33 and tetravalent or higher valent transition metals are contained in the slab of a P2-type structure. Thus, P2-type materials are also able to contain various 3*d* and 4*d* transition metals, alkali metals and alkaline earth metals in the binary, ternary and quaternary systems. Furthermore, the influence of the different oxygen stacking of O3 and P2 types on the charge–discharge voltage is very small, as shown in Figure 1.13 (Kubota *et al.* 2018a), except for the high Na content region of $0.0 \leq x \leq 1/3$ in $Na_{1-x}MO_2$. Thus, the strategy and knowledge of the multiple transition metal systems with inert-metal doping or surface coating, which are developed in O3-type materials, can be applied to P2-type materials and also from P2- to O3-type materials. Because of the specific oxygen stacking of the P2-type materials and unstable occupation of transition metal ions at the octahedral interslab sites, migration of transition metal ions is suppressed even in the high voltage region above 4 V. Thus, P2-type materials are usually charged to >4.0 V in order to increase the working voltage.

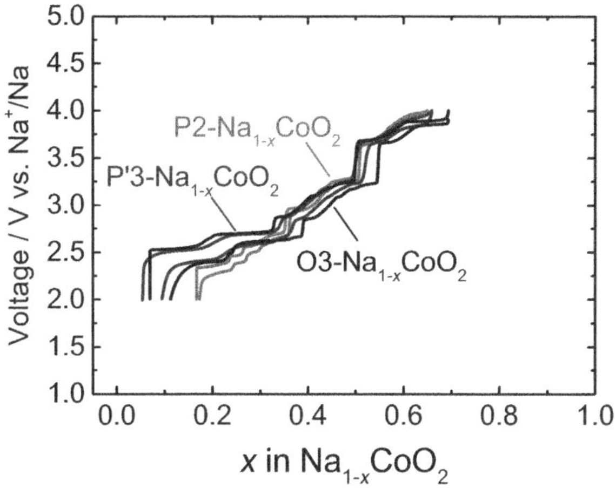

Figure 1.13. *Charge/discharge curves of P2, P'3 and O3 type Na_xCoO_2 in Na cells. Reprinted with permission from Kubota* et al. *(2018a). Copyright 2018, Wiley-VCH. For a color version of this figure, see www.iste.co.uk/monconduit/batteries.zip*

It should be noted that P2-$Na_{2/3}MnO_{2+y}$ ($Na_{4/(6+3y)}Mn_{2/(2+y)}O_2$, $0.05 \leq y \leq 0.25$) has higher oxidation state of Mn than that of P'2-$Na_{2/3}MnO_2$ and Mn defects in the slab without lattice distortion (Stoyanova *et al.* 2010; Kumakura *et al.* 2016), which significantly influences the electrode performances. Our group recently reported that P2-$Na_{0.59}Mn_{0.90}O_2$ having Mn defects in the slab without lattice distortion shows smooth charge/discharge voltage profiles and delivers an initial discharge capacity of 198 mAh g^{-1} in the range of 1.5–4.4 V. However, the capacity apparently decays during cycles, which is probably influenced by the Mn defects. On the other hand, P'2-$Na_{0.64}MnO_2$ ($Na_{0.64}Mn^{3+}{}_{0.63}Mn^{4+}{}_{0.37}O_2$) has lattice distortion caused by a higher content of Jahn–Teller active Mn^{3+} in the structure with the Mn-defect-free slab unlike P2-$Na_{0.59}Mn_{0.90}O_2$. P'2-$Na_{0.64}MnO_2$ shows stepwise voltage profiles and delivers discharge capacity of 216 mAh g^{-1} in the range of 1.5–4.4 V. The stepwise voltage behaviors are mainly derived from Na^+/vacancy ordering related with Jahn–Teller distortion (Kumakura *et al.* 2016). Jahn–Teller distortion is regarded as a detrimental factor to the battery performances of Li- and Na-ion batteries due to local strains and irreversible structural changes during charge/discharge and Jahn–Teller active Mn^{3+} is often avoided (Clement *et al.* 2016). However, P'2-$Na_{0.64}MnO_2$ having lattice distortion at initial state exhibits superior capacity retention and highly reversible phase transitions, which is quite interesting for upsetting conventional chemistry in the battery field. Li *et al.* (2016) also recently proposed beneficial assist of Jahn–Teller active Fe^{4+} ion enhancing Na diffusion at the end of charge, leading to further Na extraction. These inexpensive transition metals of Mn and Fe are available as a redox center in layered oxides for Na-ion batteries unlike Li-ion. Further new chemistry different from Li-ion is expected to appear in Na-ion chemistry.

1.4.2. P2-$Na_{2/3}[Mn,Co,M]O_2$

As mentioned above, P2-type single 3*d* transition metal systems are found in only P2-$Na_{2/3}VO_2$, P2-$Na_{0.6}Mn_{0.9}O_2$, P'2-$Na_{2/3}MnO_2$ and P2-$Na_{2/3}CoO_2$. Since Na_xMnO_2 delivers the largest reversible capacities of >200 mAh g^{-1} among single 3*d* transition metal P2-type systems (Kumakura *et al.* 2016), various solid solutions are synthesized based on P'2-$Na_{2/3}MnO_2$ and P2-$Na_{2/3}CoO_2$ as parent oxides and their Na (de)intercalation properties have been extensively investigated in Na cells.

A solid solution between P'2-$Na_{2/3}MnO_2$ and P2-$Na_{2/3}CoO_2$ is crystallized in any ratio of Mn/Co (Paulsen and Dahn 1999; Wang *et al.* 2013). From the late 1990s to the early 2000s, the solid solution was studied as positive electrode materials for Li-ion batteries after Li^+/Na^+ exchange (Paulsen and Dahn 1999; Lu and Dahn 2001b; Eriksson *et al.* 2003). Na intercalation properties of the sodium solid solution were first reported in P2-$Na_{2/3}Mn_{1/3}Co_{2/3}O_2$ by Carlier *et al.* (2011). P2-$Na_{2/3}Mn_{1/3}Co_{2/3}O_2$ delivers a reversible capacity of 112 mAh g^{-1} in the range of 1.5–4.0 V, as shown in Figure 1.14. The voltage profiles of P2-$Na_{2/3}Mn_{1/3}Co_{2/3}O_2$ are relatively smooth except for the voltage jump at x = 2/3 and 1/2 in P2-$Na_xMn_{1/3}Co_{2/3}O_2$ compared to the end-members of P'2-$Na_{2/3}MnO_2$ and P2-$Na_{2/3}CoO_2$, suggesting that the phase transitions and Na^+/vacancy ordering in the Mn/Co-disordered phase are of less significance (Beck *et al.* 2014). According to XAS data of $Na_{2/3}Mn_{1/3}Co_{2/3}O_2$ during charge/discharge in Na cells (Cheng *et al.* 2014), LS Co^{3+} is oxidized into LS Co^{4+} by Na extraction and HS Mn^{4+} is maintained during charging to 4.0 V. On discharge to 2.5 V, LS Co^{4+} is reversibly reduced into LS Co^{3+}. P2-$Na_{2/3}Mn_{1/3}Co_{2/3}O_2$ can accommodate more Na in the structure and both HS Mn^{4+} and LS Co^{4+} were reported to reduce into HS Mn^{3+} and HS Co^{2+}, respectively, with major modification of hybridization between M 3*d* and O 2p orbitals by Na insertion on discharge below 2.5 V. The electronic structures significantly change with the formation of Jahn–Teller active Mn^{3+} and the reduction from LS-Co^{3+} to HS-Co^{2+} on discharge, resulting in much lower voltage than P2-$Na_{2/3}CoO_2$ below 2.5 V in the Na cells (Cheng *et al.* 2014; Watanabe *et al.* 2019). The extra 0.16 Na insertion into $Na_{2/3}Mn_{1/3}Co_{2/3}O_2$ and the large reversible capacity on discharging from 2.5 to 1.25 V is attractive, but Na compensation is necessary for the practical use. Even for P2-$Na_{2/3}[Mn,Co,Ni]O_2$, the low working voltage and indispensableness of external Na compensation are drawbacks.

1.4.3. *P2-$Na_{2/3}[Mn,Fe,M]O_2$*

A solid solution of P2-$Na_{2/3}[Mn,Fe]O_2$ was also synthesized and the battery performances were investigated in Li cells in the early 2000s (Eriksson *et al.* 2003; Matsumura *et al.* 2003). Our group evaluated Na (de)intercalation properties of P2-$Na_{2/3}Fe_{1/2}Mn_{1/2}O_2$ and P2-$Na_{2/3}Fe_{1/2}Mn_{1/2}O_2$ deliver a reversible capacity of ca. 190 mAh g^{-1} and exhibit good rate performance in Na cells in the range of 1.5–4.3 V (Yabuuchi *et al.* 2012a), as shown in Figure 1.14. P2-$Na_{2/3}Fe_{1/2}Mn_{1/2}O_2$ consists only of the abundant elements in the Earth's crust without cobalt and shows acceptable capacity retention different from O3-$NaFeO_2$, which is probably due to less or reversible Fe migration into interslab space. The P2-$Na_{2/3}Fe_{1/2}Mn_{1/2}O_2$ phase transforms into an OP4-like phase by Na extraction on charge to 4.3 V and tetrahedral sites exist in the O-type interslab space. P2-$Na_{2/3}Fe_{1/2}Mn_{1/2}O_2$ shows capacity decay during cycles, which is thought to be caused by irreversible migration of transition metals into the interslab space. Talaie *et al.* successfully detected migrated transition metals at tetrahedral sites in the interslab space by pair distribution function (PDF) analysis on the XRD patterns of chemically oxidized $Na_{0.1}Fe_{1/2}Mn_{1/2}O_2$, $Na_{0.15}Fe_{1/2}Mn_{1/2}O_2$, $Na_{0.15}Fe_{0.3}Mn_{0.6}Ni_{0.1}O_2$, $Na_{0.15}Fe_{0.2}Mn_{0.65}Ni_{0.15}O_2$ and $Na_{0.1}Fe_{0.2}Mn_{0.66}Cu_{0.14}O_2$ (Talaie *et al.* 2015, 2017). The amount of Fe migration is higher in $Na_xFe_{1/2}Mn_{1/2}O_2$ than in Cu and Ni doped materials. Indeed, cycling stability of the doped samples is better than that of P2-$Na_{2/3}Fe_{1/2}Mn_{1/2}O_2$. The P2-$Na_{2/3}Fe_{1/2}Mn_{1/2}O_2$ sample, however, generally contains crystal domains of O3-$NaFe_{1/2}Mn_{1/2}O_2$, leading to broad 10*l* Bragg diffraction peaks in the XRD (Yabuuchi *et al.* 2012a). The difficulty of synthesizing pure P2-$Na_{2/3}Fe_{1/2}Mn_{1/2}O_2$ complicates our understanding of the reaction mechanism. As Talaie *et al.* also mentioned in the literature, reversible charge–discharge behaviors are observed for P2-type materials in Na cells even if transition metals are migrating to interslab space, which is quite different from O3-$NaFeO_2$ and O3-$NaCrO_2$ (Figure 1.8). Mortemard de Boisse *et al.* (2014) confirmed well-crystallized phases and no significant structural evolution before and after sodium extraction/insertion for P2-$Na_xFe_{1/2}Mn_{1/2}O_2$. Except for the capacity decay, P2-$Na_{2/3}Fe_{1/2}Mn_{1/2}O_2$ has some drawbacks: (1) voltage hysteresis in between charge and discharge as seen above 3 V in Figure 1.14, (2) Na deficiency in the as-prepared state and (3) hygroscopic property like O3-type materials. First, the significant voltage hysteresis appears after charging beyond 4.1 V (Talaie *et al.* 2015). Despite extensive research, the mechanism of the voltage hysteresis has not been elucidated yet. Redox reaction of oxide ion (Talaie *et al.* 2015) and P2/OP4-like phase transition accompanied by formation of stacking faults (Mortemard de Boisse *et al.* 2014) and by migration of transition metals (Talaie *et al.* 2015) are estimated to cause the voltage hysteresis. Second, Na deficiency is a considerable issue in all P2-type materials showing capacity below 2.5 V and the large reversible capacity is achieved by Na compensation from external Na sources such as Na metal counter electrode in a half-cell and pretreatment with Na metal by ball milling (Zhang *et al.* 2016), but Na metal should be avoided in the Na-ion battery full cells and synthesis process due to safety issues. Recently, the addition of sacrificial Na sources such as NaN_3 and Na_2CO_3 is proposed to compensate for Na in the full cells without Na metal (Singh *et al.* 2013; Martinez De Ilarduya *et al.* 2017; Sathiya *et al.* 2017). Third, as discussed in the hygroscopic property of O3-type materials above, P2-type materials are also generally hygroscopic except for P2-$Na_{2/3}Ni_{1/3-x}Cu_xMn_{2/3}O_2$ ($0 \leq x \leq 1$) (Lu and Dahn 2001b; Zheng *et al.* 2017). Air stable P2-$Na_{2/3}$[Mn,Fe]O_2 is achieved by Cu doping (Li *et al.* 2015; Talaie *et al.* 2017). Doping with Cu^{2+} and alkaline earth metals would be effective to enhance the water- and moisture-resistant properties of P2-type materials. Ternary and quaternary Fe–Mn-based P2-type materials such as P2-Na_x[Fe,Mn,Ti]O_2 (Han *et al.* 2016), P2-Na_x[Fe,Mn,Co]O_2 (Thorne *et al.* 2014b), P2-Na_x[Fe,Mn,Ni]O_2 (Talaie *et al.* 2015; Bai *et al.* 2016), P2-Na_x[Fe,Mn,Co,Ni]O_2 (Kaithwas and Kundu 2015; Li *et al.* 2017) and P2-Na_x[Fe,Mn,Cu]O_2 (Li *et al.* 2015; Talaie *et al.* 2017) have also been extensively studied and Fe in P2-type materials would also assist Na diffusion at the end of charge due to the flexibility of Jahn–Teller active Fe^{4+} (Liu *et al.* 2015) as observed in O3-type materials (Li *et al.* 2016). The metal doping in P2-Na_x[Mn,Fe]O_2 suppresses the voltage hysteresis, but the Na compensation and raising working voltage are challenging.

1.4.4. *P2-$Na_{2/3}$[Ni,Mn,M]O_2*

Similar to P2-Na_x[Mn, Fe]O_2, a solid solution of P2-$Na_{2/3}$[Ni,Mn]O_2 was also studied as a precursor to synthesize a positive electrode material in Li cells by means of Li^+/Na^+ exchange by Dahn's group from the late 1990s to the early 2000s (Paulsen and Dahn 2000; Paulsen *et al.* 2000). Interestingly, the Na battery performance was examined with Na cells and phase transitions were investigated using *in situ* XRD in 2001 (Lu and Dahn 2001a). P2-$Na_{2/3}Ni_{1/3}Mn_{2/3}O_2$ delivers a reversible capacity of ca. 150 mAh g^{-1} with stepwise voltage profiles in the range of 2.0–4.5 V, as shown in Figure 1.14. The average working voltage is one of the highest (ca. 3.6 V) among O3- and P2-type materials. However, the reversible capacity rapidly deteriorates during cycles (Lu and Dahn 2001a; Lee *et al.* 2013). The capacity decay mechanism is explained by phase transition accompanied by significant shrinkage of interslab distance by Na extraction (Yoshida *et al.* 2014). P2-$Na_{2/3}Ni_{1/3}Mn_{2/3}O_2$ transforms into O2-type $Ni_{1/3}Mn_{2/3}O_2$ by extraction of almost all Na on charge through the two-phase reaction above 4.1 V (Lu and Dahn 2001a) and the P2–O2 phase transition is accompanied by huge volume shrinkage of 23% mainly attributed to shrinkage of the interslab distance (Lee *et al.* 2013; Yoshida *et al.* 2014). The volume change significantly influences battery performances such as capacity retention and rate performance in Li and Na batteries and is recently reported to be suppressed by partial metal doping in the $Ni_{1/3}Mn_{2/3}O_2$ slab. For example, our group synthesized P2-$Na_{2/3}[Ni_{1/3}Mn_{1/2}Ti_{1/6}]O_2$ and the Ti-substituted phase delivers a reversible capacity of 127 mAh g^{-1} with smooth voltage curves as shown in Figure 1.14 (Yoshida *et al.* 2014). The volume change is successfully suppressed into 12–13% by Ti substitution. Our group systematically compared influence of inert-metal substitution in P2-$Na_{2/3}[Ni_{1/3}Mn_{2/3}]O_2$ and revealed that Al^{3+} or Ti^{4+} substitution successfully enhanced the cycling stability and rate performances (Kubota *et al.* 2018b). Multiple substitution by inert metals such as Cu^{2+}-Ti^{4+} coupling (Mu *et al.* 2019) is expected to further improve the electrochemical performances as found for O3-$Na[Ni_{1/2}Mn_{1/2}]O_2$.

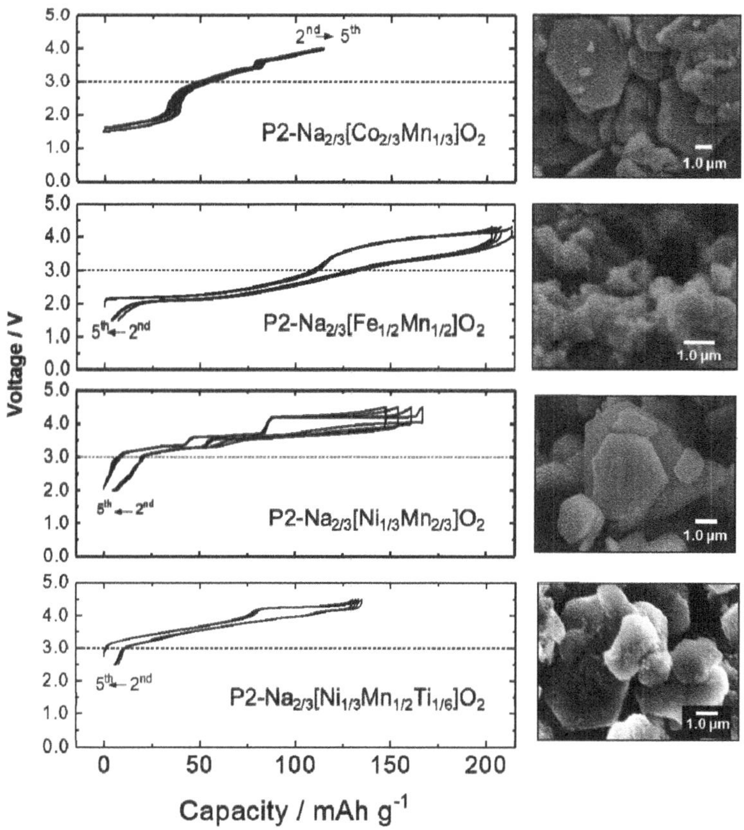

Figure 1.14. *Comparison of galvanostatic charge/discharge curves of layered P2 type binary and ternary 3d transition metal oxides (left). Morphology of particles for each sample is also compared (right)*

1.5. Summary and prospects

Partial substitution of various metals for 3*d* transition metals in sodium layered transition metal oxides is efficient to modulate their electrochemical properties. An unique variety of sodium layered oxides is attractive unlike lithium layered transition metal oxides. Actually, selection of transition metals and dopant elements controls the crystal structure, electronic/ionic conductivity, moist air–resistant property, phase transitions during Na (de)intercalation and surface reactivity with the electrolyte.

This chapter's role of mainly reviewing the 3*d* transition metal elements will help to further enhance electrode performance of layered sodium transition metal oxides as positive electrode materials for Na-ion batteries. From previous studies on Li-ion, Na-ion and K-ion batteries, not only the study on positive electrode materials but also comprehensive studies with negative electrode materials, current collectors, binders, surface coating and concentration gradient of active materials, and electrolyte salts and solvents, etc., are also required to enhance overall performances of Na-ion batteries. Even if a huge demand of lithium resources is actualized for the Li-ion batteries in transportation applications and there is a risk of undersupply of lithium resources, which are distributed unevenly in the Earth's crust, further developments to enhance not only energy density but also cycle life and safety are, of course, required for the stationary use of Na-ion batteries. Furthermore, unique and featured advantages over Li-ion have been found and will be further desired for practical performance of future Na-ion batteries.

The advantages of lower Lewis acidity, weaker Coulombic interaction and smaller crystal ion size for Na^+ than Li^+ ions are the key to maximizing the characteristics of Na-ion batteries, whereas the hygroscopic property of sodium transition metal oxides is similar to that of $LiNiO_2$-based materials. As development of high capacity $LiNiO_2$-based materials is desired in next-generation Li-ion batteries, new findings and chemistry to control the hygroscopic issue are still being developed (Bianchini *et al.* 2019).

Similar progress is also sometimes expected in layered sodium transition metal oxides such as the hygroscopic property and coverage of alkali metal carbonate on the oxides disturbing phase transitions, possibly, at least until the 2040s. Therefore, interactive research and developments between Li-ion and Na-ion (possibly K-ion) batteries are believed to lead to a positive effect on the fundamental understanding and realization of sustainable energy technology.

1.6. Acknowledgments

This study is in part supported by Japan Science and Technology Agency (JST) through Adaptable and Seamless Technology Transfer Program (A-STEP) and by MEXT program "Elements Strategy Initiative to Form Core Research Center" (No. JPMXP0112101003).

1.7. References

Abakumov, A.M., Tsirlin, A.A., Bakaimi, I., Van Tendeloo, G., and Lappas, A. (2014). Multiple twinning as a structure directing mechanism in layered rock-salt-type oxides: $NaMnO_2$ polymorphism, redox potentials, and magnetism. *Chemistry of Materials,* 26(10), 3306–3315.

Adamczyk, E. and Pralong, V. (2017). $Na_2Mn_3O_7$: A suitable electrode material for Na-ion batteries? *Chemistry of Materials,* 29(11), 4645–4648.

Ado, K., Tabuchi, M., Kobayashi, H., Kageyama, H., Nakamura, O., Inaba, Y., Kanno, R., Takagi, M., and Takeda, Y. (1997). Preparation of $LiFeO_2$ with alpha-$NaFeO_2$-type structure using a mixed-alkaline hydrothermal method. *Journal of the Electrochemical Society,* 144(7), L177–L180.

Amatucci, G.G., Tarascon, J.M., and Klein, L.C. (1996). CoO_2, the end member of the $LixCoO_2$ solid solution. *Journal of the Electrochemical Society,* 143(3), 1114–1123.

Andersson, S. and Wadsley, A.D. (1962). $Na_XTi_4O_8$, an alkali metal titanium dioxide bronze. *Acta Crystallographica,* 15(3), 201–206.

Bai, Y., Zhao, L.X., Wu, C., Li, H., Li, Y., and Wu, F. (2016). Enhanced sodium ion storage behavior of P2-type $Na_{2/3}Fe_{1/2}Mn_{1/2}O_2$ synthesized via a chelating agent assisted route. *Acs Applied Materials & Interfaces,* 8(4), 2857–2865.

Barpanda, P., Oyama, G., Nishimura, S., Chung, S.C., and Yamada, A. (2014). A 3.8-V earth-abundant sodium battery electrode. *Nat. Commun.,* 5, 4358.

Beck, F.R., Cheng, Y.Q., Bi, Z.H., Feygenson, M., Bridges, C.A., Moorhead-Rosenberg, Z., Manthiram, A., Goodenough, J.B., Paranthaman, M.P., and Manivannan, A. (2014). Neutron diffraction and electrochemical studies of $Na_{0.79}CoO_2$ and $Na_{0.79}Co_{0.7}Mn_{0.3}O_2$ cathodes for sodium-ion batteries. *Journal of the Electrochemical Society,* 161(6), A961–A967.

Berthelot, R., Carlier, D., and Delmas, C. (2011). Electrochemical investigation of the P2-$NaxCoO_2$ phase diagram. *Nat Mater,* 10(1), 74–80.

Bianchini, M., Roca-Ayats, M., Hartmann, P., Brezesinski, T., and Janek, J. (2019). There and back again–The journey of $LiNiO_2$ as a cathode active material. *Angewandte Chemie-International Edition,* 58(31), 10434–10458.

Billaud, J., Clement, R.J., Armstrong, A.R., Canales-Vazquez, J., Rozier, P., Grey, C.P., and Bruce, P.G. (2014). Beta-$NaMnO_2$: A high-performance cathode for sodium-ion batteries. *Journal of the American Chemical Society,* 136(49), 17243–17248.

Blesa, M.C., Moran, E., Menendez, N., Tornero, J.D., and Torron, C. (1993). Hydrolysis of sodium orthoferrite [alpha-$NaFeO_2$]. *Materials Research Bulletin,* 28(8), 837–847.

Bo, S.H., Li, X., Toumar, A.J., and Ceder, G. (2016). Layered-to-rock-salt transformation in desodiated Na_xCrO_2 (x 0.4). *Chemistry of Materials,* 28(5), 1419–1429.

Braconnier, J.-J., Delmas, C., Fouassier, C., and Hagenmuller, P. (1980). Comportement electrochimique des phases Na_xCoO_2. *Materials Research Bulletin,* 15(12), 1797–1804.

Braconnier, J.J., Delmas, C., and Hagenmuller, P. (1982). Etude par désintercalation électrochimique des systèmes Na_xCrO_2 et Na_xNiO_2. *Materials Research Bulletin,* 17(8), 993–1000.

Breger, J., Kang, K., Cabana, J., Ceder, G., and Grey, C.P. (2007). NMR, PDF and RMC study of the positive electrode material $Li(Ni_{0.5}Mn_{0.5})O_2$ synthesized by ion-exchange methods. *Journal of Materials Chemistry,* 17(30), 3167–3174.

Carlier, D., Cheng, J.H., Berthelot, R., Guignard, M., Yoncheva, M., Stoyanova, R., Hwang, B.J., and Delmas, C. (2011). The P2-$Na_{2/3}Co_{2/3}Mn_{1/3}O_2$ phase: Structure, physical properties and electrochemical behavior as positive electrode in sodium battery. *Dalton Transactions,* 40(36), 9306–9312.

Chang, F.M. and Jansen, M. (1985). Synthesis and crystal-structure of $Na_2Mn_3O_7$. *Zeitschrift Fur Anorganische Und Allgemeine Chemie,* 531(12), 177–182.

Chen, C.-Y., Matsumoto, K., Nohira, T., Hagiwara, R., Fukunaga, A., Sakai, S., Nitta, K., and Inazawa, S. (2013). Electrochemical and structural investigation of $NaCrO_2$ as a positive electrode for sodium secondary battery using inorganic ionic liquid NaFSA–KFSA. *Journal of Power Sources,* 237(0), 52–57.

Chen, X., Wang, Y.C., Wiaderek, K., Sang, X.H., Borkiewicz, O., Chapman, K., Lebeau, J., Lynn, J., and Li, X. (2018). Super charge separation and high voltage phase in Na_xMnO_2. *Adv. Funct. Mater.* 28(50), 1805105. https://doi.org/10.1002/adfm.201805105.

Cheng, J.H., Pan, C.J., Lee, J.F., Chen, J.M., Guignard, M., Delmas, C., Carlier, D., and Hwang, B.J. (2014). Simultaneous reduction of Co3+ and Mn4+ in P2-$Na_{2/3}Co_{2/3}Mn_{1/3}O_2$ as evidenced by X-ray absorption spectroscopy during electrochemical sodium intercalation. *Chemistry of Materials,* 26(2), 1219–1225.

Clement, R., Billaud, J., Armstrong, R., Singh, G., Rojo, T., Bruce, P.G., and Grey, C.P. (2016). Structurally stable Mg-doped P2-$Na_{2/3}Mn_{1-y}Mg_yO_2$ sodium-ion battery cathodes with high rate performance: Insights from electrochemical, NMR and diffraction studies. *Energy & Environmental Science,* 9(10), 3240–3251.

Clement, R.J., Bruce, P.G., and Grey, C.P. (2015). Review-manganese-based P2-type transition metal oxides as sodium-ion battery cathode materials. *Journal of the Electrochemical Society,* 162(14), A2589–A2604.

Croguennec, L., Pouillerie, C., Mansour, A.N., and Delmas, C. (2001). Structural characterisation of the highly deintercalated $LixNi_{1.02}O_2$ phases (with x <= 0.30). *Journal of Materials Chemistry,* 11(1), 131–141.

Delmas, C., Braconnier, J.J., Maazaz, A., and Hagenmuller, P. (1982). Soft chemistry in A_xMO_2 sheet oxides. *Revue De Chimie Minerale,* 19(4–5), 343–351.

Delmas, C., Fouassier, C., and Hagenmuller, P. (1977). Crystallochemical evolution and physical-properties of some lamellar oxides. *Materials Science and Engineering,* 31, 297–301.

Delmas, C., Fouassier, C., and Hagenmuller, P. (1980). Structural classification and properties of the layered oxides. *Physica B & C,* 99(1–4), 81–85.

Deng, J., Luo, W. B., Chou, S. L., Liu, H. K., & Dou, S. X. (2018). Sodium-Ion Batteries: From Academic Research to Practical Commercialization. *Advanced Energy Materials,* 8(4), 1701428.

Deng, J.Q., Luo, W.B., Lu, X., Yao, Q.R., Wang, Z.M., Liu, H.K., Zhou, H.Y., and Dou, S.X. (2018). High energy density sodium-ion battery with industrially feasible and air-stable O3-type layered oxide cathode. *Advanced Energy Materials,* 8(5), 1701610.

Didier, C., Guignard, M., Denage, C., Szajwaj, O., Ito, S., Saadoune, I., Darriet, J., and Delmas, C. (2011). Electrochemical Na-deintercalation from $NaVO_2$. *Electrochemical and Solid State Letters,* 14(5), A75–A78.

Dyer, L.D., Borie, B.S., and Smith, G.P. (1954). Alkali metal nickel oxides of the type $MNiO_2$. *Journal of the American Chemical Society,* 76(6), 1499–1503.

Eriksson, T.A., Lee, Y.J., Hollingsworth, J., Reimer, J.A., Cairns, E.J., Zhang, X.F., and Doeff, M.A. (2003). Influence of substitution on the structure and electrochemistry of layered manganese oxides. *Chemistry of Materials,* 15(23), 4456–4463.

Fang, C., Huang, Y.H., Zhang, W.X., Han, J.T., Deng, Z., Cao, Y.L., and Yang, H.X. (2016). Routes to high energy cathodes of sodium-ion batteries. *Advanced Energy Materials,* 6(5), 18.

Fouassier, C., Delmas, C., and Hagenmuller, P. (1975). Evolution structurale et propriétés physiques des phases $AXMO_2$ (A = Na, K; M = Cr, Mn, Co) (x,©Ω 1). *Materials Research Bulletin,* 10(6), 443–449.

Fouassier, C., Matejka, G., Reau, J.-M., and Hagenmuller, P. (1973). Sur de nouveaux bronzes oxygénés de formule $Na\chi CoO_2$ (χ1). Le système cobalt-oxygène-sodium. *Journal of Solid State Chemistry,* 6(4), 532–537.

Goldsztaub, M.S. (1935). Etude de quelques dérives de l'oxyde ferrique (FeO * OH, FeO_2 Na, FeOCl) détermination de leurs structures. *Bulletin de la Société Francaise de Minéralogie,* 58, 6.

Goodenough, J.B., Mizushima, K., and Takeda, T. (1980). Solid-solution oxides for storage-battery electrodes. *Japanese Journal of Applied Physics,* 19(3), 305–313.

Hamani, D., Ati, M., Tarascon, J.M., and Rozier, P. (2011). Na_xVO_2 as possible electrode for Na-ion batteries. *Electrochemistry Communications,* 13(9), 938–941.

Han, M.H., Gonzalo, E., Casas-Cabanas, M., and Rojo, T. (2014). Structural evolution and electrochemistry of monoclinic $NaNiO_2$ upon the first cycling process. *Journal of Power Sources,* 258, 266–271.

Han, M.H., Gonzalo, E., Sharma, N., Del Amo, J.M.L., Armand, M., Avdeev, M., Garitaonandia, J.J.S., and Rojo, T. (2016). High-performance P2-phase $Na_{2/3}Mn_{0.8}Fe_{0.1}Ti_{0.1}O_2$ cathode material for ambient-temperature sodium-ion batteries. *Chemistry of Materials,* 28(1), 106–116.

Hinuma, Y., Meng, Y. S., & Ceder, G. (2008). Temperature-concentration phase diagram of P2-Na_xCoO_2 from first-principles calculations. *Physical Review B*, 77(22), 224111.

Hoffmann, L. and Hoppe, R. (1977). Alpha- $LiFeO_2$-type of structure. *Zeitschrift für anorganische und allgemeine Chemie,* 430(3), 115–120.

Hoppe, R., Schepers, B., Rohrborn, H.-J., and Vielhaber, E. (1965). Über Oxoscandate: $LiScO_2$ und $NaScO_2$. *Zeitschrift für anorganische und allgemeine Chemie,* 339(3–4), 130–143.

Huang, J.T., Furukawa, T., and Aoto, K. (2005). High temperature behavior of Na-Fe oxides in $H_2O + CO_2$ atmosphere. *Journal of Physics and Chemistry of Solids,* 66(2–4), 388–391.

Hwang, J.-Y., Myung, S.-T., Choi, J.U., Yoon, C.S., Yashiro, H., and Sun, Y.-K. (2017a). Resolving the degradation pathways of the O3-type layered oxide cathode surface through the nano-scale aluminum oxide coating for high-energy density sodium-ion batteries. *Journal of Materials Chemistry A,* 5(45), 23671–23680.

Hwang, J.-Y., Myung, S.-T., and Sun, Y.-K. (2017b). Sodium-ion batteries: Present and future. *Chemical Society Reviews,* 46(12), 3529–3614.

Jansen, M. and Hoppe, R. (1972). New oxocobaltates. *Naturwissenschaften,* 59(5), 215.

Jian, Z.L., Zhao, L., Pan, H.L., Hu, Y.S., Li, H., Chen, W., and Chen, L.Q. (2012). Carbon coated $Na_3V_2(PO_4)_3$ as novel electrode material for sodium ion batteries. *Electrochemistry Communications,* 14(1), 86–89.

Kaithwas, C.K. and Kundu, T.K. (2015). Development of high capacity $Na_{0.7}(Ni_{0.4}Mn_{0.4}Co_{0.1}Fe_{0.1})O_2$ cathode material for sodium ion batteries. *IOP Conference Series: Materials Science and Engineering,* 75(1), 012008.

Kang, J., Baek, S., Mathew, V., Gim, J., Song, J., Park, H., Chae, E., Rai, A.K., and Kim, J. (2012). High rate performance of a $Na_3V_2(PO_4)_3$/C cathode prepared by pyro-synthesis for sodium-ion batteries. *Journal of Materials Chemistry,* 22(39), 20857–20860.

Kanno, R., Shirane, T., Inaba, Y., and Kawamoto, Y. (1997). Synthesis and electrochemical properties of lithium iron oxides with layer-related structures. *Journal of Power Sources,* 68(1), 145–152.

Kaufman, J. L., & Van der Ven, A. (2019). Na_xCoO_2 phase stability and hierarchical orderings in the O3/P3 structure family. *Physical Review Materials*, 3(1), 015402.

Kim, D., Lee, E., Slater, M., Lu, W.Q., Rood, S., and Johnson, C.S. (2012a). Layered $Na[Ni_{1/3}Fe_{1/3}Mn_{1/3}]O_2$ cathodes for Na-ion battery application. *Electrochemistry Communications,* 18, 66–69.

Kim, H., Kim, H., Ding, Z., Lee, M.H., Lim, K., Yoon, G., and Kang, K. (2016). Recent progress in electrode materials for sodium-ion batteries. *Advanced Energy Materials,* 6(19), 1600943.

Kim, S., Ma, X.H., Ong, S.P., and Ceder, G. (2012b). A comparison of destabilization mechanisms of the layered Na_xMO_2 and $LixMO_2$ compounds upon alkali de-intercalation. *Physical Chemistry Chemical Physics,* 14(44), 15571–15578.

Komaba, S. (2019). Systematic study on materials for lithium-, sodium-, and potassium-ion batteries. *Electrochemistry,* 87(6), 312–320.

Komaba, S., Hasegawa, T., Dahbi, M., and Kubota, K. (2015). Potassium intercalation into graphite to realize high-voltage/high-power potassium-ion batteries and potassium-ion capacitors. *Electrochemistry Communications,* 60, 172–175.

Komaba, S., Murata, W., Ishikawa, T., Yabuuchi, N., Ozeki, T., Nakayama, T., Ogata, A., Gotoh, K., and Fujiwara, K. (2011). Electrochemical Na insertion and solid electrolyte interphase for hard-carbon electrodes and application to Na-ion batteries. *Advanced Functional Materials,* 21(20), 3859–3867.

Komaba, S., Nakayama, T., Ogata, A., Shimizu, T., Takei, C., Takada, S., Hokura, A., and Nakai, I. (2009). Electrochemically reversible sodium intercalation of layered $NaNi_{0.5}Mn_{0.5}O_2$ and $NaCrO_2$. *ECS Transactions,* 16(42), 43–55.

Komaba, S., Takei, C., Nakayama, T., Ogata, A., and Yabuuchi, N. (2010). Electrochemical intercalation activity of layered $NaCrO_2$ vs. $LiCrO_2$. *Electrochemistry Communications,* 12(3), 355–358.

Komaba, S., Yabuuchi, N., Nakayama, T., Ogata, A., Ishikawa, T., and Nakai, I. (2012). Study on the reversible electrode reaction of $Na_{1-x}Ni_{0.5}Mn_{0.5}O_2$ for a rechargeable sodium-ion battery. *Inorganic Chemistry,* 51(11), 6211–6220.

Komaba, S., Yabuuchi, N., Yano, M., and Kuze, S. (2014). Positive electrode active substance for sodium secondary cell, positive electrode for sodium secondary cell, and sodium secondary cell. Japan patent application PCT/JP2014/057122

Kubota, K., Asari, T., Yoshida, H., Yabuuchi, N., Shiiba, H., Nakayama, M., and Komaba, S. (2016). Understanding the structural evolution and redox mechanism of a $NaFeO_2$-$NaCoO_2$ solid solution for sodium-ion batteries. *Advanced Functional Materials,* 26(33), 6047–6059.

Kubota, K., Dahbi, M., Hosaka, T., Kumakura, S., and Komaba, S. (2018a). Towards K-ion and Na-ion batteries as "beyond Li-ion". *Chem. Rec.,* 18(4), 459–479.

Kubota, K., Ikeuchi, I., Nakayama, T., Takei, C., Yabuuchi, N., Shiiba, H., Nakayama, M., and Komaba, S. (2015a). New insight into structural evolution in layered $NaCrO_2$ during electrochemical sodium extraction. *Journal of Physical Chemistry C,* 119(1), 166–175.

Kubota, K. and Komaba, S. (2015). Review-practical issues and future perspective for Na-ion batteries. *Journal of the Electrochemical Society,* 162(14), A2538–A2550.

Kubota, K., Kumakura, S., Yoda, Y., Kuroki, K., and Komaba, S. (2018). Electrochemistry and Solid-State Chemistry of $NaMeO_2$ (Me= 3d Transition Metals). *Advanced Energy Materials*, 8(17), 1703415.

Kubota, K., Miyazaki, M., and Komaba, S. (eds) (2015b). Structural and electrochemical studies on $NaMnO_2$ for Na-ion batteries. *228th ECS Meeting,* Phoenix, AZ.

Kubota, K., Yabuuchi, N., Yoshida, H., Dahbi, M., and Komaba, S. (2014). Layered oxides as positive electrode materials for Na-ion batteries. *Mrs Bulletin,* 39(5), 416–422.

Kubota, K., Yoda, Y., and Komaba, S. (2017). Origin of enhanced capacity retention of P2-Type $Na_{2/3}Ni_{1/3-x}Mn_{2/3}CuxO_2$ for Na-ion batteries. *Journal of The Electrochemical Society,* 164(12), A2368–A2373.

Kumakura, S., Tahara, Y., Kubota, K., Chihara, K., and Komaba, S. (2016). Sodium and manganese stoichiometry of P2-type $Na_{2/3}MnO_2$. *Angewandte Chemie International Edition,* 55(41), 12760–12763.

Kundu, D., Talaie, E., Duffort, V., and Nazar, L.F. (2015). The emerging chemistry of sodium ion batteries for electrochemical energy storage. *Angewandte Chemie International Edition,* 54(11), 3431–3448.

Lee, D.H., Xu, J., and Meng, Y.S. (2013). An advanced cathode for Na-ion batteries with high rate and excellent structural stability. *Physical Chemistry Chemical Physics,* 15(9), 3304–3312.

Lee, E., Brown, D.E., Alp, E.E., Ren, Y., Lu, J., Woo, J.-J., and Johnson, C.S. (2015). New insights into the performance degradation of Fe-based layered oxides in sodium-ion batteries: Instability of Fe^{3+}/Fe^{4+} redox in α-$NaFeO_2$. *Chemistry of Materials,* 27(19), 6755–6764.

Legoff, P., Baffier, N., Bach, S., Pereiraramos, J.P., and Messina, R. (1993). Structural and electrochemical characteristics of a lamellar sodium manganese oxide synthesized via a sol-gel process. *Solid State Ionics,* 61(4), 309–315.

Lei, Y.C., Li, X., Liu, L., and Ceder, G. (2014). Synthesis and stoichiometry of different layered sodium cobalt oxides. *Chemistry of Materials,* 26(18), 5288–5296.

Li, X., Wang, Y., Wu, D., Liu, L., Bo, S.-H., and Ceder, G. (2016). Jahn–Teller assisted Na diffusion for high performance Na ion batteries. *Chemistry of Materials,* 28(18), 6575–6583.

Li, X., Wu, D., Zhou, Y.N., Liu, L., Yang, X.Q., and Ceder, G. (2014). O3-type $Na(Mn_{0.25}Fe_{0.25}Co_{0.25}Ni_{0.25})O_2$: A quaternary layered cathode compound for rechargeable Na ion batteries. *Electrochemistry Communications,* 49, 51–54.

Li, Y., Yang, Z., Xu, S., Mu, L., Gu, L., Hu, Y.S., Li, H., and Chen, L. (2015). Air-stable copper-based P2-$Na_{7/9}Cu_{2/9}Fe_{1/9}Mn_{2/3}O_2$ as a new positive electrode material for sodium-ion batteries. *Adv. Sci. (Weinh),* 2(6), 1500031.

Li, Y.J., Gao, Y.R., Wang, X.F., Shen, X., Kong, Q.Y., Yu, R.C., Lu, G., Wang, Z.X., and Chen, L.Q. (2018). Iron migration and oxygen oxidation during sodium extraction from $NaFeO_2$. *Nano Energy,* 47, 519–526.

Li, Z.-Y., Gao, R., Sun, L., Hu, Z., and Liu, X. (2017). Zr-doped P2-$Na_{0.75}Mn_{0.55}Ni_{0.25}Co_{0.05}Fe_{0.10}Zr_{0.05}O_2$ as high-rate performance cathode material for sodium ion batteries. *Electrochimica Acta,* 223, 92–99.

Lim, S.Y., Kim, H., Chung, J., Lee, J.H., Kim, B.G., Choi, J.J., Chung, K.Y., Cho, W., Kim, S.J., Goddard, W.A., Jung, Y., and Choi, J.W. (2014). Role of intermediate phase for stable cycling of $Na_7V_4(P_2O_7)_4PO_4$ in sodium ion battery. *Proceedings of the National Academy of Sciences of the United States of America,* 111(2), 599–604.

Liu, L., Li, X., Bo, S.-H., Wang, Y., Chen, H., Twu, N., Wu, D., and Ceder, G. (2015). High-performance P2-type $Na_{2/3}(Mn_{1/2}Fe_{1/4}Co_{1/4})O_2$ cathode material with superior rate capability for Na-ion batteries. *Advanced Energy Materials,* 5(22), 1500944.

Lu, Z.H. and Dahn, J.R. (2001a). In situ X-ray diffraction study of P2-$Na_{2/3}[Ni_{1/3}Mn_{2/3}]O_2$. *Journal of the Electrochemical Society,* 148(11), A1225–A1229.

Lu, Z.H. and Dahn, J.R. (2001b). Intercalation of water in P2, T2 and O2 structure $A(z)[COxNi_{1/3-x}Mn_{2/3}]O_2$. *Chemistry of Materials,* 13(4), 1252–1257.

Ma, C.Z., Alvarado, J., Xu, J., Clement, R.J., Kodur, M., Tong, W., Grey, C.P., and Meng, Y.S. (2017). Exploring oxygen activity in the high energy P2-type $Na_{0.78}Ni_{0.23}Mn_{0.69}O_2$ cathode material for Na-ion batteries. *Journal of the American Chemical Society,* 139(13), 4835–4845.

Ma, X.H., Chen, H.L., and Ceder, G. (2011). Electrochemical properties of monoclinic $NaMnO_2$. *Journal of the Electrochemical Society,* 158(12), A1307–A1312.

Maazaz, A. and Delmas, C. (1982). On new phases with formula Na_xTiO_2. *Comptes Rendus De L Académie Des Sciences Série Ii,* 295(8), 759–760.

Maazaz, A., Delmas, C., and Hagenmuller, P. (1983). A study of the Na_xTiO_2 system by electrochemical deintercalation. *Journal of Inclusion Phenomena,* 1(1), 45–51.

Marcus, Y. (1985). Thermodynamic functions of transfer of single ions from water to nonaqueous and mixed-solvents. 3. Standard potentials of selected electrodes. *Pure and Applied Chemistry,* 57(8), 1129–1132.

Mariyappan, S., Hemalatha, K., Ramesha, K., Tarascon, J.M., and Prakash, A.S. (2012). Synthesis, structure, and electrochemical properties of the layered sodium insertion cathode material: $NaNi_{1/3}Mn_{1/3}Co_{1/3}O_2$. *Chemistry of Materials,* 24(10), 1846–1853.

Mariyappan, S., Thomas, J., Batuk, D., Pimenta, V., Gopalan, R., and Tarascon, J.-M. (2017). Dual stabilization and sacrificial effect of Na_2CO_3 for increasing capacities of Na-ion cells based on P2-Na_xMO_2 electrodes. *Chemistry of Materials,* 29(14), 5948–5956.

Mariyappan, S., Wang, Q., and Tarascon, J.M. (2018b). Will sodium layered oxides ever be competitive for sodium ion battery applications? *Journal of The Electrochemical Society,* 165(16), A3714–A3722.

Mariyappan, S., Marchandier, T., Rabuel, F., Iadecola, A., Rousse, G., Morozov, A. V., and Tarascon, J. M. (2020). The role of divalent ($Zn^{2+}/Mg^{2+}/Cu^{2+}$) substituents in achieving full capacity of sodium layered oxides for Na-ion battery applications. *Chemistry of Materials*, 32(4), 1657–1666.

Martinez De Ilarduya, J., Otaegui, L., López Del Amo, J.M., Armand, M., and Singh, G. (2017). NaN3 addition, a strategy to overcome the problem of sodium deficiency in P2-$Na_{0.67}[Fe_{0.5}Mn_{0.5}]O_2$ cathode for sodium-ion battery. *Journal of Power Sources,* 337, 197–203.

Matsumura, T., Sonoyama, N., and Kanno, R. (2003). Synthesis, structure and electrochemical properties of layered material, $Li_{2/3}[Mn_{1/3}Fe_{2/3}]O_2$, with mixed stacking states. *Solid State Ionics,* 161(1–2), 31–39.

Mendiboure, A., Delmas, C., and Hagenmuller, P. (1985). Electrochemical intercalation and deintercalation of Na_xMnO_2 bronzes. *Journal of Solid State Chemistry,* 57(3), 323–331.

Mishra, S.K. and Ceder, G. (1999). Structural stability of lithium manganese oxides. *Physical Review B,* 59(9), 6120–6130.

Mizushima, K., Jones, P.C., Wiseman, P.J., and Goodenough, J.B. (1980). Lithium cobalt oxide(LixCoO_2) ($0<x<1$): A new cathode material for batteries of high energy density. *Materials Research Bulletin,* 15(6), 783–789.

Momma, K. and Izumi, F. (2011). VESTA 3 for three-dimensional visualization of crystal, volumetric and morphology data. *Journal of Applied Crystallography,* 44(6), 1272–1276.

Monyoncho, E. and Bissessur, R. (2013). Unique properties of alpha-$NaFeO_2$: De-intercalation of sodium via hydrolysis and the intercalation of guest molecules into the extract solution. *Materials Research Bulletin,* 48(7), 2678–2686.

Mortemard De Boisse, B., Carlier, D., Guignard, M., Bourgeois, L., and Delmas, C. (2014). P2-$Na_xMn_{1/2}Fe_{1/2}O_2$ phase used as positive electrode in Na batteries: Structural changes induced by the electrochemical (de)intercalation process. *Inorganic Chemistry,* 53(20), 11197–11205.

Mortemard de Boisse, B. M., Liu, G., Ma, J., Nishimura, S. I., Chung, S. C., Kiuchi, H., and Yamada, A. (2016). Intermediate honeycomb ordering to trigger oxygen redox chemistry in layered battery electrode. *Nature communications*, 7(1), 1–9

Mortemard de Boisse, B., Nishimura, S. I., Watanabe, E., Lander, L., Tsuchimoto, A., Kikkawa, J., and Yamada, A. (2018). Highly Reversible Oxygen-Redox Chemistry at 4.1 V in $Na_{4/7-x}[\square_{1/7}Mn_{6/7}]O_2$ (□: Mn Vacancy). *Advanced Energy Materials*, 8(20), 1800409.

Mortemard de Boisse, B. M., Reynaud, M., Ma, J., Kikkawa, J., Nishimura, S. I., Casas-Cabanas, M., and Yamada, A. (2019). Coulombic self-ordering upon charging a large-capacity layered cathode material for rechargeable batteries. *Nature Communications*, 10(1), 1–7.

Mu, L., Xu, S., Li, Y., Hu, Y.-S., Li, H., Chen, L., and Huang, X. (2015). Prototype sodium-ion batteries using an air-stable and Co/Ni-free O3-layered metal oxide cathode. *Advanced Materials,* 27(43), 6928–6933.

Mu, L.Q., Hou, Q.P., Yang, Z.Z., Zhang, Y., Rahman, M.M., Kautz, D.J., Sun, E., Du, X.W., Du, Y.G., Nordlund, D., and Lin, F. (2019). Water-processable P2-$Na_{0.67}Ni_{0.22}Cu_{0.11}Mn_{0.56}Ti_{0.11}O_2$ cathode material for sodium ion batteries. *Journal of the Electrochemical Society,* 166(2), A251–A257.

Nanba, Y., Iwao, T., De Boisse, B.M., Zhao, W.W., Hosono, E., Asakura, D., Niwa, H., Kiuchi, H., Miyawaki, J., Harada, Y., Okubo, M., and Yamada, A. (2016). Redox potential paradox in Na_xMO_2 for sodium-ion battery cathodes. *Chemistry of Materials,* 28(4), 1058–1065.

Newman, G.H. and Klemann, L.P. (1980). Ambient-temperature cycling of an Na-TiS_2 cell. *Journal of the Electrochemical Society,* 127(10), 2097–2099.

Nitta, K., Inazawa, S., Sakai, S., Fukunaga, A., Itani, E., Numata, K., Hagiwara, R., and Nohira, T. (2013). Development of molten salt electrolyte battery. *SEI Tech. Rev.,* 76, 27–33.

Nose, M., Shiotani, S., Nakayama, H., Nobuhara, K., Nakanishi, S., and Iba, H. (2013). $Na_4Co_{2.4}Mn_{0.3}Ni_{0.3}(PO_4)_2P_2O_7$: High potential and high capacity electrode material for sodium-ion batteries. *Electrochemistry Communications,* 34, 266–269.

Okada, S., Takahashi, Y., Kiyabu, T., Doi, T., Yamaki, J.-I., and Nishida, T. (eds) (2006). Layered transition metal oxides as cathodes for sodium secondary battery. *210th ECS Meeting*, Cancun, Mexico. The Electrochemical Society.

Orlandi, F., Aza, E., Bakaimi, I., Kiefer, K., Klemke, B., Zorko, A., and Manuel, P. (2018). Incommensurate atomic and magnetic modulations in the spin-frustrated β– $NaMnO_2$ triangular lattice. *Physical Review Materials*, 2(7), 074407.

Ortiz-Vitoriano, N., Drewett, N.E., Gonzalo, E., and Rojo, T. (2017). High performance manganese-based layered oxide cathodes: Overcoming the challenges of sodium ion batteries. *Energy & Environmental Science,* 10(5), 1051–1074.

Pan, H.L., Hu, Y.S., and Chen, L.Q. (2013). Room-temperature stationary sodium-ion batteries for large-scale electric energy storage. *Energy & Environmental Science,* 6(8), 2338–2360.

Parant, J.-P., Olazcuaga, R., Devalette, M., Fouassier, C., and Hagenmuller, P. (1971). Sur quelques nouvelles phases de formule Na_xMnO_2 ($x \leqslant 1$). *Journal of Solid State Chemistry,* 3(1), 1–11.

Park, S., Yoon, W.S., and Vogt, T. (2007). Structure and magnetism of the mono-layer hydrate $Na_{0.3}NiO_2{\cdot}0.7H_2O$. *Solid State Communications,* 142(1–2), 75–79.

Paulsen, J.M. and Dahn, J.R. (1999). Studies of the layered manganese bronzes, $Na_{2/3}[Mn_{1-x}M_x]O_2$ with M = Co, Ni, Li, and $Li_{2/3}[Mn_{1-x}M_x]O_2$ prepared by ion-exchange. *Solid State Ionics,* 126(1-2), 3–24.

Paulsen, J.M. and Dahn, J.R. (2000). O-2-Type Li-2/3[Ni1/3Mn2/3]O-2: A new layered cathode material for rechargeable lithium batteries–II. Structure, composition, and properties. *Journal of the Electrochemical Society,* 147(7), 2478–2485.

Paulsen, J.M., Thomas, C.L., and Dahn, J.R. (2000). O2 structure $Li_{2/3}[Ni_{1/3}Mn_{2/3}]O_2$: A new layered cathode material for rechargeable lithium batteries I. Electrochemical properties. *Journal of the Electrochemical Society,* 147(3), 861–868.

Qi, X., Wang, Y., Jiang, L., Mu, L., Zhao, C., Liu, L., Hu, Y.-S., Chen, L., and Huang, X. (2016). Sodium-deficient O3-$Na_{0.9}[Ni_{0.4}MnxTi_{0.6-x}]O_2$ layered-oxide cathode materials for sodium-ion batteries. *Particle & Particle Systems Characterization,* 33(8), 538–544.

Rozier, P., Sathiya, M., Paulraj, A.-R., Foix, D., Desaunay, T., Taberna, P.-L., Simon, P., and Tarascon, J.-M. (2015). Anionic redox chemistry in Na-rich $Na_2Ru_{1-y}SnyO_3$ positive electrode material for Na-ion batteries. *Electrochemistry Communications,* 53(0), 29–32.

Rudnick, R.L. and Gao, S. (2014). 4.1—Composition of the continental crust. In *Treatise on Geochemistry (Second Edition)*, Holland, H.D., Turekian, K.K. (eds). Elsevier, Oxford, 1–51.

Rüdorff, W. and Becker, H. (1954). Die Strukturen von $LiVO_2$, $NaVO_2$, $LiCrO_2$ und $NaCrO_2$. *Zeitschrift für Naturforschung B,* 9(9), 614–615.

Sathiya, M., Jacquet, Q., Doublet, M. L., Karakulina, O. M., Hadermann, J., and Tarascon, J. M. (2018). A chemical approach to raise cell voltage and suppress phase transition in O3 sodium layered oxide electrodes. *Advanced Energy Materials*, 8(11), 1702599.

Scholder, R. and Kyri, H. (1952). Über die Oxydation von Mangan(II)-hydroxyd mit Sauerstoff in konzentrierten Laugen. *Zeitschrift für anorganische und allgemeine Chemie,* 270(1-4), 56–68.

Shacklette, L.W., Jow, T.R., and Townsend, L. (1988). Rechargeable electrodes from sodium cobalt bronzes. *Journal of The Electrochemical Society,* 135(11), 2669–2674.

Shacklette, L.W., Toth, J.E., and Elsenbaumer, R.L. (1985). Conjugated polymer as substrate for the plating of alkali metal in a nonaqueous secondary battery. EP patent application US 1985-749325

Shannon, R. (1976). Revised effective ionic radii and systematic studies of interatomic distances in halides and chalcogenides. *Acta Crystallographica Section A,* 32(5), 751–767.

Shirane, T., Kanno, R., Kawamoto, Y., Takeda, Y., Takano, M., Kamiyama, T., and Izumi, F. (1995). Structure and physical-properties of lithium iron-oxide, $LiFeO_2$, synthesized by ionic exchange-reaction. *Solid State Ionics,* 79, 227–233.

Shishikura, T., Takeuchi, M., Murakoshi, Y., Konuma, H., and Kameyama, M. (1989). Secondary cobalt sodium oxide-sodium alloy battery. EP patent application.

Shishkin, M., Kumakura, S., Sato, S., Kubota, K., Komaba, S., and Sato, H. (2018). Unraveling the role of doping in selective stabilization of $NaMnO_2$ polymorphs: Combined theoretical and experimental study. *Chemistry of Materials,* 30(4), 1257–1264.

Singer, A., Zhang, M., Hy, S., Cela, D., Fang, C., Wynn, T.A., Qiu, B., Xia, Y., Liu, Z., Ulvestad, A., Hua, N., Wingert, J., Liu, H., Sprung, M., Zozulya, A.V., Maxey, E., Harder, R., Meng, Y.S., and Shpyrko, O.G. (2018). Nucleation of dislocations and their dynamics in layered oxide cathode materials during battery charging. *Nature Energy,* 3(8), 641–647.

Singh, G., Acebedo, B., Cabanas, M.C., Shanmukaraj, D., Armand, M., and Rojo, T. (2013). An approach to overcome first cycle irreversible capacity in P2-$Na_{2/3}Fe_{1/2}Mn_{1/2}O_2$. *Electrochemistry Communications,* 37, 61–63.

Slater, M.D., Kim, D., Lee, E., and Johnson, C.S. (2013). Sodium-ion batteries. *Advanced Functional Materials,* 23(8), 947–958.

Stevens, D.A. and Dahn, J.R. (2000). High capacity anode materials for rechargeable sodium-ion batteries. *Journal of the Electrochemical Society,* 147(4), 1271–1273.

Stoyanova, R., Carlier, D., Sendova-Vassileva, M., Yoncheva, M., Zhecheva, E., Nihtianova, D., and Delmas, C. (2010). Stabilization of over-stoichiometric Mn^{4+} in layered $Na_{2/3}MnO_2$. *Journal of Solid State Chemistry,* 183(6), 1372–1379.

Sun, X., Jin, Y., Zhang, C.Y., Wen, J.W., Shao, Y., Zang, Y., and Chen, C.H. (2014). Na[Ni0.4Fe0.2Mn0.4-xTix]O-2: A cathode of high capacity and superior cyclability for Na-ion batteries. *Journal of Materials Chemistry A,* 2(41), 17268–17271.

Takahashi, Y., Kiyabu, T., Okada, S., Yamaki, J., and Nakane, K. (eds) (2004). *The 45th Battery Symposium in Japan.* Abstr., 3B23 (2004) [in Japanese].

Takeda, Y., Nakahara, K., Nishijima, M., Imanishi, N., Yamamoto, O., Takano, M., and Kanno, R. (1994). Sodium deintercalation from sodium iron-oxide. *Materials Research Bulletin,* 29(6), 659–666.

Talaie, E., Duffort, V., Smith, H.L., Fultz, B., and Nazar, L.F. (2015). Structure of the high voltage phase of layered P2-$Na_{2/3-z}[Mn_{1/2}Fe_{1/2}]O_2$ and the positive effect of Ni substitution on its stability. *Energy & Environmental Science,* 8(8), 2512–2523.

Talaie, E., Kim, S.Y., Chen, N., and Nazar, L.F. (2017). Structural evolution and redox processes involved in the electrochemical cycling of P2-$Na_{0.67}[Mn_{0.66}Fe_{0.20}Cu_{0.14}]O_2$. *Chemistry of Materials,* 29(16), 6684–6697.

Thorne, J.S., Chowdhury, S., Dunlap, R.A., and Obrovac, M.N. (2014a). Structure and electrochemistry of $Na_xFe_xTi1\text{-}xO_2$ ($1.0 \geq x \geq 0.75$) for Na-ion battery positive electrodes. *Journal of The Electrochemical Society,* 161(12), A1801–A1805.

Thorne, J.S., Dunlap, R.A., and Obrovac, M.N. (2014b). Investigation of P2-$Na_{2/3}Mn_{1/3}Fe_{1/3}Co_{1/3}O_2$ for Na-ion battery positive electrodes. *Journal of the Electrochemical Society,* 161(14), A2232–A2236.

Toumar, A. J., Ong, S. P., Richards, W. D., Dacek, S., and Ceder, G. (2015). Vacancy Ordering in O 3-Type Layered Metal Oxide Sodium-Ion Battery Cathodes. *Physical Review Applied*, 4(6), 064002.

Treacy, M.M.J., Newsam, J.M., and Deem, M.W. (1991). A general recursion method for calculating diffracted intensities from crystals containing planar faults. *Proceedings of the Royal Society-Mathematical and Physical Sciences,* 433(1889), 499–520.

Vassilaras, P., Kwon, D.H., Dacek, S.T., Shi, T., Seo, D.H., Ceder, G., and Kim, J.C. (2017). Electrochemical properties and structural evolution of O3-type layered sodium mixed transition metal oxides with trivalent nickel. *Journal of Materials Chemistry A,* 5(9), 4596–4606.

Vassilaras, P., Ma, X.H., Li, X., and Ceder, G. (2013). Electrochemical properties of monoclinic $NaNiO_2$. *Journal of the Electrochemical Society,* 160(2), A207–A211.

Vassilaras, P., Toumar, A.J., and Ceder, G. (2014). Electrochemical properties of $NaNi_{1/3}Co_{1/3}Fe_{1/3}O_2$ as a cathode material for Na-ion batteries. *Electrochemistry Communications,* 38, 79–81.

Vinckeviciute, J., Radin, M.D., and Van Der Ven, A. (2016). Stacking-sequence changes and Na ordering in layered intercalation materials. *Chemistry of Materials,* 28(23), 8640–8650.

Wang, H., Liao, X.Z., Yang, Y., Yan, X.M., He, Y.S., and Ma, Z.F. (2016). Large-scale synthesis of $NaNi_{1/3}Fe_{1/3}Mn_{1/3}O_2$ as high performance cathode materials for sodium ion batteries. *Journal of the Electrochemical Society,* 163(3), A565–A570.

Wang, J., Zhou, Z.F., Li, Y.S., Li, M., Wang, F., Yao, Q.R., Wang, Z.M., Zhou, H.Y., and Deng, J.Q. (2019a). High-rate performance O3-$NaNi_{0.4}Mn_{0.4}Cu_{0.1}Ti_{0.1}O_2$ as a cathode for sodium ion batteries. *Journal of Alloys and Compounds,* 792, 1054–1060.

Wang, L.G., Wang, J.J., Zhang, X.Y., Ren, Y., Zuo, P.J., Yin, G.P., and Wang, J. (2017a). Unravelling the origin of irreversible capacity loss in $NaNiO_2$ for high voltage sodium ion batteries. *Nano Energy*, 34, 215–223.

Wang, P.F., Yao, H.R., Liu, X.Y., Zhang, J.N., Gu, L., Yu, X.Q., Yin, Y.X., and Guo, Y.G. (2017b). Ti-substituted $NaNi_{0.5}Mn_{0.5-x}TixO_2$ cathodes with reversible O3-P3 phase transition for high-performance sodium-ion batteries. *Advanced Materials*, 29(19).

Wang, Q., Mariyappan, S., Vergnet, J., Abakumov, A. M., Rousse, G., Rabuel, F., and Tarascon, J. M. (2019). Reaching the Energy Density Limit of Layered O3-$NaNi_{0.5}Mn_{0.5}O_2$ Electrodes via Dual Cu and Ti Substitution. *Advanced Energy Materials*, 9(36), 1901785.

Wang, X.F., Liu, G.D., Iwao, T., Okubo, M., and Yamada, A. (2014). Role of ligand-to-metal charge transfer in O3-type $NaFeO_2$-$NaNiO_2$ solid solution for enhanced electrochemical properties. *Journal of Physical Chemistry C,* 118(6), 2970–2976.

Wang, X.F., Tamaru, M., Okubo, M., and Yamada, A. (2013). Electrode properties of P2-$Na_{2/3}Mn_yCo_{1-y}O_2$ as cathode materials for sodium-ion batteries. *Journal of Physical Chemistry C,* 117(30), 15545–15551.

Wang, Y., Li, W., Hu, G.R., Peng, Z.D., Cao, Y.B., Gao, H.C., Du, K., and Goodenough, J.B. (2019c). Electrochemical performance of large-grained $NaCrO_2$ cathode materials for Na-ion batteries synthesized by decomposition of $Na_2Cr_2O_7{\cdot}2H_2O$. *Chemistry of Materials*, 31(14), 5214–5223.

Wang, Y., Xiao, R., Hu, Y.S., Avdeev, M., and Chen, L. (2015). P2-$Na_{0.6}[Cr_{0.6}Ti_{0.4}]O_2$ cation-disordered electrode for high-rate symmetric rechargeable sodium-ion batteries. *Nat. Commun.*, 6, 7954.

Watanabe, E., Zhao, W.W., Sugahara, A., De Boisse, B.M., Lander, L., Asakura, D., Okamoto, Y., Mizokawa, T., Okubo, M., and Yamada, A. (2019). Redox-driven spin transition in a layered battery cathode material. *Chemistry of Materials*, 31(7), 2358–2365.

Wu, D., Li, X., Xu, B., Twu, N., Liu, L., and Ceder, G. (2015). $NaTiO_2$: A layered anode material for sodium-ion batteries. *Energy & Environmental Science*, 8(1), 195–202.

Xia, X. and Dahn, J.R. (2012). $NaCrO_2$ is a fundamentally safe positive electrode material for sodium-ion batteries with liquid electrolytes. *Electrochemical and Solid State Letters*, 15(1), A1–A4.

Xie, Y., Wang, H., Xu, G., Wang, J., Sheng, H., Chen, Z., Ren, Y., Sun, C.-J., Wen, J., Wang, J., Miller, D.J., Lu, J., Amine, K., and Ma, Z.-F. (2016). In operando XRD and TXM study on the metastable structure change of $NaNi_{1/3}Fe_{1/3}Mn_{1/3}O_2$ under electrochemical sodium-ion intercalation. *Advanced Energy Materials*, 6(24), 1601306.

Xu, J., Lee, D.H., and Meng, Y.S. (2013). Recent advances in sodium intercalation positive electrode materials for sodium ion batteries. *Functional Materials Letters*, 6(1), 1330001.

Yabuuchi, N., Ikeuchi, I., Kubota, K., and Komaba, S. (2016). Thermal stability of Na_xCrO_2 for rechargeable sodium batteries; studies by high-temperature synchrotron X-ray diffraction. *Acs Applied Materials & Interfaces*, 8(47), 32292–32299.

Yabuuchi, N., Kajiyama, M., Iwatate, J., Nishikawa, H., Hitomi, S., Okuyama, R., Usui, R., Yamada, Y., and Komaba, S. (2012a). P2-type $Na_x[Fe_{1/2}Mn_{1/2}]O_2$ made from earth-abundant elements for rechargeable Na batteries. *Nature Materials*, 11(6), 512–517.

Yabuuchi, N., Yoshida, H., and Komaba, S. (2012b). Crystal structures and electrode performance of alpha-$NaFeO_2$ for rechargeable sodium batteries. *Electrochemistry*, 80(10), 716–719.

Yabuuchi, N., Hara, R., Kajiyama, M., Kubota, K., Ishigaki, T., Hoshikawa, A., and Komaba, S. (2014). New O2/P2-type Li-Excess Layered Manganese Oxides as Promising Multi-Functional Electrode Materials for Rechargeable Li/Na Batteries. *Advanced Energy Materials*, 4(13), 1301453.

Yabuuchi, N., Kubota, K., Dahbi, M., and Komaba, S. (2014). Research development on sodium-ion batteries. *Chemical Reviews*, 114(23), 11636–11682.

Yabuuchi, N., Yano, M., Yoshida, H., Kuze, S., and Komaba, S. (2013). Synthesis and electrode performance of O3-type $NaFeO_2$-$NaNi_{1/2}Mn_{1/2}O_2$ solid solution for rechargeable sodium batteries. *Journal of The Electrochemical Society*, 160(5), A3131–A3137.

Yao, H.R., Wang, P.F., Gong, Y., Zhang, J.N., Yu, X.Q., Gu, L., Ouyang, C.Y., Yin, Y.X., Hu, E.Y., Yang, X.Q., Stavitski, E., Guo, Y.G., and Wan, L.J. (2017). Designing air-stable O3-type cathode materials by combined structure modulation for Na-ion batteries. *Journal of the American Chemical Society,* 139(25), 8440–8443.

Yoshida, H., Yabuuchi, N., and Komaba, S. (2013). $NaFe_{0.5}Co_{0.5}O_2$ as high energy and power positive electrode for Na-ion batteries. *Electrochemistry Communications,* 34, 60–63.

Yoshida, H., Yabuuchi, N., Kubota, K., Ikeuchi, I., Garsuch, A., Schulz-Dobrick, M., and Komaba, S. (2014). P2-type $Na_{2/3}Ni_{1/3}Mn_{2/3-x}Ti_xO_2$ as a new positive electrode for higher energy Na-ion batteries. *Chemical Communications,* 50(28), 3677–3680.

You, Y., Dolocan, A., Li, W.D., and Manthiram, A. (2019). Understanding the air-exposure degradation chemistry at a nanoscale of layered oxide cathodes for sodium-ion batteries. *Nano Letters,* 19(1), 182–188.

Yu, C.-Y., Park, J.-S., Jung, H.-G., Chung, K.-Y., Aurbach, D., Sun, Y.-K., and Myung, S.-T. (2015). $NaCrO_2$ cathode for high-rate sodium-ion batteries. *Energy & Environmental Science,* 8(7), 2019–2026.

Yue, J.L., Yin, W.W., Cao, M.H., Zulipiya, S., Zhou, Y.N., and Fu, Z.W. (2015). A quinary layer transition metal oxide of $NaNi_{1/4}Co_{1/4}Fe_{1/4}Mn_{1/8}Ti_{1/8}O_2$ as a high-rate-capability and long-cycle-life cathode material for rechargeable sodium ion batteries. *Chemical Communications,* 51(86), 15712–15715.

Zandbergen, H. W., Foo, M., Xu, Q., Kumar, V., and Cava, R. J. (2004). Sodium ion ordering in Na_xCoO_2: Electron diffraction study. *Physical Review B*, 70(2), 024101.

Zhang, B., Dugas, R., Rousse, G., Rozier, P., Abakumov, A. M., and Tarascon, J. M. (2016). Insertion compounds and composites made by ball milling for advanced sodium-ion batteries. *Nature Communications*, 7(1), 1-9.

Zhang, Y., Zhang, R., and Huang, Y. (2019). Air-Stable Na_xTMO_2 Cathodes for Sodium Storage. *Frontiers in Chemistry*, 7, 335.

Zhao, J., Zhao, L.W., Dimov, N., Okada, S., and Nishida, T. (2013). Electrochemical and thermal properties of alpha-$NaFeO_2$ cathode for Na-ion batteries. *Journal of the Electrochemical Society,* 160(5), A3077–A3081.

Zheng, L.T., Li, J.R., and Obrovac, M.N. (2017). Crystal structures and electrochemical performance of air-stable $Na_{2/3}Ni_{1/3-x}$,$CuxMn_{2/3}O_2$ in sodium cells. *Chemistry of Materials,* 29(4), 1623–1631.

Zheng, L.T. and Obrovac, M.N. (2017). Investigation of O3-type $Na_{0.9}Ni_{0.45}MnxTi_{0.55-x}O_2$ (0 <= x <= 0.55) as positive electrode materials for sodium-ion batteries. *Electrochimica Acta,* 233, 284–291.

Zhou, Y.-N., Ding, J.-J., Nam, K.-W., Yu, X., Bak, S.-M., Hu, E., Liu, J., Bai, J., Li, H., Fu, Z.-W., and Yang, X.-Q. (2013). Phase transition behavior of $NaCrO_2$ during sodium extraction studied by synchrotron-based X-ray diffraction and absorption spectroscopy. *Journal of Materials Chemistry A,* 1(37), 11130–11134.

2

Polyanionic-Type Compounds as Positive Electrodes for Na-ion batteries

Long H. B. NGUYEN,[1,2,3] Fan CHEN,[2,3] Christian MASQUELIER[2,3,4] and Laurence CROGUENNEC[1,3,4]

[1] *CNRS, University of Bordeaux, Bordeaux INP, ICMCB UMR 5026, Pessac, France*

[2] *Laboratoire de Réactivité et de Chimie des Solides, CNRS-UMR 7314, University of Picardie Jules Verne, Amiens, France*

[3] *RS2E, Réseau Français sur le Stockage Electrochimique de l'Energie, FR CNRS 3459, Amiens, France*

[4] *ALISTORE-ERI European Research Institute, FR CNRS 3104, Amiens, France*

2.1. Introduction

2.1.1. *Oxides and polyanionic frameworks as positive electrodes for sodium ion-batteries*

The booming of electric and hybrid vehicles' market conjugated to the development of renewable energy technologies in recent years have generated a strong interest in alternative electrochemical energy storage technologies to lithium-ion batteries (LIBs). Among them, Na-ion batteries (NIBs) are considered as the most promising one (Larcher and Tarascon 2015). Several materials have been intensively studied as prospective positive electrodes for NIBs, and depending on their composition and crystal structure, they are commonly classified into layered oxides, polyanionics, Prussian blue analogues and organic compounds (Hwang *et al.* 2017).

Na-ion Batteries,
coordinated by Laure MONCONDUIT and Laurence CROGUENNEC.

Layered transition metal oxides are probably the most promising ones because of their high theoretical capacity and high ionic mobility, with three possible structures (P2, P3 and O3-Na_xMO_2) as defined by Delmas *et al.* (1981) and a great diversity in their compositions. The up-to-date performances of these layered oxides as well as the challenges that still need to be tackled have been addressed in the previous chapter of this book and thus will not be recalled here.

Besides layered oxides, polyanionic compounds form an important class of materials for positive electrodes for NIBs. They are highly attractive due to their high operating voltage and their robust structural framework, which can support long-term cycling. The electrochemical potential of these polyanionic compounds can be tailored by varying the nature of the involved redox couple as well as the composition and the structure of the anion building the framework (PO_4, P_2O_7, PO_4F, etc.) (Figure 2.1). The theoretical specific capacity associated with these materials is usually lower than for layered oxides, and the high molar mass of the polyanionic group is an obvious penalty for their gravimetric capacity. On the other hand, the inductive effect occurring in polyanionic structures provides higher operating voltages for a given redox couple and hence polyanionic compounds are still considered as promising materials for NIBs, especially for stationary applications for which the weight is not a limiting criterion.

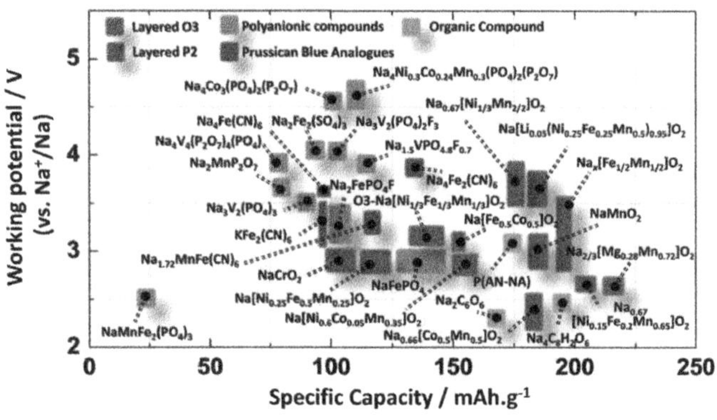

Figure 2.1. *A gathering of the working potential and the specific capacity of different types of materials developed as positive electrodes in Na-ion batteries (Hwang* et al. *2017). Reprinted with permission from Hwang J.-Y., Myung S.-T., Sun Y.-K. (2017). Creative Commons Licence - Published by The Royal Society of Chemistry. For a color version of this figure, see www.iste.co.uk/monconduit/batteries.zip*

The crystal structures of polyanionic materials are diverse. Depending on the anionic composition as well as the connectivity between the polyhedral units in the structure, these three-dimensional frameworks can be classified into different families, according to:

– the nature of the anionic group: they can thus be classified into sulfates, phosphates, silicates, borates, fluorinated oxy-phosphates, oxalates;

– the adopted crystal structure type: they can be Na Super Ionic CONductor (NASICON)-type, Tavorite-type, Olivine-type, Alluaudite-type, etc.;

– the dimension of the ionic diffusion paths that can be 1D for olivine $NaFePO_4$, for instance, 2D for $Na_3V_2(PO_4)_2F_3$ and 3D for NASICON $Na_3V_2(PO_4)_3$.

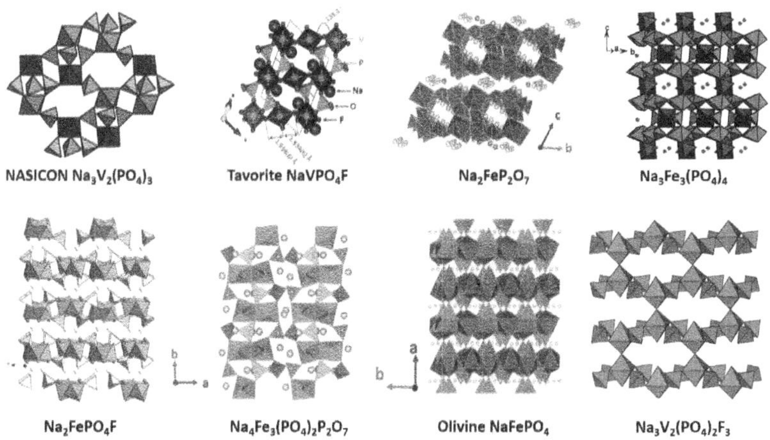

Figure 2.2. *Illustration of the structural frameworks of the main families of polyanionic materials studied as positive electrodes for Na-ion batteries (Adapted with permission from Masquelier C., Croguennec L. Polyanionic (2013). Copyright (2020) American Chemical Society. Adapted with permission from Trad K., Carlier D., Croguennec L., Wattiaux A., Lajmi B., Ben Amara M., Delmas C. (2010). Copyright (2020) American Chemical Society. Adapted with permission from Boivin E., Chotard J.-N., Bamine Th., Carlier D., Serras P., Paomares V., Rojo T., Iadecola A., Dupont L., Bourgeois L., Fauth F., Masquelier C., Croguennec L. (2017). Copyright (2020) The Royal Society of Chemistry). For a color version of this figure, see www.iste.co.uk/monconduit/batteries.zip*

The structural frameworks of some typical polyanionic materials are given in Figure 2.2. Among them, $Na_3V_2(PO_4)_3$ and $Na_3V_2(PO_4)_2F_3$ are the two most promising active materials for positive electrodes in NIBs as they both show Na^+-extractions at relatively high voltage with two Na^+ ions per formula unit that can be cycled reversibly in the stable electrochemical window of all the currently available electrolytes. Despite being really different in terms of crystal structures and electrochemical properties, many reports improperly classified these two compounds into the NASICON-type family. In this review, we will highlight the main characteristics of their crystal structures to show that they actually belong to two different structural types; furthermore, we will highlight recent achievements in the development of new electrode materials for NIBs based on these two structures.

2.1.2. *NASICONs and $Na_3V_2(PO_4)_2F_3$*

2.1.2.1. *Structural descriptions and differences*

The NASICON structure, with the general formula of $A_xMM'(XO_4)_3$, can be described as a 3D framework of MO_6 octahedra and XO_4 tetrahedra connected to each other by corners sharing. The basic repeating unit "$MM'(XO_4)_3$", which is made of two MO_6 octahedra connected by three XO_4 tetrahedra, is named as "lantern unit" (Figure 2.3(a), inset). Each lantern is connected to six other lanterns. Normally, these lanterns are well aligned along the *c* direction of the structure to build a rhombohedral cell. Depending on the synthesis routes, these lanterns may display a zig-zag alignment building a monoclinic cell known as "anti-NASICON". This zig-zag arrangement, leads to a less open structure than in the NASICON form, and the anti-NASICON structure is less favorable for alkali ion transportation (Figure 2.3(a)).

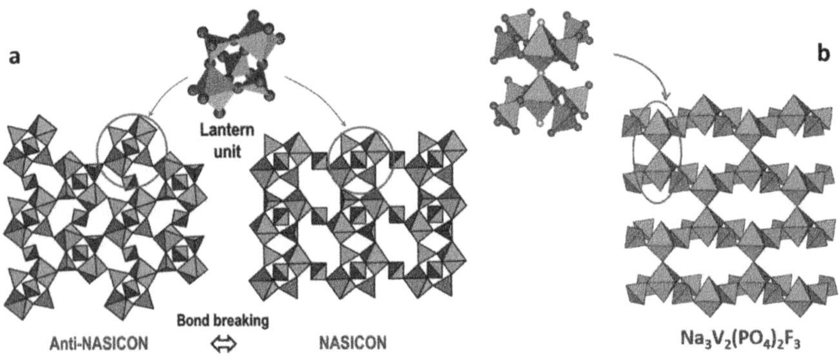

Figure 2.3. *(a) NASICON (generally rhombohedral) and anti-NASICON (generally monoclinic) frameworks of the general formula $A_xMM'(XO_4)_3$. (b) $Na_3V_2(PO_4)_2F_3$ structural framework. Reprinted with permission from Masquelier C., Croguennec L. (2013). Copyright (2020) American Chemical Society. For a color version of this figure, see www.iste.co.uk/monconduit/batteries.zip*

In the structure of $Na_3V_2(PO_4)_2F_3$, each V^{3+} is coordinated by four oxygen atoms in the equatorial plane and two fluorine atoms on the axial positions forming the VO_4F_2 subunit. Two VO_4F_2 octahedra share one common fluorine vertex to form a bi-octahedral unit $V_2O_8F_3$, F—VO_4—F—VO_4—F (Figure 2.3(b), inset). These bi-octahedral units are connected by PO_4 groups to generate a 3D network (Figure 2.3(b)). The Na^+ ions occupy and can only diffuse in the channels locating at $z = 0$ and $z = ½$. Furthermore, there is no ion cross-over between two planes localized at different coordinates along the z-axis, indicating that $Na_3V_2(PO_4)_2F_3$ and all isostructural materials are 2D ionic conductors (Jalem *et al.* 2016).

IMPORTANT.– $Na_3V_2(PO_4)_2F_3$ does not belong to the NASICON family. It is totally wrong to describe $Na_3V_2(PO_4)_2F_3$ structure like "replacing one $(PO_4)^{3-}$ from $Na_3V_2(PO_4)_3$ by 3 F^-".

2.1.2.2. *Traditional synthesis methods*

Solid state reaction and wet method are two traditional synthesis routes for NASICON compounds. In the solid-state synthesis method, precursors are first well mixed in a mortar or by ball-milling, and then transferred into a furnace. Depending on the oxidation state of precursors, a reducing atmosphere (H_2/Ar gas mixture flow, mixed with carbon source, etc.) may be applied during the high temperature annealing procedure (Jian *et al.* 2012). In the wet method, precursors are dissolved or well dispersed in the dilute water, then the recovered powder is annealed at high temperature (Song *et al.* 2014).

The most direct way to synthesize $Na_3V_2(PO_4)_2F_3$ is a "ceramic" high temperature reaction between NaF and VPO_4 (Gover *et al.* 2006); carbothermal synthesis (Jian *et al.* 2013) and wet synthesis methods (Serras *et al.* 2012) were also reported. For these latter methods, most of the obtained materials are in fact $Na_3V^{3+}{}_{2-y}V^{4+}{}_y(PO_4)_2F_{3-y}O_y$ ($0 < y \leq 2$), whose crystal structure and electrochemical properties are very close to those of $Na_3V_2(PO_4)_2F_3$. For wet synthesis routes, a vanadium-containing source (usually V^{3+} or V^{5+}) is dissolved in an aqueous or organic medium and a reducing agent, such as citric acid, is added to the solution to reduce all the V^{5+} ions into V^{3+}. Nevertheless, in most cases, the presence of V^{4+} residue is observed after the synthesis leading to the formation of mixed valence materials $Na_3V^{3+}{}_{2-y}V^{4+}{}_y(PO_4)_2F_{3-y}O_y$.

As for most of the polyanionic compounds, the electronic conductivity of NASICON and $Na_3V_2(PO_4)_2F_3$ is rather low due to the presence of highly covalent phosphate groups that isolate the transition metal centers from each other's within the structure. The low electronic conductivity of these polyanionic materials appears to be detrimental to their electrochemical performance, especially at high cycling rates. Nanocomposites and surface-conducting modification (e.g., carbon coating) are often formed during the synthesis process or in a posttreatment step. Nanostructures provide a huge electrode/electrolyte contact interface, shorten the sodium diffusion distance, and thus, enhance the kinetics of the electrochemical reactions. On the other hand, the incorporation of carbon sources covers the surface of the material by a conductive layer to form a composite with an enhanced electronic conductivity. For example, Deng *et al.* (2019) developed a facile one-step carbothermal reduction synthesis of $Na_3V_2(PO_4)_2F_3/C$ by an *in situ* decomposition of glucose during the synthesis step and the as-obtained materials' surface was covered by a thin layer (10 nm) of carbon. An optimized sample with 1.7 wt.% of carbon-coating showed 89.7% of capacity retention after 100 cycles at 1 C rate. The use of carbon nanotubes (CNTs) or graphene encapsulation were also proposed to enhance the cycling rate and the long-term cycling of $Na_3V_2(PO_4)_3$, $Na_3V^{3+}{}_{2-y}V^{4+}{}_y(PO_4)_2F_{3-y}O_y$ and isostructural materials. Shen *et al.* (2018b) claimed that the charge transfer resistance of $Na_3V_2(PO_4)_2F_3$@C dispersed within the CNTs frameworks was significantly lower than that of conventional $Na_3V_2(PO_4)_2F_3$@C.

2.1.2.3. *Electrochemical signatures*

Flat voltage-composition plateaus can be observed from the cycling profile of $Na_3V_2(PO_4)_3$. During the charging process, two Na^+ ions can be extracted from the structure at ~3.4 V versus Na^+/Na, reaching the $NaV_2(PO_4)_3$ composition (Figure 2.4(a)). During the discharging process, one Na^+ ions can be inserted into the structure at ~1.6 V versus Na^+/Na, reaching the $Na_4V_2(PO_4)_3$ composition (Figure 2.4(a)).

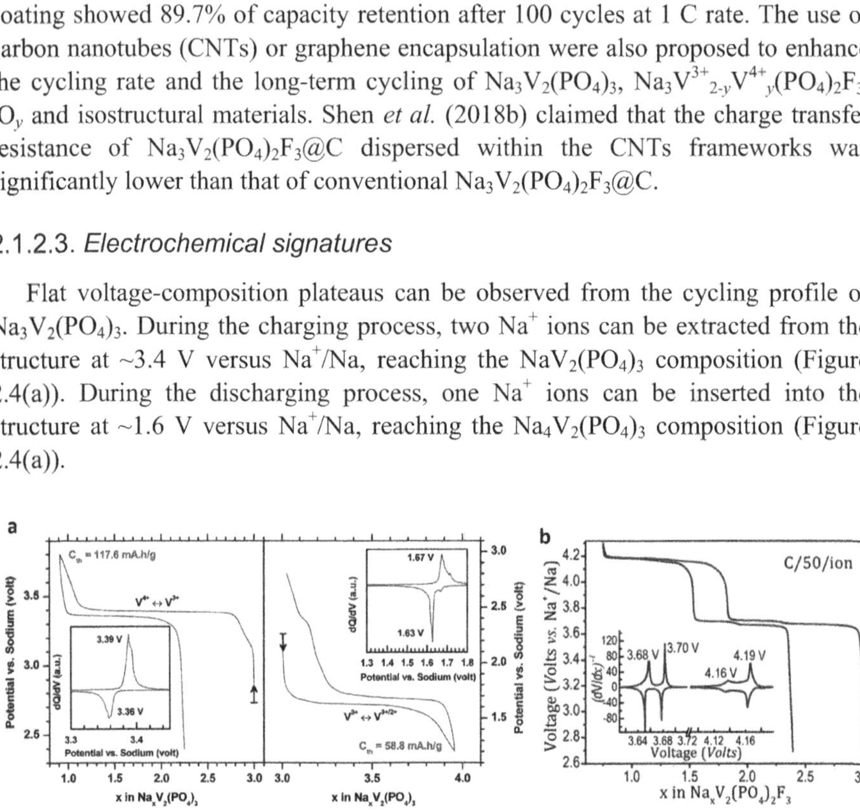

Figure 2.4. *Galvanostatic electrochemical first cycles obtained for (a) $Na_3V_2(PO_4)_3$ and (b) $Na_3V_2(PO_4)_2F_3$. The corresponding derivative curves are given in inset. Reprinted with permission from Lalère F., Seznec V., Courty M., David R., Chotard J.N., Masquelier C. (2015). Copyright (2020). The Royal Society of Chemistry. For a color version of this figure, see www.iste.co.uk/monconduit/batteries.zip*

Due to the presence of fluorine, the highest electronegative element, in the first coordination sphere of V^{3+}, its 3*d* orbitals are strongly polarized giving rise to an increasing voltage of 0.5 V for the V^{4+}/V^{3+} redox couple involved during the Na^+-extraction from $Na_3V_2(PO_4)_2F_3$ versus what is observed for $Na_3V_2(PO_4)_3$. Two Na^+ ions can be extracted and reinserted reversibly from/into $Na_3V_2(PO_4)_2F_3$ at ~3.70 and 4.20 V versus Na^+/Na (Figure 2.4(b)). Unlike what is observed for $Na_3V_2(PO_4)_3$, the V^{3+}/V^{2+} redox couple cannot be activated in $Na_3V_2(PO_4)_2F_3$ by an electrochemical reaction involving Na^+ insertion (Bianchini *et al.* 2017).

2.2. NASICON structures as model frameworks in sodium-ion battery applications

2.2.1. *Compositional diversity from solid electrolytes to electrodes*

NASICON-type materials constitute an important family of polyanionic compounds. They also hold an essential role as solid electrolytes and electrode materials for both LIBs and NIBs. The attractiveness of these materials lies in their structural stability as well as their excellent ionic conductivity, as identified in the early days by Goodenough *et al.* (1976). The general formula of NASICON compositions can be written as $A_xMM'(XO_4)_3$ and possible compositions that can be found in the literature are as follows:

– A = alkali ions (Li^+, Na^+, K^+, Rb^+ and Cs^+), alkaline earth ions (Mg^{2+}, Ca^{2+}, Sr^{2+} and Ba^{2+}), H^+, Cu^+, Cu^{2+}, Ag^+. The crystallographic positions occupied by A can also be vacant;

– M or M' = di- (Zn^{2+}, Ni^{2+}, Mn^{2+}, Co^{2+}), tri- (Fe^{3+}, Sc^{3+}, Ti^{3+}, V^{3+}, Cr^{3+}, Al^{3+}, In^{3+}, Ga^{3+}, Y^{3+}, Lu^{3+}), tetra- (Ti^{4+}, Zr^{4+}, Hf^{4+}, Sn^{4+}, Ge^{4+}, Mo^{4+}), and penta- (V^{5+}, Nb^{5+}, Sb^{5+}) valent transition metal ions;

– $(XO_4)^{n-}$ = $(SO_4)^{2-}$, $(WO_4)^{2-}$, $(MoO_4)^{2-}$, $(PO_4)^{3-}$, $(AsO_4)^{3-}$ and $(SiO_4)^{4-}$.

The number of alkali cations per $A_xMM'(XO_4)_3$ formula, x, can be adjusted depending on the nature and oxidation state of M and M', and of the X element. For instance, niobium, titanium and iron, mostly stable in air in their oxidation states +5, +4 and +3, respectively, when combined with $(PO_4)^{3-}$ anion will lead to the formation of $Nb^{5+}Ti^{4+}(PO_4)_3$, $LiTi^{4+}{}_2(PO_4)_3$, $Na_2Fe^{3+}Ti^{4+}(PO_4)_3$ and $Na_3Fe^{3+}{}_2(PO_4)_3$ that all adopt the NASICON structure and can be obtained easily as pure phases. Other examples can be $Fe^{3+}{}_2(SO_4)_3$ and $Na_4Zr^{4+}{}_2(SiO_4)_3$, with the anionic groups $(SO_4)^{2-}$ and $(SiO_4)^{4-}$, respectively. Depending on the composition (especially on the content in alkali ions) and on the temperature, the crystal structure of NASICON materials can be rhombohedral, orthorhombic, monoclinic or triclinic. The stable and flexible three-dimensional framework generates a large interstitial space and offers excellent interconnecting conduction pathways, resulting in a fast ion transport in the NASICON structure (Susman *et al.* 1983).

The Zr-containing NASICON compounds of general formula $Na_{1+x}Zr_2(PO_4)_{3-x}$ $(SiO_4)_x$ ($0 \leq x \leq 3$), generally called NZPS where $NaZr_2(PO_4)_3$ (NZP) and $Na_4Zr_2(SiO_4)_3$ (NZS) are the two end members, were first proposed as solid electrolytes for NIBs by Goodenough *et al.* (1976). The crystal structure of this series of compounds is commonly described in the *R*-3*c* space group (rhombohedral structure) except in the range $1.8 \leq x \leq 2.2$, where a monoclinic distortion generally occurs due to the Na^+–vacancy ordering, leading to a symmetry reduction to *C*2/*c*. The highest ionic conductivity is obtained in $Na_3Zr_2(PO_4)(SiO_4)_2$, which can reach 2.0×10^{-1} $S{\cdot}cm^{-1}$ at 300°C and 6.7×10^{-4} $S{\cdot}cm^{-1}$ at room temperature, even though the two end members NZP and NZS are poor ionic conductors. For $x = 0$ (NZP), all the Na(1) positions are fully occupied, while Na(2) positions are vacant. For $x = 3$ (NZS), all the Na(1) and Na(2) positions are fully occupied. For compositions with $x > 0$, Na^+ ions are introduced into Na(2) sites, which are interconnected between them via Na(1) sites, forming zigzag conduction pathways of sodium ions (Na(2) → Na(1) → Na(2) sites) (Boilot *et al.* 1986).

One interesting feature of the NASICON series of compounds $Na_xMM'(PO_4)_{3-y}(SiO_4)_y$ is their compositional diversity allowing the presence of several metal ions with different valence states in the structure, and thus for charge compensation, the variation of x from 0 to 4. The M and M' can be di-, tri-, tetra- or pentavalent transition metal cations, and the resulted compounds can be used for ion recognition, gas sensors and sodium ion conductive electrolyte membranes, but they are nowadays widely reinvestigated as the solid electrolyte for sodium-based all-solid-state batteries (ASSBs) (Noguchi *et al.* 2013). Among all the various compositions, Sc^{3+}-containing compounds $Na_3Sc_yZr_{2-y}(PO_4)_{1+y}(SiO_4)_{2-y}$ and $Na_2Sc_yZr_{2-y}(PO_4)_{2+y}(SiO_4)_{1-y}$ ($0 \leq y \leq 1$) are the most promising material families, which have been insightfully investigated by combining experimental and theoretical approaches (Kaus *et al.* 2017). The as-mentioned materials can form unlimited solid solution, as Sc^{3+} and Zr^{4+} are characterized by similar ionic radii ($R_{Sc^{3+}} = 75$ pm versus $R_{Zr^{4+}} = 72$ pm), which can further be used as solid electrolytes since the *d* orbitals of Sc^{3+} and Zr^{4+} are completely empty. A hard carbon/hybrid $Na_3Zr_2(PO_4)(SiO_4)_2$–PVDF–ether based electrolyte/$NaFePO_4$ ASSB was shown to deliver a discharge capacity of ~120 $mAh{\cdot}g^{-1}$ at *C*/5 (Kim *et al.* 2015).

Regarding lithium-based ASSBs using NASICON-type solid electrolytes, $Li_{1+x}Al_xTi_{2-x}(PO_4)_3$ (LATP) and $Li_{1+x}Al_xGe_{2-x}(PO_4)_3$ (LAG) have been intensively investigated and used as the solid electrolyte. A symmetric $Li_3V_2(PO_4)_3/Li_{1.5}Al_{0.5}Ge_{1.5}(PO_4)_3/Li_3V_2(PO_4)_3$ ASSB, with $Li_3V_2(PO_4)_3$ at the two electrodes, can deliver a discharge capacity of 101 $mAh.g^{-1}$ at 11 $\mu A.cm^{-2}$ (corresponding to *C*/20 rate) at 80°C (Kobayashi *et al.* 2010).

Nevertheless, the presence of several solid-solid interfaces in ASSBs induces a large interfacial resistance due to interdiffusion and reactivity between the electrode materials and the electrolyte, which lowers the ionic conductivity and the cell capacity despite the superionic conductivity in the bulk electrolyte. In order to obtain perfect interfaces between electrode and electrolyte, a battery made from a single material might be a better choice, as proposed recently by Inoishi *et al.* (2017).

With the increasing demand for large-scale energy storage applications, lithium and sodium containing NASICON materials have been, and still are, greatly investigated as positive (or negative in some cases) electrodes due to their specific 3D framework, stable long-term cyclability and high ion mobility. Indeed, Li^+ or Na^+ can be easily intercalated or de-intercalated into/from the NASICON structure within a wide range of operating voltages.

2.2.2. *NASICON-typed materials as electrodes for Na batteries*

2.2.2.1. *One transition-metal-based $Na_xM_2(XO_4)_3$ NASICONs*

The history of reversible alkali ions insertion/extraction into/from NASICON materials went back to the 1980s with the works of Torardi *et al.* (1986) and Delmas *et al.* (1987) on the alkali ions insertion into $Fe_2(MoO_4)_3$ and $(Li,Na)Ti_2(PO_4)_3$ structures, respectively. At that time, the intercalation of Na^+ ions into $NaTi_2(PO_4)_3$ (NTP) resulted in the formation of $Na_3Ti_2(PO_4)_3$ through a biphasic electrochemical reaction with the reduction of Ti^{4+} into Ti^{3+} at 2.2 V versus Na^+/Na and delivered a gravimetric capacity of 133 mAh g^{-1}. This can be considered as one of the first examples of NASICON materials used as positive electrodes for NIBs. The activity of the Ti^{3+}/Ti^{2+} redox couple upon further discharging $Na_3Ti_2(PO_4)_3$ at lower voltage (0.4 V versus Na^+/Na), leading to the formation of $Na_4Ti_2(PO_4)_3$, was later reported by Senguttuvan *et al.* (2013). These preliminary works have opened an avenue toward the development of symmetric full cells with the use of $Na_3Ti_2(PO_4)_3$ as both positive and negative electrodes.

An overall illustration of some accessible $Na_xM_2(XO_4)_3$ compositions and their corresponding operating voltages involved the Na^+ intercalation and de-intercalation processes is given in Figure 2.5.

By comparing the potential values of the Fe^{3+}/Fe^{2+} redox couple in different $Li_xFe_2(XO_4)_3$ compositions, Manthiram and Goodenough (1987) introduced the so-called inductive effect concept in order to qualitatively illustrate the impact of the nature of the tetrahedral XO_4 groups on the position of the Fe^{3+}/Fe^{2+} redox couple in these materials. In all these $Li_xFe_2(XO_4)_3$ compositions, there exist the Fe^{3+}—O—X bond chains. When the X element bears a high electronegativity or a high formal charge value, the O–X chemical bond is more covalent, thus the opposite Fe^{3+}–O bond is more ionic and a higher redox potential value is observed for the Fe^{3+}/Fe^{2+} redox couple. For instance, the redox potential value of the Fe^{3+}/Fe^{2+} redox couple (versus Li^+/Li) varies from ~2.8 V for $Li_3Fe_2(PO_4)_3$, to 3.0 V for $Fe_2(MoO_4)_3$ and $Fe_2(WO_4)_3$, and to 3.6 V for $Fe_2(SO_4)_3$ when X is replaced by P^{+5}, Mo^{+6} or W^{+6} and S^{+6}, respectively. The same phenomena were also observed for Na-containing compounds: isostructural $Na_3Fe_2(PO_4)_3$ and $Fe_2(SO_4)_3$ exhibit operating voltages that differ by 0.7 V upon Na^+ intercalation, where the latter being higher than the former (Shiva *et al.* 2016).

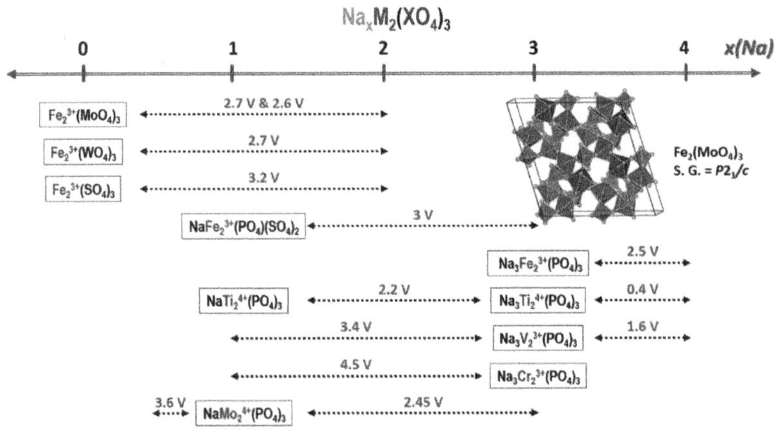

Figure 2.5. *Accessible compositions and the corresponding voltages of the $M^{n+}/M^{(n-1)+}$ redox couples (versus Na^+/Na) involved during Na^+ intercalation and de-intercalation in/from NASICON or anti-NASICON $Na_xM_2(XO_4)_3$ structures. Operating voltages are given for each redox couple (Delmas* et al. *1987; Bruce and Miln 1990; Saravanan* et al. *2013; Senguttuvan* et al. *2013; Zhou* et al. *2016a; Liu* et al. *2017b; Ben Yahia* et al. *2018; Chung* et al. *2018; Kawai* et al. *2018; Panin* et al. *2018). For a color version of this figure, see www.iste.co.uk/monconduit/batteries.zip*

NASICON-type $Na_3V_2(PO_4)_3$ (NVP) has so far been identified as the most promising positive electrode material from the polyanionic family as it possesses satisfactory energy density, high Na^+-extraction voltages and high stability, which are suitable for long-term cycling (Jian *et al.* 2012). The electrochemical Na^+-extraction from $Na_3V_2(PO_4)_3$, through the activation of the V^{4+}/V^{3+} redox couple, occurs at 3.4 V versus Na^+/Na, can deliver a theoretical capacity of 118 $mAh{\cdot}g^{-1}$. The V^{3+} ions in $Na_3V_2(PO_4)_3$ can also be reduced into V^{2+} through the formation of $Na_4V_2(PO_4)_3$ at 1.8 V versus Na^+/Na leading to a theoretical capacity of 56 $mAh{\cdot}g^{-1}$.

Because of the possible involvement of two different redox couples, V^{3+}/V^{2+} and V^{4+}/V^{3+}, NVP can practically be used at both negative and positive electrodes in NIBs with a mass balance of 2/1 for the negative/positive weight ratio (Noguchi *et al.* 2013). The electrochemical activities of $Na_3Cr_2(PO_4)_3$ (Kawai *et al.* 2018) and $NaMo_2(PO_4)_3$ (Panin *et al.* 2018) have recently been reported at 4.5 V and 3.45 V versus Na^+/Na, respectively, making them more attractive than NVP in terms of delivered energy. Nevertheless, the presence of potentially toxic Cr^{4+} ions in $Na_{3-x}Cr_2(PO_4)_3$ and the heavy element, Mo, in $NaMo_2(PO_4)_3$ result in little interest of these compositions for practical applications.

2.2.2.2. *Multitransition-metal-based $Na_xMM'(PO_4)_3$ NASICONs*

As described above, the NASICON framework can accommodate a variety of transition metal ions on the M and M' positions and depending on their nature and oxidation state, the operating voltage of the resulting phases can vary in a wide range (Figure 2.6). Furthermore, the number x of Na^+ in $Na_xMM'(PO_4)_3$ can also be monitored. It can vary from $x_{min} = 0$ in $Nb^{5+}Ti^{4+}(PO_4)_3$ to $x = 2$ in $Na_2Fe^{3+}Ti^{4+}(PO_4)_3$ and up to $x_{max} = 4$ in $Na_4Fe^{2+}Fe^{3+}(PO_4)_3$. Maximizing the number of reversible Na^+ ions involved in the electrochemical reaction is an essential approach to design new positive electrode materials with superior capacity.

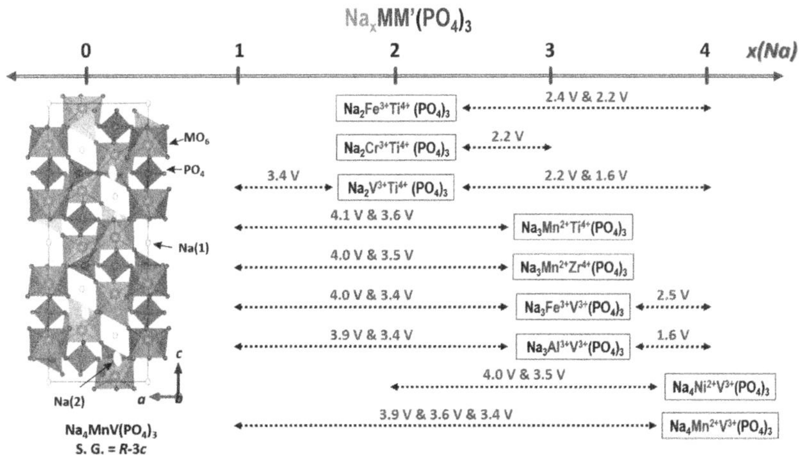

Figure 2.6. *Accessible compositions and the corresponding voltages of the $M^{n+}/M^{(n-1)+}$ redox couples (versus Na^+/Na) involved during Na^+ intercalation and de-intercalation in/from NASICON $Na_xMM'(PO_4)_3$ structures. Operating voltages are given for each redox couple (Patoux* et al. *2003; Aragón* et al. *2015; Lalère* et al. *2015, 2018; Gao and Goodenough 2016; Zhou* et al. *2016b; Chen* et al. *2018; Gao* et al. *2018). For a color version of this figure, see www.iste.co.uk/monconduit/batteries.zip*

Noticeably, due to the absence of direct connectivity (no face, edge or corner sharing) between two (M,M')O_6 octahedra in this structure, the voltage value of a given $M^{n+}/M^{(n-1)+}$ redox couple is independent of the nature of the neighboring transition metal element M' and vice versa. For instance, the position of the Ti^{4+}/Ti^{3+} redox couple in $Na_{2+x}FeTi(PO_4)_3$ (Tillement *et al.* 1991a), $Na_xTiNb(PO_4)_3$ (Tillement *et al.* 1991b) and $Na_2MTi^{3+}(PO_4)_3$ (M = Fe, Cr) (Patoux *et al.* 2003) is the same as in $NaTi_2(PO_4)_3$, i.e. ~2.2 V versus Na^+/Na.

Among all the multi-transition-metal-based NASICON compositions, $Na_4Mn^{2+}V^{3+}(PO_4)_3$ and $Na_4Ni^{2+}V^{3+}(PO_4)_3$, which were recently reported by Zhou *et al.* (2016b), are the two most noticeable because of their high operating voltages and excellent reversibility even at high cycling rates. Furthermore, this was the first time that Ni^{2+}- and Mn^{2+}-containing NASICONs were reported, which allowed to stabilize NASICONs with four Na^+ ions per formula unit, thus providing the possibility of extracting more than one Na^+ ion per transition metal. The electrochemical performance of $Na_4MnV(PO_4)_3$ will be discussed later dedicated to this compound.

2.2.3. *$Na_3V_2(PO_4)_3$ (NVP)*

2.2.3.1. *Structural modifications as a function of temperature*

The synthesis of $Na_3V_2(PO_4)_3$ (NVP) was first reported by Delmas *et al.* (1978), but different solid-state synthesis approaches were quickly developed. The precursors containing sodium (e.g. Na_2CO_3, NaH_2PO_4, Na_3PO_4), phosphorus (e.g. $NH_4H_2PO_4$) and vanadium (e.g. V_2O_3, VPO_4, V_2O_5, NH_4VO_3) are first mixed in the right stoichiometry, then transferred into a tubular furnace and annealed at high temperature (~800°C), and depending on the vanadium oxidation state, an inert (N_2 or Ar) or reducing (5–10% vol. H_2) atmosphere can be used. NVP can also be obtained by sol-gel method by dissolving a stoichiometric amount of precursors in water together with reducing carbon-containing sources (e.g. citric acid, oxalic acid, sugar) and maintaining it under continuous stirring. After the sol and then the gel formation, the recovered gel is transferred into a tubular furnace and annealed at high temperature in an inert atmosphere.

	T (°C)	S.G.	*a* (Å)	*b* (Å)		β (°)	*Z*	*V/Z* (Å³)
α	–10	*C*2/*c*	15.1244(6)	8.7287(3)	21.6143(8)	90.163(2)	12	237.786
β	60	*C*2/*c*	15.113(2)	8.7297(6)	8.8346(11)	124.568(9)	4	239.647
β'	130	*C*2/*c*	15.119(2)	8.7347(4)	8.8543(7)	124.557(6)	4	240.756
γ	200	*R*-3c	8.73382(4)	–	21.9144(2)	–	6	241.277

Table 2.1. *Cell parameters of different $Na_3V_2(PO_4)_3$ polymorphs observed at different temperatures*

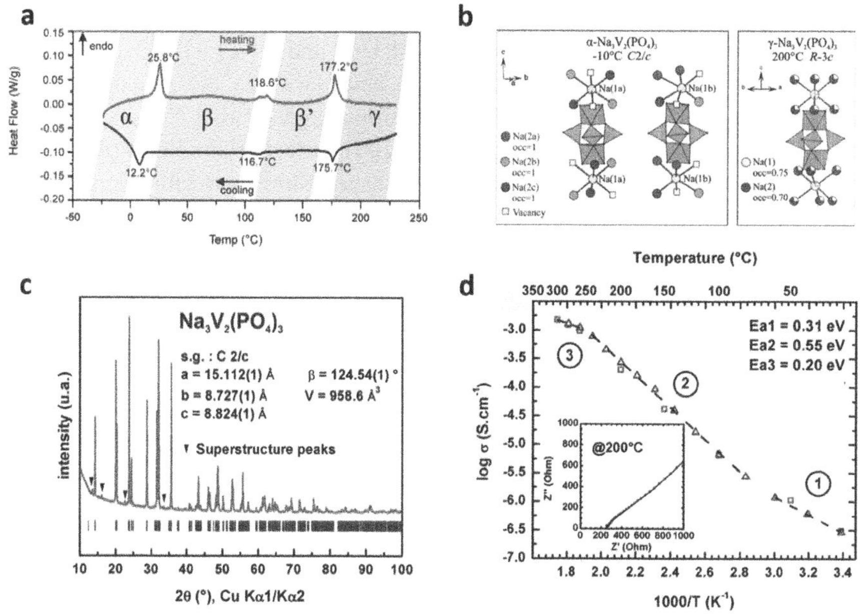

Figure 2.7. *(a) DSC curve recorded for $Na_3V_2(PO_4)_3$ from –30 to 225°C showing the phase transition between four different crystal structures (α, β, β' and γ). (b) Na^+ ordering around the "lantern units" in the low temperature phase, α-$Na_3V_2(PO_4)_3$, and in the high temperature phase, γ-$Na_3V_2(PO_4)_3$. (c) X-ray diffraction patterns of $Na_3V_2(PO_4)_3$ at 25°C. (d) Conductivity measurements of NVP on heating (red square) and cooling (blue triangle). Inset shows Nyquist plot obtained at 200°C Reprinted with permission from Chotard J.-N., Rousse G., David R., Mentré O., Courty M., Masquelier C. (2015). Copyright (2020). American Chemical Society. Reprinted with permission Lalère F., Leriche J.B., Courty M., Boulineau S., Viallet V., Masquelier C., Seznec V. (2014). Copyright (2020). Elsevier. For a color version of this figure, see www.iste.co.uk/monconduit/batteries.zip*

Until recently, $Na_3V_2(PO_4)_3$ was described in a hexagonal unit cell with the "classical" *R*-3*c* space group as adopted by many NASICON compositions. Chotard *et al.* (2015) performed several differential scanning calorimetry (DSC) measurements in the temperature range of –30 to 225°C and discovered that NVP underwent three distinct reversible phase transitions (Figure 2.7(a)): α ↔ β at 25.8°C, β ↔ β' at 118.6°C, and finally β' ↔ γ at 177.2°C. These reversible phase transitions are the results of changes in the Na^+ ions distribution. The unit cell parameters corresponding to these different polymorphs are presented in Table 2.1.

The high temperature γ-NVP form is described in the space group *R*-3*c* (Figure 2.7(b)) with the lattice parameters *a* = 8.7338(1) Å and *c* = 21.9144(2) Å at 200°C. There are two independent Na sites (Na(1) and Na(2)) in the structure and the anisotropic thermal factors of Na^+, obtained from Single crystal X-Ray diffraction, indicate a great mobility perpendicular to the [001] direction for Na(1) (disc shape) and slightly tilted from [001] for Na(2) (cigar shape). Similar structures have already been reported for other NASICON compounds, including α-$Na_3Ti_2(PO_4)_3$ and α-$Na_3Fe_2(PO_4)_3$ (Kabbour *et al.* 2011).

The low temperature polymorph, α-NVP, crystallizes in a monoclinic unit cell described in the *C*2/*c* space group (Figure 2.7(b)). At first sight, the high and low temperature X-ray diffraction (XRD) patterns look quite similar except for the presence at low temperature of extra diffraction peaks with low intensity and the splitting of the (116) and (211) reflections. It is clear that the rhombohedral unit cell (*R*-3*c*) cannot fully explain all the observed reflections and these superstructure peaks can be successfully indexed in a monoclinic supercell (*Z* = 12), which is two times larger than that of the high temperature rhombohedral cell (*Z* = 6). The refined lattice parameters are *a* = 15.1244(6) Å, *b* = 8.7287(3) Å, *c* = 21.6143(8) Å, β = 90.163(2)° and *V*/*Z* = 237.78(1) $Å^3$ at −10°C. A similar monoclinic distortion had already been reported for $Na_3Ti_2(PO_4)_3$ and $Na_3Fe_2(PO_4)_3$.

The monoclinic supercell generates 31 fully occupied independent atomic positions including five for sodium atoms. The general NASICON framework is retained (VO_6 octahedra and PO_4 tetrahedra forming the "lantern units" along the *c* direction) in the monoclinic structure, while the lowering of symmetry implies an increasing number of independent Wyckoff sites. The original Na(1) (6*b*) site in the γ phase splits into two fully occupied sodium sites Na(1a) (4*e*) and Na(1b) (8*f*). The Na(2) site in the γ phase splits into five sites, including three fully occupied sodium sites Na(2a) (8*f*), Na(2b) (8*f*) and Na(2c) (8*f*), and two vacant sites v1 (4*e*) and v2 (8*f*). This low temperature α-NVP (*C*2/*c*) can be considered as an ordered form of the Na-disordered high temperature phase of NVP (γ, *R*-3*c*).

The DSC measurement also shows the existence of two other forms: β-NVP and β'-NVP (Lalere *et al.* 2014). The structure of β-NVP is indexed in a monoclinic sub-cell (space group *C*2/*c*, *Z* = 4) with *a* = 15.112(1) Å, *b* = 8.727(1) Å, *c* = 8.824(1) Å, β = 124.54(1)°, and *V* = 958.6 $Å^3$ (Figure 2.7(c)). Nevertheless, from high-quality synchrotron data, a great number of tiny extra superstructure peaks are clearly observed, and they cannot be fully indexed by the monoclinic cell except considering an incommensurate modulated structure. Considering the β-NVP, the modulation can be fully explained by using two vectors: $\mathbf{q_1}$ = (0.573(1), 0, 0.7143(7)) and $\mathbf{q_2}$ = (0.210(3), 0.787(2), 0.502(2)). For the β' phase, the situation is more complex since three different vectors are needed to index all the reflections: $\mathbf{q_1}$ = (0.567(4), 0, 0.732(3)), $\mathbf{q_2}$ = (0.219(2), 0.780(1), 0.516(2)) and $\mathbf{q_3}$ = (0.784(2), 0.780(1), 0.484(2)).

The conductivity measurement of NVP for both heating and cooling sequences is reported in Figure 2.7(d). The data follow some sort of diffuse transition profile that can be approximated by three linear domains with activation energies of 0.31, 0.55 and 0.20 eV, respectively, which is in good agreement with the DSC results. The significant decrease in the activation energy from low temperature to high temperature is typical of order-disorder transitions toward a superionic conducting phase into which Na^+ ions are disordered.

2.2.3.2. *Electrochemical properties and applications of NVP in Na batteries*

Gopalakrishnan and Kasthuri Rangan (1992) reported the possible extraction of three Na^+ ions and the formation of the novel sodium-free $V^{4+}V^{5+}(PO_4)_3$ composition by chemical oxidation, but this result has never been reproduced by other groups. Afterwards, the electrochemical extraction and insertion of Na^+ from/into $Na_3V_2(PO_4)_3$ were extensively investigated.

The electrochemical performance of carbon coated NVP was first reported by Jian *et al.* (2012). An Na//C-NVP half-cell cycled at a cycling rate of *C*/20 at the voltage of 2.7–3.8 V versus Na^+/Na can deliver a reversible capacity of 92 $mAh{\cdot}g^{-1}$ after 10 cycles. A flat plateau is observed at 3.4 V versus Na^+/Na with a polarization of 0.07 V, corresponding to the (de)intercalation of two Na^+ between $Na_3V_2(PO_4)_3$ and $NaV_2(PO_4)_3$ (Figure 2.8(a)). When used as negative electrode, C-NVP//Na half-cell can be cycled at a current rate of *C*/40 at the voltage of 1.0–3.0 V versus Na^+/Na to deliver a reversible capacity of 59 $mAh{\cdot}g^{-1}$ after 50 cycles. A flat voltage-composition plateau is observed at 1.6 V versus Na^+/Na with nearly no polarization, corresponding to the insertion of one Na^+ into $Na_3V_2(PO_4)_3$ (Figure 2.8(b)).

In order to understand the structural evolution behind this electrochemical reactivity, *in situ* XRD was performed on a Na//NVP half-cell cycled between 2.7 and 3.7 V at a current rate of *C*/10 (Figure 2.8(c)) (Jian *et al.* 2013). The reaction is clearly biphasic during the Na^+-extraction process and it is highly reversible in the full charge/discharge cycle. In NVP, there are three sodium ions per formula unit, however only two of them locating on the Na(2) (18*e*) sites can be extracted. The third sodium on Na(1) (6*b*) site remains in the NVP structure even at the end of charge. Indeed, even though the sodium-free $V_2(PO_4)_3$ structure was reported by

Gopalakrishnan and Kasthuri Rangan (1992), it has never been obtained by electrochemical methods.

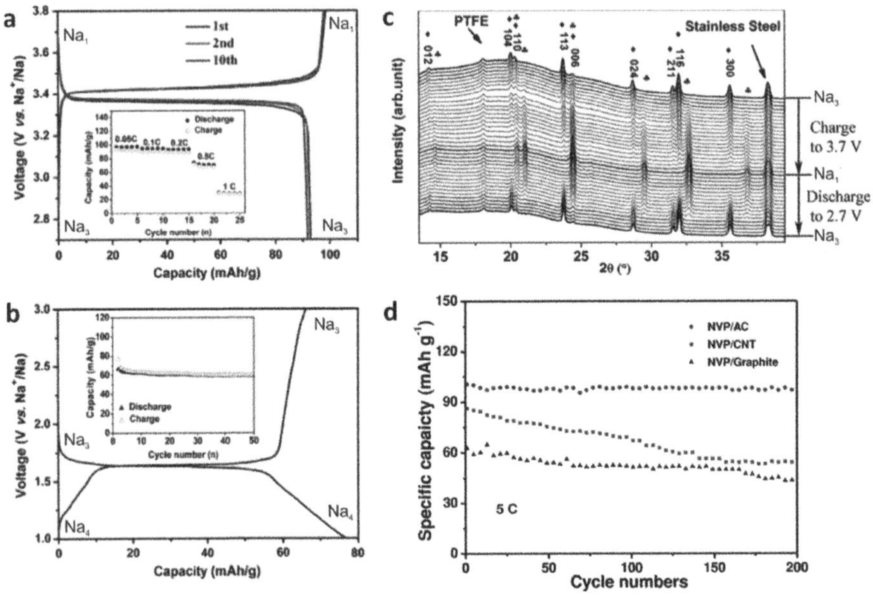

Figure 2.8. *(a) Charge/discharge profiles for the C-coated $Na_3V_2(PO_4)_3$ sample at the cycling rate of C/20 at the voltage of 2.7–3.8 V versus Na^+/Na. Inset shows the rate performance. (b) Charge/discharge profiles for the C-coated $Na_3V_2(PO_4)_3$ sample recorded at the cycling rate of C/40 at the voltage of 1.0–3.0 V versus Na^+/Na. Inset shows the cycling performance. (c) In situ XRD patterns recorded on a Na//$Na_3V_2(PO_4)_3$ half-cell cycled between 3.7 and 2.7 V at a current rate of C/10. (d) Cycling performance of the NVP/AC, NVP/CNT and NVP/graphite electrodes for 200 cycles at 5 C Reprinted with permission Jian Z., Zhao L., Pan H., Hu Y.-S., Li H., Chen W., Chen L. (2012). Copyright (2020). Elsevier. Reprinted with permission from Jian Z., Han W., Lu X., Yang H., Hu Y.-S., Zhou J., Li J., Chen W., Chen D., Chen L. (2013). Copyright (2020). Wiley. Reprinted with permission from Li S., Dong Y., Xu L., Xu X., He L., Mai L. (2014). Copyright (2020). Wiley. For a color version of this figure, see www.iste.co.uk/monconduit/batteries.zip*

The presence of the highly covalent P–O bonds in the structure and a lack of electron exchange between vanadium centers are responsible for the poor electronic conductivity of NVP, which adversely affects the overall electrochemical performance, especially at high cycling rates. In order to improve the electronic conductivity and to obtain better cycling performance, nanocomposites with electronically conductive additives (such as graphene, CNT, interconnected frameworks, etc.) were reported, not only to enhance the grain conductivity but also to prevent the particle agglomeration during the long-term cycling at high current rates, and these carbon matrices are proposed to have indeed a significant influence on the performance of $Na_3V_2(PO_4)_3$. "Double Carbon Wrapped" NVP (NVP first wrapped by a layer of amorphous carbon, and then graphitized carbon) was reported to possess outstanding stability upon long-term cycling (up to 10,000 cycles at 100*C*) in half-cells using Na at the negative electrode (Li *et al.* 2017b). Figure 2.8(d) presents the cycling performance of NVP coated by acetylene carbon (AC) nanospheres, CNT and graphite nanosheets. Among them, NVP/AC composites are reported to possess outstanding properties.

Compared to the classical organic-based liquid electrolytes, aqueous electrolytes may present significant advantages in terms of lower cost, higher safety and environmental footprint. NVP was thus also investigated as positive electrode material with 1 M Na_2SO_4 electrolyte in a three-electrode system. Two symmetric redox peaks were observed from the cyclic voltammetry (CV) curves at various scan rates as a signature of a highly reversible Na^+ intercalation/de-intercalation reaction (Song *et al.* 2014). The Na^+ diffusion coefficient was determined to be as high as 10^{-10} cm^2 s^{-1}. The formation of the carbon coating and nanocomposites was shown to improve the cycling performance of NVP in aqueous-based system. For instance, Zhang *et al.* (2015) prepared carbon-coated NVP nanocomposites by sol-gel method to be used as positive electrode between 0 and 0.9 V versus saturated calomel electrode (SCE) in 1 M Na_2SO_4 electrolyte. The battery was shown to deliver a specific discharge capacity of 94.5 mAh·g^{-1} at 10*C*, 90.5 mAh·g^{-1} at 15*C* and 71.7 mAh·g^{-1} at 20*C*. Furthermore, an all NASICON-structured aqueous NIB with $NaTi_2(PO_4)_3$ at the negative electrode and $Na_3V_2(PO_4)_3$ at the positive electrode demonstrated a flat discharge plateau at 1.2 V, corresponding to the voltage difference between the potential of NVP in discharge (0.45 V versus SCE) and the potential of NTP in charge (–0.80 V versus SCE). When cycled between 0.5 and 1.6 V, the cell delivered an energy density of 29 Wh kg^{-1}. Other NASICON materials were also investigated as materials for aqueous-based systems, such as $Na_3MnTi(PO_4)_3$ (Gao and Goodenough 2016) and $Na_2VTi(PO_4)_3$ (Wang *et al.* 2018).

2.2.3.2.1. Symmetric NASICON NVP//NVP full cells

The use of NVP in "symmetric cells" was reported by several groups. Indeed, the V^{3+} ions in NVP can undergo either oxidation ($V^{3+} \rightarrow V^{4+} + e^-$) or reduction ($V^{3+} + e^- \rightarrow V^{2+}$) at 3.4 V and 1.6 V versus Na^+/Na, respectively. The key reactions can be described as follows:

Negative electrode:

$$Na_3V^{3+}{}_2(PO_4)_3 + x\,Na^+ + x\,e^- \rightarrow Na_{3+x}V^{2+}{}_xV^{3+}{}_{2-x}(PO_4)_3$$

Positive electrode:

$$Na_3V^{3+}{}_2(PO_4)_3 \rightarrow Na_{3-x}V^{4+}{}_xV^{3+}{}_{2-x}(PO_4)_3 + x\,Na^+ + x\,e^-$$

Overall cell reaction:

$$2\,Na_3V^{3+}{}_2(PO_4)_3 \rightarrow Na_{3+x}V^{2+}{}_xV^{3+}{}_{2-x}(PO_4)_3 + Na_{3-x}V^{4+}{}_xV^{3+}{}_{2-x}(PO_4)_3$$

Box 2.1. *Key reactions in NVP symmetric cells*

For instance, Zhang *et al.* (2016c) assembled a symmetric NVP//NVP full cell using self-supported NVP: carbon foam electrodes. The cell was cycled in an electrolyte containing 1 M $NaClO_4$ in ethylene carbonate: dimethyl carbonate (EC/DMC = 1/1 by weight) with 5 wt.% fluoroethylene carbonate (FEC) additive. The output voltage of the cell is 1.8 V, delivering a capacity of 90.2 $mAh \cdot g^{-1}$ at *C*/4 and a high energy density of 162 $Wh \cdot kg^{-1}$. This symmetric cell exhibits long cycling stability with 81% of capacity retention at 2*C* after 280 cycles. Duan *et al.* (2014) prepared NVP@C core-shell nanocomposites through a sol-gel assisted method. Ascorbic acid and polyethylene glycol 400 (PEG-400) were used in order to provide a coating surface of conductive carbon and to control/limit the particle growth. A symmetric full cell was assembled with the weight ratio of 3/5 between the positive and negative electrodes. The output voltage of the cell was 1.7 V with a capacity of 90.9 $mAh \cdot g^{-1}$ at 2*C* and a high energy density of 154.5 $Wh \cdot kg^{-1}$, which is in good agreement with earlier results.

ASSBs have been the subject of intense research in recent years due to safety concerns for many applications. Non-flammable ion-conducting solid electrolytes have been considered as promising candidates for alternative energy storage devices. Noguchi *et al.* (2013) fabricated sodium-based ASSBs by a combination of screen printing and hot pressing, using NASICON $Na_3Zr_2(SiO_4)_2PO_4$ (NZSP) as a solid electrolyte. A conventional non-symmetric Na-metal containing Na/NZSP/NVP all-solid-state half-cell was first designed, which demonstrated reversible capacities of 30 $mAh \cdot g^{-1}$ and 80 $mAh \cdot g^{-1}$ at 1.6 V and 3.4 V, respectively. The capacity of NVP positive electrodes was found to decay over 50 cycles while such a tendency was not observed for the negative electrodes. Then, a symmetric NVP/NZSP/NVP ASSB was fabricated and cycled at 25°C between 0.01 V and 1.9 V at current densities of 1.2 and 10 $\mu A \cdot cm^{-2}$ (*C* rate was not mentioned). The discharge capacity for the first cycle was 68 $mAh \cdot g^{-1}$. The authors mentioned that, although disappointing, their preliminary results indicated that the cycle performance of the all-solid-state symmetric cell could be significantly improved by decreasing the internal porosity of the NASICON solid electrolyte as well as by precise control of the contact at the NVP/NASICON interface.

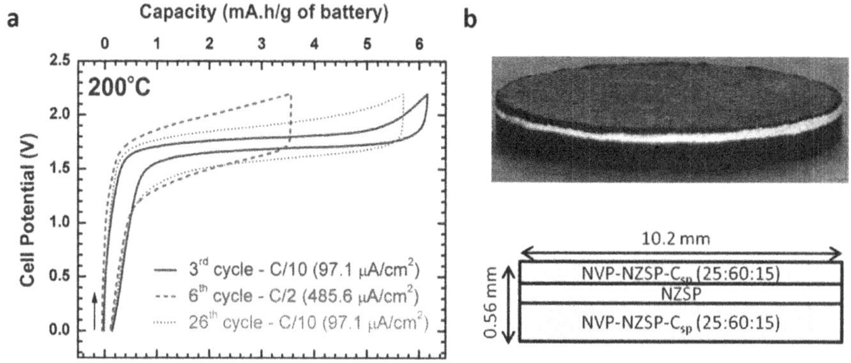

Figure 2.9. *(a) Galvanostatic cycling at current rate of C/2 or C/10 at 200°C, with potential limitation of 2.2 V for the third, sixth and 26th cycles. (b) Photo of the battery and battery stack composition. Reprinted with permission from Chotard J.-N., Rousse G., David R., Mentré O., Courty M., Masquelier C. (2015). Copyright (2020). American Chemical Society. For a color version of this figure, see www.iste.co.uk/monconduit/ batteries.zip*

Lalere *et al.* (2014) successfully developed a symmetric monolithic all-solid-state NVP/NZSP/NVP battery by using the spark plasma sintering (SPS) technique (Figure 2.9). Positive and negative electrodes were fabricated using 40 mg and 80 mg mixtures of 25% of NVP, 60% of NZSP and 15% of carbon super P. Between the two electrodes, 30 mg of NZSP solid electrolyte were added. The battery was first assembled layer by layer in a 10 mm diameter graphite die, and then the sandwiched cell was sintered at 900°C by SPS for 10 min. The cell was cycled at 200°C with 85% of the theoretical capacity of NVP at a current rate of *C*/10. The whole NASICON battery offered good chemical stability and good interface between electrodes and electrolyte, which was shown to greatly improve the electrochemical performance of ASSBs.

2.2.3.2.2. Recent achievements in NIBs using $Na_3V_2(PO_4)_3$ as positive electrodes

NVP can be assembled as positive electrode material with a large variety of negative electrode materials (such as carbonaceous materials, polyanionic compounds, metal oxides, alloys, metal sulfides) to fabricate sodium full cells with high electrochemical performance.

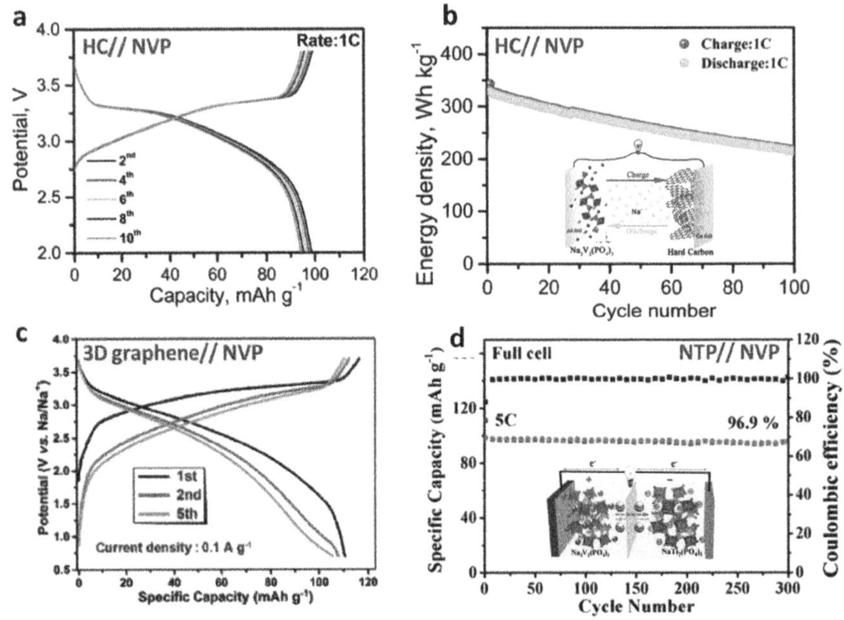

Figure 2.10. *(a) Typical charge–discharge profiles of a liquid-based hard carbon//NVP@NSC full cell for the first 10 cycles at 1 C cycling rate. (b) Energy density of a hard carbon//NVP@NSC full cell for 100 cycles at 1 C. Inset illustrates the structure of the hard carbon// NVP@NSC full cell. (c) Typical charge–discharge profile of a 3D graphene//NVP full cell in its first five cycles at a current density of 0.1 A·g^{-1}. (d) The cycling performance and the corresponding coulombic efficiency of a NTP//NVP full cell. Inset shows a schematic illustration of the NTP//NVP full cell. Reprinted with permission from Liang X., Ou X., Zheng F., Pan Q., Xiong X., Hu R., Yang C., Liu M. (2017). Copyright (2020). American Chemical Society. Reprinted with permission from Cao X., Pan A., Liu S., Zhou J., Li S., Cao G., Liu J., Liang S. (2017). Copyright (2020). Wiley. Reprinted with permission from Ren W., Zheng Z., Xu C., Niu C., Wei Q., An Q., Zhao K., Yan M., Qin M. Mai L. (2016). Copyright (2020). Elsevier. For a color version of this figure, see www.iste.co.uk/monconduit/batteries.zip*

Carbonaceous materials (including graphite, hard carbon, graphene, etc.) are promising and cost-effective negative electrodes for NIBs. Recently, a sodium full cell based on nitrogen and sulfur dual-doped carbon layer (NSC) wrapped $Na_3V_2(PO_4)_3$ nanoparticles (NVP@NSC, 96.47 wt.% NVP) and hard carbon was reported by Liang *et al.* (2017) (Figures 2.10(a) and (b)). The battery can deliver an impressive discharge capacity of 99.0 mAh·g^{-1} at 1*C* with an energy density of 329 Wh·kg^{-1} with a satisfactory capacity retention. In addition to hard carbon, graphene negative electrodes were proposed. Graphene//NVP full cells displayed a sloping profile with an average operating voltage of 2.7 V and a high specific capacity of 109 $mAh \cdot g^{-1}$ at a current rate of 0.1 $A \cdot g^{-1}$ (Zhu *et al.* 2015) (Figure 2.10(c)).

NASICON-type NTP is also a promising negative electrode for NIBs. A NTP//NVP full cell has been developed by Ren *et al.* (2016) (Figure 2.10(d)). The 3D NVP nanofiber network was synthesized by self-sacrificed route. In this self-sacrificed route, the fast nucleation driven by antisolvent *n*-dimethylformamide led to the formation of NVP microspheres. In the following 3 h, the surface of microspheres was converted into nanosheet-assembled microflowers via Ostwald ripening. With the increasing time, the metastable nanosheet presenting on the surface served as the self-sacrificed template, which was dissolved and recrystallized into the nanofibers, forming a 3D network. The full cell exhibited excellent cycling stability, with nearly 100% coulombic efficiency and 96.9% capacity retention over 300 cycles at 5*C*.

Only 2 Na^+ per formula unit can be extracted from $Na_3V_2(PO_4)_3$: The oxidation state of vanadium is limited to +4 ($Na_3V^{3+}{}_2(PO_4)_3 \rightarrow 2Na^+ + 2e^- + NaV^{4+}{}_2(PO_4)_3$), while +5 could be theoretically achieved. By cationic substitution, a possible activation of the V^{5+}/V^{4+} redox couple at higher voltages may occur and contribute to a substantial increase in the energy density of NVP-based compounds. This hypothesis was first proposed by Lalère *et al.* (2015) by introducing a spectator ion such as Al^{3+} (0.5 per formula unit) into the NVP structure. By this way, 0.5 V^{4+} per formula unit was activated and oxidized to V^{5+} at 3.9 V versus Na^+/Na (Figure 2.11(a)), increasing the theoretical energy density up to 425 Wh kg^{-1}. However, the observed substantial irreversibility is one of the greatest challenges for these substituted materials, as also reported for other Al-substituted polyanionic systems (Kovrugin *et al.* 2017).

$Na_2TiV(PO_4)_3$ and $Na_3FeV(PO_4)_3$ have recently been reported and described in the rhombohedral (*R*-3*c*) and monoclinic (*C*2/*c*) cells, respectively. Na^+ deintercalation from $Na_2TiV(PO_4)_3$, with the formation of $NaTiV(PO_4)_3$, occurs through a biphasic mechanism at 3.33 V versus Na^+/Na for a theoretical capacity of 62.4 $mAh \cdot g^{-1}$. Na^+ intercalation into $Na_2Ti^{4+}V^{3+}(PO_4)_3$, to form $Na_4Ti^{3+}V^{2+}(PO_4)_3$, occurs, first through a biphasic reaction at 2.17 V versus Na^+/Na for the Ti^{4+}/Ti^{3+} redox couple, and then through a monophasic solid solution type reaction at 1.60 V versus Na^+/Na for the V^{3+}/V^{2+} redox couple, with the formation of the intermediate phase $Na_3Ti^{3+}V^{3+}(PO_4)_3$. $Na_3FeV(PO_4)_3$ delivers a reversible capacity of 103 mAh g^{-1} between 2 and 3.8 V at a current rate of 1*C*. The two charge/discharge plateaus observed at 3.3 V and 2.5 V versus Na^+/Na correspond to the V^{4+}/V^{3+} and Fe^{3+}/Fe^{2+} redox couples, respectively (Figures 2.11(b) and (c)).

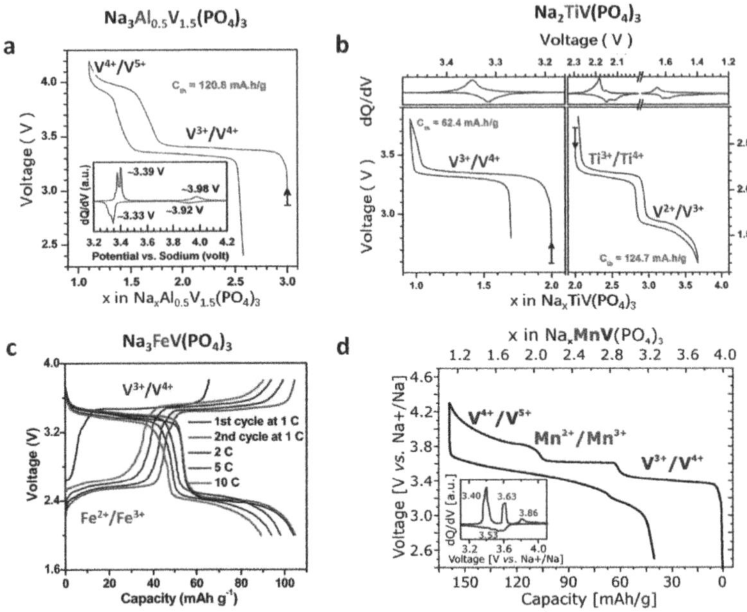

Figure 2.11. *Electrochemical signatures of (a) $Na_3Al_{0.5}V_{1.5}(PO_4)_3$, (b) $Na_2TiV(PO_4)_3$, (c) $Na_3FeV(PO_4)_3$ and (d) $Na_4MnV(PO_4)_3$ electrode materials upon Na^+ extraction or Na^+ insertion. Reprinted with permission from Lalère F., Seznec V., Courty M., David R., Chotard J.N., Masquelier C. (2015). Copyright (2020). The Royal Society of Chemistry. Reprinted with permission from Lalère F., Seznec V., Courty M., Chotard J.N., Masquelier C. (2018). Copyright (2020). The Royal Society of Chemistry. Reprinted with permission from Zhou W., Xue L. Lü X., Gao H., Li Y., Xin S., Fu G., Cui Z., Zhu Y., Goodenough J.B. (2016b). Copyright (2020) American Chemical Society. Reprinted with permission from Chen F., Kovrugin V. M., David R., Mentré O., Fauth F., Chotard J.-N., Masquelier C. (2018). Copyright (2020). Wiley. For a color version of this figure, see www.iste.co.uk/monconduit/batteries.zip*

Recently, Zhou *et al.* (2016b) reported a quite interesting compound, $Na_4MnV(PO_4)_3$ (NMnVP), which generates two reversible charge/discharge voltage-composition "plateaus" at 3.3 V and 3.6 V versus Na^+/Na, delivering a reversible capacity of 101 mAh·g^{-1} at a cycling rate of 1*C*. $Na_4MnV(PO_4)_3$ was described in the rhombohedral cell (*R*-3*c*), with the two sodium sites fully occupied. Na(1) exhibited almost isotropic thermal oscillations, whereas the shape of thermal ellipsoids around Na(2) could be described as a flat plate slightly tilted along [001]. This thermal-motion behavior of Na^+ ions was quite different from that observed within the non-substituted $Na_3V_2(PO_4)_3$ compound where both Na(1) and Na(2) sites are partially occupied. Through galvanostatic electrochemical charge of the Na//NMnVP cell, three Na^+ per formula unit can be extracted between 2.5 V and 4.3 V versus Na^+/Na, thus delivering a high gravimetric capacity of 156 $mAh \cdot g^{-1}$ (Figure 2.11(d)). Three redox phenomena can be observed during the first charging process at 3.4, 3.6 and 3.9 V versus Na^+/Na, corresponding to the V^{4+}/V^{3+}, Mn^{3+}/Mn^{2+} and V^{5+}/V^{4+} redox couples, respectively. Unlike the biphasic mechanism in NVP, the V^{4+}/V^{3+} and V^{5+}/V^{4+} plateaus in NMnVP occurred through a solid solution mechanism, while the Mn^{3+}/Mn^{2+} process is associated with a biphasic reaction, as detected by XRD analysis performed in *operando* conditions during the cycling of the battery. Mn substitution into NVP structure promoted the electrochemical oxidation of V^{4+} into V^{5+}, bringing an important increase in the gravimetric capacity at a higher operating voltage. During the first discharge, an irreversible structural evolution is observed, then leading to a solid solution mechanism all over the composition range, which is a common drawback of such materials upon the activation of the V^{5+}/V^{4+} couple, like recently identified for the third sodium extraction from $Na_3V_2(PO_4)_2F_3$ (Yan *et al.* 2019).

There is no doubt that $Na_3V_2(PO_4)_3$ and its iso-structural compounds are some of the most promising materials for positive electrodes in NIBs. Nevertheless, there are still several unanswered questions as well as further developments that have to be addressed in order to improve the performance and the stability of these materials:

– In partial substituted NVP phases, such as $Na_3V_{1.5}Al_{0.5}(PO_4)_3$ and $Na_4VMn(PO_4)_3$, the V^{5+}/V^{4+} redox couple can be activated at a high cut-off voltage; nonetheless, this behavior has never been observed in the pristine $Na_3V_2(PO_4)_3$ composition. The presence of a foreign cation in the NVP framework seems to be an essential factor in the activation of V^{5+}/V^{4+} redox couple, but this has never been discussed in details elsewhere.

– An irreversible structural modification as well as a significant capacity loss were observed for $Na_3V_{1.5}Al_{0.5}(PO_4)_3$ and $Na_4VMn(PO_4)_3$ upon the activation of the V^{5+}/V^{4+} redox couple at high voltage. However, the reason behind this peculiar behavior has never been addressed in the literature. Therefore, the future development of NVP would be identifying the nature as well as the content of the cationic substituent required to activate the V^{5+}/V^{4+} redox couple and stabilize the NVP framework at a high cut-off voltage.

2.3. $Na_3V_2(PO_4)_2F_3$ used as a model framework in sodium-ion battery applications

2.3.1. *Structural description and compositional diversity*

Le Meins *et al.* (1999) discovered a family of compounds with the general chemical formula $Na_3M_2(PO_4)_2F_3$, where M can be V^{3+}, Ga^{3+}, Al^{3+}, Fe^{3+} or Cr^{3+}. The crystal structure of $Na_3M_2(PO_4)_2F_3$ is built from bi-octahedral units $M_2O_8F_3$, which are linked together by highly covalent phosphate groups, and the Na^+ ions occupy the empty spaces in the structural framework. The Na^+ ions distribution depends strongly on the nature of the M metal and the temperature. At ambient temperature, $Na_3V_2(PO_4)_2F_3$ crystallizes in the tetragonal system within the *P4₂/mnm* space group with $a = b = 9.047(2)$ Å and $c = 10.705(2)$ Å. The Na^+ ions were announced to be distributed on two different crystallographic sites, the Na(1) site fully occupied and the Na(2) site half occupied (Figure 2.12(b)). Recently, by using high-resolution synchrotron XRD, Bianchini *et al.* (2014) revealed an obvious systematic splitting of the (200)/(020) and (400)/(040) diffraction lines (Figure 2.12(c)) for the stoichiometric $Na_3V_2(PO_4)_2F_3$ and thus described its crystal structure in the *Amam* space group with a subtle orthorhombic distortion ($a \neq b$, $b/a = 1.002$). This distortion is correlated to the distribution of the Na^+ ions over three different crystallographic sites: Na(1) is fully occupied while Na(2) and Na(3) are partially occupied (Figure 2.12(d)). This subtle orthorhombic distortion and the triangular distribution of Na^+ ions in $Na_3V_2(PO_4)_2F_3$ allow to minimize the Na^+-Na^+ electrostatic repulsions, as later on confirmed by the theoretical work of Dacek *et al.* (2016).

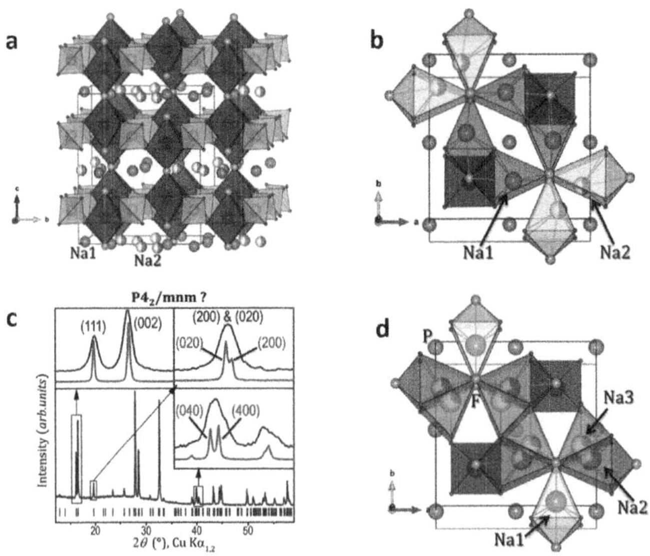

Figure 2.12. *(a) Structural framework of $Na_3V_2(PO_4)_2F_3$. (b) Distribution of the Na^+ ions in the crystallographic plane localized at z = 0 in $Na_3V_2(PO_4)_2F_3$ as proposed by Le Meins* et al. *(1999). (c) Comparison of the XRD patterns recorded for $Na_3V_2(PO_4)_2F_3$ using a laboratory Cu $K\alpha_{1,2}$ diffractometer (blue) and the high-resolution synchrotron X-ray radiation at $\lambda = 0.6687$ Å (Red). (d) Distribution of the Na^+ ions in the crystallographic plane localized at z = 0 in $Na_3V_2(PO_4)_2F_3$ as proposed by Bianchini* et al. *(2014). For a color version of this figure, see www.iste.co.uk/monconduit/batteries.zip*

Upon heating up to 127°C, an order–disorder phase transition occurs. The structure of $Na_3V_2(PO_4)_2F_3$ above this temperature is thus described in the tetragonal *I4/mmm* space group to take into account the circle distribution of the Na^+ ions around the fluorine atoms localized at the extrema of the bi-octahedral units $V_2O_8F_3$. The observation of the "cationic rings" reveals the increased mobility of the alkali ions at high temperature.

In recent years, $Na_3V_2(PO_4)_2F_3$ was intensively investigated in particular through cationic and/or anionic substitution in order to tailor its redox potential, its diffusion properties and thus its electrochemical performance. Depending on the nature of the substitution, different goals could be targeted:

– *Exchange of the mobile ion*: The great size of the cavities within the crystal structure, together with the high mobility of Na^+ ions, makes $Na_3V_2(PO_4)_2F_3$ an excellent ionic conductor. The Na^+ ions can be easily replaced by other alkali ions, either by a chemical or by an electrochemical ion exchange reaction. The ion exchange has been reported for Li^+, K^+ and Ag^+, and the obtained compositions can (or could) be used as positive electrode materials in Li^+, K^+ and Ag^+-ion batteries, respectively. In $K_3V_2(PO_4)_2F_3$ and $Ag_3V_2(PO_4)_2F_3$, the K^+ and Ag^+ ions simply replace the Na^+ ions in the same crystallographic sites of the 3D structural framework and the resulting materials can also be described in the *Amam* space group.

– *Modulation of the redox potential*, and thus of the number of Na^+ ions participating into the electrochemical reaction *through ionic substitution*: $Na_3V_2(PO_4)_2F_3$ contains V^{3+}, which can be oxidized into V^{4+} upon Na^+ extraction, and the electrochemical potential depends on the redox value of the V^{4+}/V^{3+} redox couple. The V^{3+} ions can be fully or partially substituted by V^{4+}, Al^{3+} and Ga^{3+}. This cationic substitution modifies the crystal structure, the electronic structure of the active material, and thus its redox potential and possibly the number of Na^+ ions to be extracted. The other alternative to tailor the redox potential is to employ the inductive effect originated from the anions belonging to the bi-octahedral units $V_2O_8F_3$. By the use of density functional theory (DFT), Xu *et al.* (2014) predicted that the terminal fluorine anions could be substituted by oxygen or by other halogen elements, such as chlorine and bromine. Depending on the nature of the substituent, the redox potential of the V^{4+}/V^{3+} redox couple in $Na_3V_2(PO_4)_2F_3$ is lowered when the terminal fluorine is replaced by chlorine or bromine.

2.3.2. $Na_3V_2(PO_4)_2F_3$: a promising active material for positive electrodes in NIBs

2.3.2.1. *A versatile promising material: from LIBs to NIBs*

Barker *et al.* (2006) were the first to employ $Na_3V_2(PO_4)_2F_3$ as a positive electrode material in rechargeable LIBs and they have successfully recorded the first electrochemical signature of a graphite/1M $LiPF_6$ in EC/DMC (2/1 by weight)/$Na_3V_2(PO_4)_2F_3$ full cell. Two alkali ions were extracted and re-inserted reversibly into the structure of $Na_3V_2(PO_4)_2F_3$ at 3.88 and 4.43 V versus Li^+/Li and the full cell showed very limited capacity fading after 40 cycles performed at a current rate of 2*C*. Similar full cells were cycled in a larger potential window, with a cut-off voltage of 5.0 V versus Li^+/Li, in order to extract the third Na^+ ion. A capacity corresponding to the extraction of three Na^+ ions could be observed in charge, but only two alkali ions could be re-inserted in discharge (Gover *et al.* 2006). They speculated that only a fraction of the third Na^+ could be cycled reversibly. The XRD analysis performed *ex situ* on the fully charged sample recovered at 5.0 V revealed that "the full extraction of sodium ion from the $Na_3V_2(PO_4)_2F_3$ material induced some irreversible structural degradation".

With the recent emergence of NIBs, the use of $Na_3V_2(PO_4)_2F_3$ as positive electrode materials was reported by several groups. The charge–discharge profile obtained for $Na_3V_2(PO_4)_2F_3$ versus Na metal in a Na^+-containing organic electrolyte is characterized by two electrochemical voltage domains corresponding to the reversible extraction of two Na^+ ions at 3.70 and 4.20 V versus Na^+/Na (Figure 2.4(b)) resulting in a reversible capacity of 125 $mAh \cdot g^{-1}$.

Ponrouch *et al.* (2013) first reported the electrochemical performance of a hard carbon//$Na_3V_2(PO_4)_2F_3$ coin cell-type full cell, cycled in an electrolyte containing 1 M $NaPF_6$ in a mixture of EC/PC/DMC = 0.45/0.45/0.10. The full cell could deliver a capacity of 100 $mAh \cdot g^{-1}$ at the cycling rate of *C*/5. Despite an increased polarization observed at 5*C*, a capacity of 70 $mAh \cdot g^{-1}$ could still be achieved (Figure 2.13(a)). Broux *et al.* (2019) have recently reported that an optimized hard carbon// $Na_3V_2(PO_4)_2F_3$ 18650 full cell prepared by CEA (France) and cycled at 1*C* at ambient temperature could maintain more than 80% of its initial capacity after 4,000 cycles (Figure 2.13(b)). Furthermore, the electrochemical signatures of the cell obtained at *C*/10 at 25°C and at 0°C and their corresponding *operando* XRD patterns were very similar, indicating that Na^+ can maintain high mobility in $Na_3V_2(PO_4)_2F_3$ even at low temperatures.

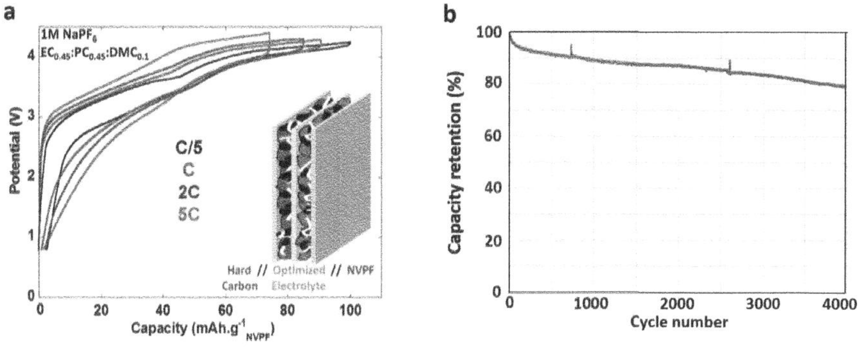

Figure 2.13. *(a) Voltage versus capacity electrochemical curves obtained for a hard carbon//$Na_3V_2(PO_4)_2F_3$ full cell cycled in an electrolyte containing 1 M $NaPF_6$ in EC/PC/DMC (0.45/0.45/0.10) at different current rates (Ponrouch* et al. *2013). (b) Long-term cyclability of a Hard carbon//$Na_3V_2(PO_4)_2F_3$ 18,650 75Wh·kg^{-1} full cell cycled at ambient temperature and at a 1 C cycling rate upon charge and discharge (Broux* et al. *2019). For a color version of this figure, see www.iste.co.uk/monconduit/batteries.zip*

2.3.2.2. *A complex phase diagram observed upon cycling, controlled by the charge ordering*

As already mentioned, the electrochemical signature obtained upon charge and discharge of $Na_3V_2(PO_4)_2F_3$ versus Na^+/Na is characterized by the presence of two voltage domains at 3.70 and 4.20 V versus Na^+/Na (Figure 2.4(b)), suggesting the formation of intermediate phases especially at the intermediate mixed valence composition $Na_2V^{3+}V^{4+}(PO_4)_2F_3$ corresponding to the potential jump.

By the use of *operando* synchrotron XRD, Bianchini *et al.* (2015) identified a succession of several biphasic and monophasic (solid solution) reactions during the desodiation process from $Na_3V_2(PO_4)_2F_3$. The monophasic mechanism was only observed in the composition range $1.3 \leq x \leq 1.8$ in $Na_xV_2(PO_4)_2F_3$, while biphasic reactions were identified in all the other composition ranges (Figure 2.14). Several intermediate phases were detected at x = 2.4, 2.2, 2.0, 1.8, 1.3 and 1.0. For all of them, the structural framework remains similar to that of the pristine phase, but a reorganization of the Na^+ ions distribution occurs in order to minimize their mutual electrostatic repulsions. In the pristine composition $Na_3V_2(PO_4)_2F_3$, the two vanadium sites of the bi-octahedral unit are highly symmetric (6 x $d_{V–F/\,V–O}$ = ~2.0 Å) and crystallographically equivalent (Bianchini *et al.* 2014). At the end of charge, in $Na_1V_2(PO_4)_2F_3$, one vanadium site of the bi-octahedral unit becomes highly distorted (4 x $d_{V–O}$ = ~ 1.9 Å, $d_{V–F}$ = 2.00 Å, $d_{V–F}$ = 1.67 Å), while the other remains symmetric (4 x $d_{V–O}$ = ~ 2.0 Å, $d_{V–F}$ = 2.02 Å, $d_{V–F}$ = 1.88 Å). Combining this observation with bond valence sum analysis, the distorted site was assigned to V^{5+} and the symmetric one to V^{3+}.

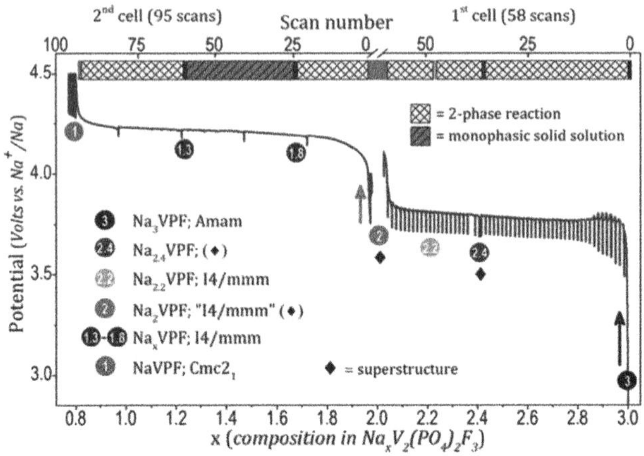

Figure 2.14. *Potential–composition electrochemical curves obtained upon Na^+ extraction from $Na_3V_2(PO_4)_2F_3$. Several biphasic and solid solution reactions were observed during the Na^+ extraction process. The regions where biphasic or solid solution mechanisms occur are indicated in the upper bar of the figure. The Na-content of the pure intermediate phases isolated during cycling are given in the small circles on the electrochemical profile (Bianchini* et al. *2015). For a color version of this figure, see www.iste.co.uk/monconduit/batteries.zip*

2.3.2.3. *A paradox related to vanadium redox couples*

The evolution of vanadium oxidation state in $Na_3V_2(PO_4)_2F_3$ during charge is followed by *operando* X-ray absorption spectroscopy (XAS) at vanadium *K* edge (5,465 eV).

Figure 2.15(a) shows the evolution of the *operando* X-ray absorption near edge structure (XANES) spectra recorded at vanadium *K* edge on an operating $Na//Na_3V_2(PO_4)_2F_3$ half-cell. Each XANES spectrum can be divided into the pre-edge and edge regions (Figure 2.15(a)). The pre-edge signal is related to the forbidden $1s \rightarrow 3d$ transition and its intensity can be used to evaluate the local symmetry of the vanadium site while the edge signal is associated with the allowed $1s \rightarrow 4p$ transition and its position and shape are related to the average vanadium oxidation state and to the metal–ligand bond lengths around vanadium.

In the initial state, the intensity of the pre-edge signal is very weak as all the V^{3+} ions in $Na_3V_2(PO_4)_2F_3$ are in highly symmetric octahedral sites and the $1s \rightarrow 3d$

transition is forbidden. Upon electrochemical oxidation, the intensity of the pre-edge signal increases gradually and shifts toward higher energy (Figure 2.15(a), inset), figuring the oxidation of the V^{3+} ions together with a distortion in their local environment. This oxidation is also obvious through the linear shift in the edge energy of the *operando* XANES spectra (Figure 2.15(a)).

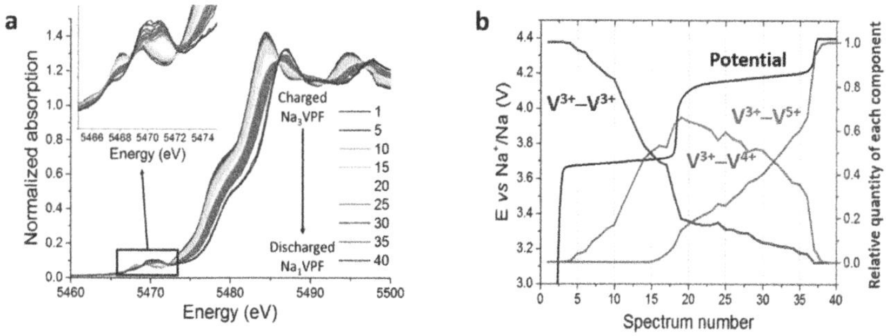

Figure 2.15. *(a) Vanadium K edge XANES spectra collected operando during charge of an Na//$Na_3V_2(PO_4)_2F_3$ half cell cycled at a rate of C/10 per Na^+. Inset shows the evolution of the pre-edge signal; (b) evolution in the concentration of the three principal components required to describe the whole series of Vanadium K edge XANES data recorded operando during the first charge of the battery (Broux* et al. *2017). For a color version of this figure, see www.iste.co.uk/monconduit/batteries.zip*

Three independent components are required to describe the whole system and the concentration profile of each component is given in Figure 2.15(b). Broux *et al.* (2017) did the hypothesis that the first component corresponds to the V^{3+}—F—V^{3+} pairs, as the pristine phase contains only V^{3+} in $V_2O_8F_3$ bi-octahedra. Upon charge in the first potential domain, one V^{3+} per pair is oxidized into V^{4+} leading to the formation of V^{3+}—F—V^{4+} pairs (Figure 2.15(b)). In the second potential domain, these V^{3+}—F—V^{4+} pairs are continuously replaced by V^{3+}—F—V^{5+}, either through the formation of the transient intermediate state V^{4+}—F—V^{4+} and then followed by a disproportionation ($2V^{4+} \rightarrow V^{3+} + V^{5+}$), or through the direct oxidation of V^{3+}—F—V^{4+} into V^{3+}—F—V^{5+}. Whatever the mechanism, the formation of V^{5+} at the end of charge was detected by ^{51}V solid-state nuclear magnetic resonant (NMR) spectroscopy (Broux *et al.* 2017) and by *operando* synchrotron XRD (Bianchini *et al.* 2015).

2.3.3. *Oxygen substitution in $Na_3V_2(PO_4)_2F_3$ and its effects on the electrochemical performance of substituted phases*

Massa *et al.* (2002) reported on the discovery of a new sodium vanadyl fluorite phosphate compound with the chemical formula $Na_3(VO)_2(PO_4)_2F$. The structural framework of this new material is topologically identical to that of $Na_3V_2(PO_4)_2F_3$ where the two terminal fluorine atoms of the bi-octahedral units are replaced by two terminal oxygen atoms. The structure is indexed in the *P4₂/mnm* space group at ambient temperature, and in the *I4/mmm* space group at temperatures greater than 227°C (Tsirlin *et al.* 2011). As in the case of $Na_3V_2(PO_4)_2F_3$, this phase transition is due to the redistribution of the Na^+ ions in the channels of the 3D framework. Sauvage *et al.* (2006) demonstrated that two Na^+ ions per formula unit could be extracted electrochemically in a Na//$Na_3(VO)_2(PO_4)_2F$ half-cell with a poor capacity retention in the high-voltage region. This capacity fading was associated with an increase in the cell polarization due to the formation of a surface layer resulting from the electrolyte degradation; more recently, it was attributed to the presence of impurities in the sample. Indeed, the electrochemical properties of $Na_3(VO)_2(PO_4)_2F$ have recently been revisited by several groups, and an outstanding capacity retention has been demonstrated (Serras *et al.* 2013).

Recent studies have shown that $Na_3V_2(PO_4)_2F_3$ and $Na_3(VO)_2(PO_4)_2F$ are indeed the two end members of a family of materials with the general formula $Na_3V^{3+}{}_{2-y}V^{4+}{}_y(PO_4)_2F_{3-y}O_y$ with $0 \leq y \leq 2$ (Park *et al.* 2014). The partial substitution of oxygen on the terminal fluorine of the bi-octahedral units $V_2O_8F_3$ leads to the formation of $Na_3V^{3+}{}_{2-y}V^{4+}{}_y(PO_4)_2F_{3-y}O_y$ phases, the vanadium ions bonded to these terminal oxygen atoms being oxidized into V^{4+} with the formation of very covalent vanadyl bonds (V=O).

The crystal structures of all the compositions in the $Na_3V^{3+}{}_{2-y}V^{4+}{}_y(PO_4)_2F_{3-y}O_y$ family are very similar as they are all constructed by the interconnection between $V_2O_{8+y}F_{3-y}$ bi-octahedra and PO_4 groups. Park *et al.* (2014) described the structures of these materials in a tetragonal *P4₂/mnm* space group. In fact, a subtle splitting of the (200)/(020) and (400)/(040) diffraction lines is observed for compositions with $y \leq 0.6$ and their structures should then be described in orthorhombic symmetry, using the *Amam* space group. For other compositions, descriptions in the *Amam* or *P4₂/mnm* space groups lead to the same apparent quality of fit. However, the orthorhombic description is required in order to minimize the electrostatic repulsion between the Na^+ ions in the diffusion channels (Broux *et al.* 2016). For the specific case of $Na_3(VO)_2(PO_4)_2F$, several low-intensity diffraction peaks could not be indexed in the *Amam* space group nor *P4₂/mnm*. We proposed a structural description in the monoclinic space group *P2₁/m* with a modulation of occupancies among the three Na^+ sites (Nguyen *et al.* 2019a).

A linear evolution of cell parameters with y is reported for $Na_3V^{3+}{}_{2-y}V^{4+}{}_y(PO_4)_2F_{3-y}O_y$ ($0 \leq y \leq 2$). Nonetheless, the difference in the cell parameters of these phases is so small that XRD is actually not the most appropriate technique to clearly identify each member of this family of compounds. Instead, spectroscopic techniques can be used to precisely characterize these $Na_3V^{3+}_{2-y}V^{4+}_y(PO_4)_2F_{3-y}O_y$ compositions.

Infrared (IR) spectroscopy can identify V=O stretching modes in the composition of interest, and then provide information on the composition of the phase because of the relative intensity and the position of the V=O vibrational bands (Nguyen *et al.* 2019a). The characteristic vibration of V=O bonds can be found at ~920 cm^{-1} and its corresponding overtone at ~1,850 cm^{-1}. As expected, no V=O bond is observed in $Na_3V_2(PO_4)_2F_3$ (Figure 2.16(a)). On the other hand, in $Na_3(VO)_2(PO_4)_2F$, all the vanadium ions form V=O bonds with the apical oxygen atom as confirmed by intense signals at ~920 cm^{-1} and ~1,840 cm^{-1} (Figure 2.16(a)). In the intermediate phases ($0 < y < 2$), there are two signals corresponding to V=O stretching modes at ~920 cm^{-1} (Figure 2.16(a)) and they correspond to the $O{=}V^{4+}{-}F{-}V^{4+}{=}O$ and $O{=}V^{4+}{-}F{-}V^{3+}{-}F$ groups, respectively. The relative intensity between these two signals varies with the y value.

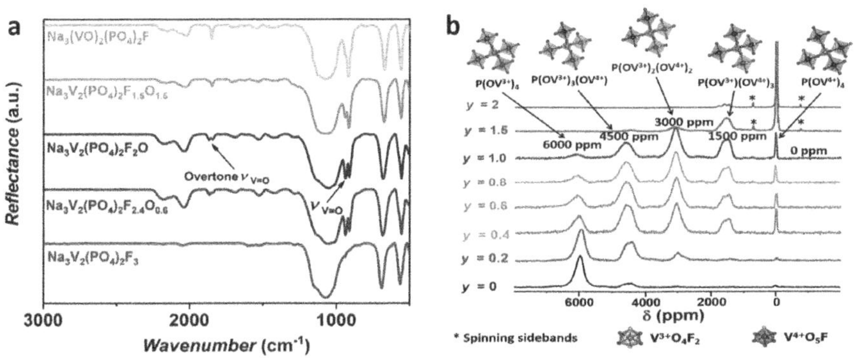

Figure 2.16. *(a) Comparison of the IR spectra of $Na_3V_2(PO_4)_2F_3$, $Na_3V_2(PO_4)_2F_{2.4}O_{0.6}$, $Na_3V_2(PO_4)_2F_2O$, $Na_3V_2(PO_4)_2F_{1.5}O_{1.5}$ and $Na_3(VO)_2(PO_4)_2F$ powders in the wavenumber range of 500–3,000 cm^{-1}; (b) ^{31}P solid-state NMR spectra of different members of the $Na_3V^{3+}_{2-y}V^{4+}_y(PO_4)_2F_{3-y}O_y$ family collected at 100 MHz, MAS = 30 kHz (Nguyen* et al. *2019c). For a color version of this figure, see www.iste.co.uk/monconduit/batteries.zip*

Another technique that can directly distinguish V^{3+} and V^{4+} is *electron paramagnetic resonance (EPR)* where the signal can be recorded in perpendicular (⊥) or parallel (‖) polarization mode. The V^{4+} ions give rise to a featureless first-differential signal at $g = 1.93$ that can be observed in both perpendicular and parallel polarization modes while the V^{3+} ions generate a broad line shape centered at

$g = 4.01$, which is only active in the parallel polarization mode. Li *et al.* (2018) have recently developed the use of *ex situ* EPR spectroscopy to study the evolution of V^{3+} and V^{4+} concentrations in $Na_3V_2(PO_4)_2O_{1.6}F_{1.4}$ during cycling; however, this technique could also be applied to investigate the mixed valence state of vanadium in $Na_3V^{3+}{}_{2-y}V^{4+}{}_y(PO_4)_2F_{3-y}O_y$.

Another essential technique is *^{23}Na and ^{31}P solid-state NMR* spectroscopy. In this technique, V^{3+} and V^{4+} are not directly probed, but indirectly through their electronic interaction with the ^{23}Na and ^{31}P nuclei present in their neighborhoods (Park *et al.* 2014). The ^{31}P NMR spectra recorded for $Na_3V^{3+}{}_{2-y}V^{4+}{}_y(PO_4)_2F_{3-y}O_y$ compositions are given in Figure 2.16(b). A "single" resonance is recorded at 6,000 ppm for $Na_3V_2(PO_4)_2F_3$, and is attributed to the $P(OV^{3+})_4$ local environment, while the single resonance detected at ~0 ppm for $Na_3(VO)_2(PO_4)_2F$ is attributed to the $P(OV^{4+})_4$ local environment. In the mixed valence phases, five ^{31}P resonances are recorded at 6,000, 4,500, 3,000, 1,500, and 0 ppm, which are attributed to the $P(OV^{3+})_4$, $P(OV^{3+})_3(OV^{4+})$, $P(OV^{3+})_2(OV^{4+})_2$, $P(OV^{3+})(OV^{4+})_3$ and $P(OV^{3+})_4$ local environments, respectively. Each V^{3+} in the proximity of the phosphorus nucleus contributes to a paramagnetic shift of 1,500 ppm (Nguyen *et al.* 2019c). The relative intensity of the five resonances recorded on the ^{31}P NMR spectrum is controlled by the ratio of V^{3+}/V^{4+} in the structure. ^{23}Na solid-state NMR can also be used to determine the V^{3+}/V^{4+} ratio in the structure as the Fermi contact shift of ^{23}Na nuclei in $Na_3V^{3+}{}_{2-y}V^{4+}{}_y(PO_4)_2F_{3-y}O_y$ depends on the number of V^{3+} ions in its closest bi-octahedron.

All the phases belonging to the $Na_3V^{3+}{}_{2-y}V^{4+}{}_y(PO_4)_2F_{3-y}O_y$ family can be used as positive electrode materials in NIBs. In the potential range of 2.5–4.3 V versus Na^+/Na, two Na^+ ions can be extracted and re-inserted reversibly for all the compositions, and the voltage at which the Na^+-extraction occurs depends strongly on the oxygen-content of the phase. In this series of compositions, $Na_3V_2(PO_4)_2F_3$ has the highest desodiation voltage versus Na^+/Na and this voltage shifts to a lower value in the mixed valance compositions (Figure 2.17). The more V^{4+} (or more precisely $\{V{=}O\}^{2+}$ vanadyl bond) presents in the structure, the lower is the desodiation voltage value. The minimum desodiation voltage is observed for $Na_3(VO)_2(PO_4)_2F$. In the mixed valence phases, the cycling profile is sloping as the Na^+ extraction/insertion reaction occurs through a solid solution mechanism without the formation of intermediate phases. For the two end members, $Na_3V^{3+}{}_2(PO_4)_2F_3$ and $Na_3(V^{4+}O)_2(PO_4)_2F$, the charge/discharge profiles show the presence of several composition voltage "plateaus" due to the formation of intermediate phases whose structures were determined using high resolution synchrotron XRD data (Sharma *et al.* 2014; Bianchini *et al.* 2015). Furthermore, analyses by electrochemical impedance spectroscopy revealed that a small amount of oxygen substituted for fluorine helps to increase the ionic conductivity of the material in comparison to that of $Na_3V_2(PO_4)_2F_3$ (Broux *et al.* 2018), which could be beneficial to the electrochemical performance.

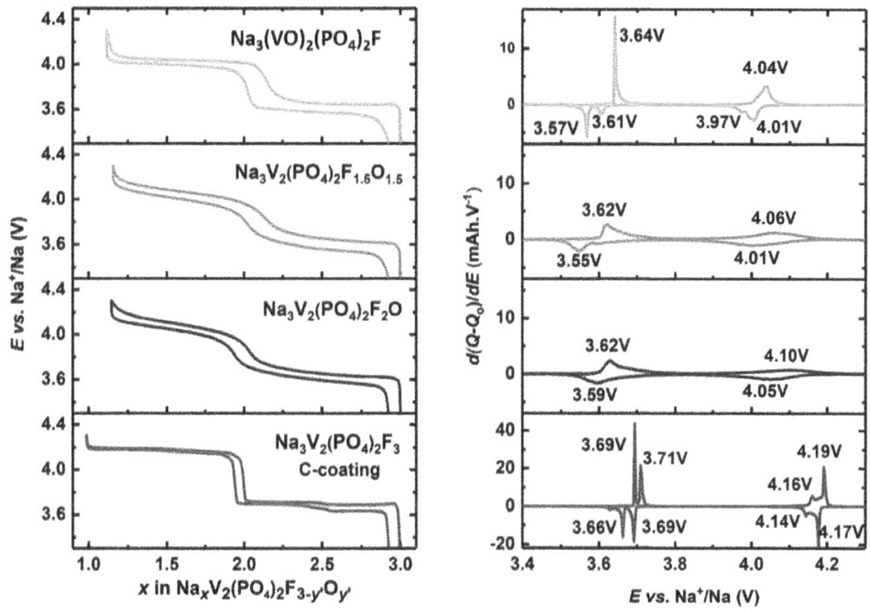

Figure 2.17. *(Left) Comparison of the electrochemical charge/discharge data obtained for carbon-coated $Na_3V_2(PO_4)_2F_3$, $Na_3V_2(PO_4)_2F_2O$, $Na_3V_2(PO_4)_2F_{1.5}O_{1.5}$ and $Na_3(VO)_2(PO_4)_2F$ electrodes in half -cells at the cycling rate of C/10 per Na^+ ion. (Right) The first derivatives of the corresponding electrochemical voltage composition profiles (Nguyen* et al. *2019a). Reprinted with permission from Kumar P.R., Jung Y.H., Lim C.H., Kim D.K. (2015). Copyright (2020). The Royal Society of Chemistry. For a color version of this figure, see www.iste.co.uk/monconduit/batteries.zip*

Importantly, $Na_3V_2(PO_4)_2F_3$ and all the related phases are stable in aqueous solution. A long immersion in water does not induce any structural degradation as revealed by combining XRD, IR and NMR analyses (Nguyen *et al.* 2019a). Therefore, they can be used to develop aqueous electrode formulation processes or active materials for aqueous batteries. Kumar *et al.* (2015) reported on the use of $Na_3V^{3+}{}_{2-y}V^{4+}{}_y(PO_4)_2F_{3-y}O_y$ phases embedded on multiwalled carbon nanotubes (MWCNT) (7.8 wt.%) as positive electrode in aqueous batteries. The material can be cycled versus Zn or $NaTi_2(PO_4)_3$ as negative electrode and the full cells exhibit excellent capacity retention upon long-term cycling (Figure 2.18). Similar results were also reported for the $NaTi_2(PO_4)_3//Na_3V_2(PO_4)_2F_3$ full cell system (Liu *et al.* 2019).

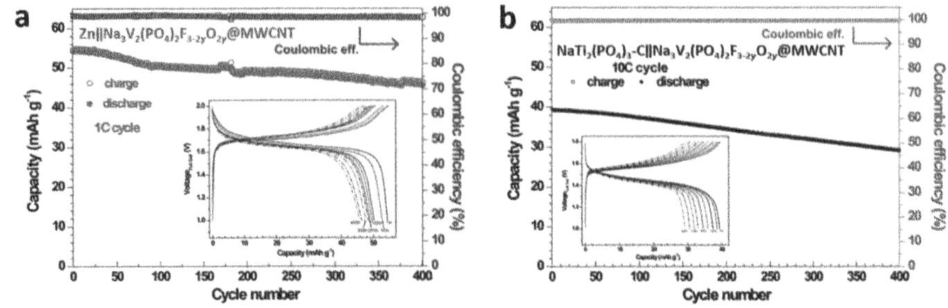

Figure 2.18. *Electrochemical cycling performance and coulombic efficiencies of (a) Zn//$Na_3V_2(PO_4)_2F_{3-y}O_y$@MWCNT full cell and (b) C-coated $NaTi_2(PO_4)_3$//$Na_3V_2(PO_4)_2F_{3-y}O_y$@MWCNT full cell cycled in aqueous 10 M $NaClO_4$ electrolyte. Insets show the corresponding charge–discharge profiles of the given full cell. Reprinted with permission from Kumar P.R., Jung Y.H., Lim C.H., Kim D.K. (2015). Copyright (2020). The Royal Society of Chemistry. For a color version of this figure, see www.iste.co.uk/monconduit/batteries.zip*

2.3.4. *Paving the way toward $Na_3V_2(PO_4)_2F_3$ with superior performance*

Despite the presence of three Na^+ ions in the chemical formula, only two Na^+ ions can apparently be extracted/re-inserted reversibly for all the $Na_3V^{3+}_{2-y}V^{4+}_y(PO_4)_2F_{3-y}O_y$ compositions. The capacity of these materials could be improved significantly if (1) the third Na^+ ion in the material could be cycled reversibly in a practical potential window or (2) extra Na^+ ion could be inserted/extracted reversibly in the structure at lower voltage. Several studies have been made in recent years in order to reach these targets and they will be summarized hereafter.

2.3.4.1. *Three Na^+ ions extraction at the extreme cycling conditions*

Figure 2.19(a) shows the computed voltage curves in the composition range of $1 \leq x \leq 4$ for $Na_xV_2(PO_4)_2F_3$ and $Na_x(VO)_2(PO_4)_2F$.

A recent theoretical study revealed that one extra Na^+ ion can be inserted into the structure of $Na_3V_2(PO_4)_2F_3$ and $Na_3(VO)_2(PO_4)_2F$ at 1.67 V and 1.85 V versus Na^+/Na, respectively (Bianchini *et al.* 2017). An extra capacity corresponding to the insertion of one Na^+ into the structure of $Na_3(VO)_2(PO_4)_2F$ can be observed experimentally at ~1.6 V versus Na^+/Na when an Na//$Na_3(VO)_2(PO_4)_2F$ half-cell is cycled between 1.1 V and 4.4 V versus Na^+/Na (Figure 2.19(b)). The same authors revealed that no electrochemical Na^+ insertion is observed in the low-voltage region for $Na_3V_2(PO_4)_2F_3$. The insertion of the fourth Na^+ into $Na_3(VO)_2(PO_4)_2F$ together with a reduction of V^{4+} into V^{3+} is possible due to the strong screening effect of the oxygen atoms localized on the apical positions of the bi-octahedral units $V_2O_8O_2F$, which is contrary to what is observed with fluorine in these positions in $Na_3V_2(PO_4)_2F_3$. However, the voltage at which this process occurs is too low to hardly find any practical application. Peng *et al.* (2019) recently reported long-term cycling of Na//$Na_3(VO)_2(PO_4)_2F$ half-cells at the voltage 1.0-4.5 V versus Na^+/Na and revealed a continuous capacity loss, that is most probably associated with the electrolyte decomposition process at low voltage.

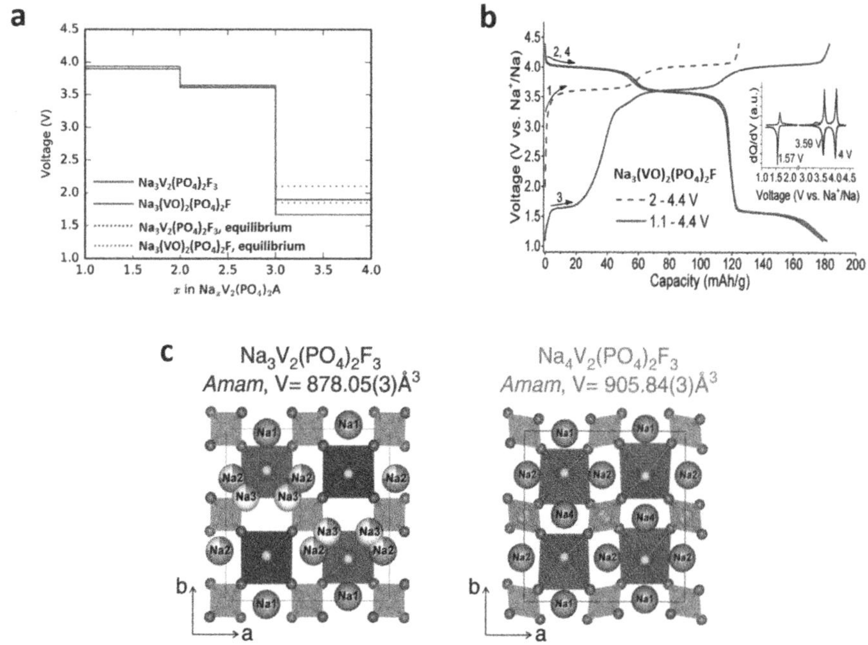

Figure 2.19. *(a) The computed voltage curves in the range of 1 ≤ x ≤ 4 for $Na_xV_2(PO_4)_2F_3$ (red) and $Na_x(VO)_2(PO_4)_2F$ (blue). The dotted curves refer to the average equilibrium potential calculated for the compositions with 3 ≤ x ≤ 4, while the full lines are calculated in the region of 3 ≤ x ≤ 3.125. (b) Electrochemical charge and discharge curves of $Na_3(VO)_2(PO_4)_2F$ cycled in an electrolyte containing 1 M $NaPF_6$ in EC/DEC (1/1) at the cycling rate of C/20, in the potential range of 1.1–4.4 V versus Na^+/Na (full line) and in the potential range of 2.0–4.4 V (dash line). Inset shows the first derivative of the cycling curve between 1.1 and 4.4 V. (c) Comparison between the crystal structure of $Na_3V_2(PO_4)_2F_3$ and $Na_4V_2(PO_4)_2F_3$ synthesized by Na reductive ball milling method. Reprinted with permission Bianchini M., Xiao P., Ceder G. (2017). Copyright (2020). Wiley. Reprinted with permission from Yan G., Mariyappan S., Rousse G., Jacquet Q., Deschamps M., David R., Mirvaux B., Freeland J.W., Tarascon J.-M. (2019). Creative Commons Licence - Published by SpringerNature. For a color version of this figure, see www.iste.co.uk/monconduit/batteries.zip*

Zhang *et al.* (2016a) succeeded in synthesizing a novel material, $Na_4V_2(PO_4)_2F_3$, by ball milling $Na_3V_2(PO_4)_2F_3$ directly with Na metal. Half of the V^{3+} ions in $Na_3V_2(PO_4)_2F_3$ are reduced into V^{2+} to compensate for the insertion of the fourth Na^+ ion into the diffusion channels of the structure. The crystal structure of $Na_4V_2(PO_4)_2F_3$ is similar to that of $Na_3V_2(PO_4)_2F_3$. The Na^+ ions are distributed on the four corners of a square so as to minimize the electrostatic repulsions between them, but the PO_4 groups are highly distorted (Figure 2.19(c)). The highly distorted structure of $Na_4V_2(PO_4)_2F_3$ might be a penalizing factor, preventing the reversible electrochemical extraction and insertion of the fourth Na^+ into the structure at room temperature. For instance, three Na^+ can be extracted from $Na_4V_2(PO_4)_2F_3$ during the first charge versus Na metal but only two can then be cycled reversibly. Nevertheless, this irreversible capacity can be – at least partially – sacrificed in the solid-electrolyte interphase (SEI) formation process that always happens at the hard carbon negative electrode during the first cycle. Indeed, it was demonstrated that the reversible capacity of a "hard carbon//$Na_3V_2(PO_4)_2F_3$ + $Na_4V_2(PO_4)_2F_3$" full cell is higher than that of a "hard carbon//$Na_3V_2(PO_4)_2F_3$" full cell.

Park *et al.* (2013) reported that "ultrahigh energy" can be obtained from $Na_3V_2(PO_4)_2F_{1.4}O_{1.6}$ with the exchange of 1.2 electrons per vanadium ion when the material is cycled at the extended voltage of 1.5–4.6 V versus Na^+/Na. More recently, it was also reported that extra Na^+ can be extracted from $Na_3V_2(PO_4)_2F_3$ when the upper cut-off voltage is increased up to 4.75 V versus Na^+/Na together with a floating period at the end of charge (Yan *et al.* 2019). The longer the floating time, the more Na^+ – up to three Na^+ ions in total – can be extracted (Figure 2.20). These three Na^+ ions can be reinserted into the host structure if the lower cut-off voltage is set at 1.0 V versus Na^+/Na. A reversible capacity of 200 $mAh{\cdot}g^{-1}$ was hence announced for $Na_3V_2(PO_4)_2F_3$ based on the reversible extraction and reinsertion of three Na^+ ions, i.e. corresponding to an increase in more than 50% versus the reported results. Once the material is de-intercalated up to 4.75 V, the discharge profile is completely modified, and the initial electrochemical signature is never recovered regardless the cycling conditions.

Indeed, the electrochemical profile of $Na_3V_2(PO_4)_2F_3$ in the first charge, performed up to 4.75 V, is characterized by the succession of three voltage domains, whereas its profile is very sloping in the next discharge and in the successive cycles. The structure undergoes an irreversible modification at high voltage, from the pristine structure described in the space group *Amam* with the Na^+ ions distributed among three sites to a structure similar to the high temperature disordered one described in the *I4/mmm* space group with the Na^+ ions distributed within a ring. The XRD patterns of the samples charged up to 4.75 V and then discharged to 1.0 V show a large broadening of the diffraction lines and a decrease in the peaks' intensity.

Furthermore, ^{31}P solid-state NMR spectra recorded on the same samples also reveal the presence of very broad resonances (Yan *et al.* 2019). In our opinion, the structure would suffer some irreversible degradations, during the charge up to ~4.75 V, as it was also suggested earlier in the work of Gover *et al.* (2006). Furthermore, a contribution of the electrolyte degradation to the capacity is expected at low voltage, as the number of exchanged electrons in a charge up to ~4.75 V is smaller than what is detected in a discharge below ~1.6 V (Figure 2.20, inset).

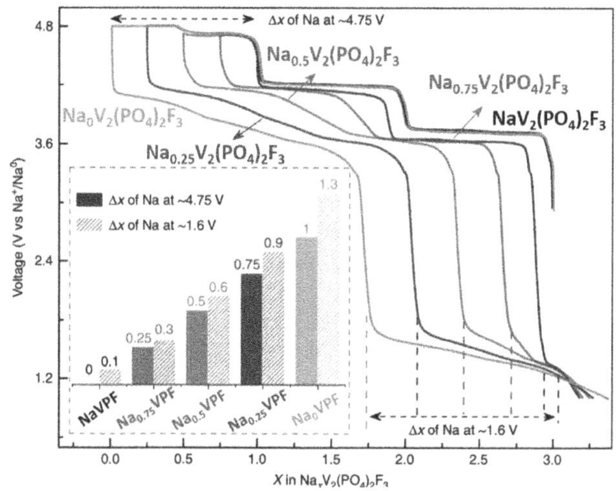

Figure 2.20. *The charge/discharge profile of $Na_3V_2(PO_4)_2F_3$ with a high cut-off voltage fixed at 4.75 V versus Na^+/Na with different floating durations. Inset compares the number of Na ions extracted at ~4.75 V versus the number of Na ions reinserted into the structure at 1.6 V. Reprinted with permission from Yan G., Mariyappan S., Rousse G., Jacquet Q., Deschamps M., David R., Mirvaux B., Freeland J.W., Tarascon J.-M. (2019). Creative Commons Licence - Published by SpringerNature. For a color version of this figure, see www.iste.co.uk/monconduit/batteries.zip*

The results announcing the extraction of the third Na^+ ion from $Na_3V_2(PO_4)_2F_3$ is really intriguing as all the theoretical works predicted that its extraction would occur at ~4.95 V versus Na^+/Na (Xu *et al.* 2014), and that the fully desodiated phase, $V_2(PO_4)_2F_3$, would be thermodynamically unstable (Dacek *et al.* 2016). A recent XAS experiment performed *operando* on a $Na//Na_3V_2(PO_4)_2F_3$ half-cell charged up to ~4.75 V versus Na^+/Na reveals a linear evolution of vanadium oxidation state from 2.0 to 4.5 V versus Na^+/Na, i.e. in the first two potential domains (Nguyen *et al.* 2019a). In the third potential domain, no change in the oxidation state of vanadium was observed. Only the local environment of the vanadium ions seems to be modified and becomes much more distorted.

2.3.4.2. *Going beyond the extraction of two Na^+ ions through ionic substitution*

As just discussed above, the voltage required to remove the third Na^+ ion is really high (Xu *et al.* 2014), and this high voltage value was attributed to the oxidation of V^{4+} to V^{5+} but also to the great activation energy required to break the Na^+ ordering formed at the end of charge. Indeed, Na^+ extraction from $Na_3V_2(PO_4)_2F_3$ leads to the formation of several intermediate phases with different and quite stable Na^+ orderings, which can slow down the diffusion process.

Figure 2.21 compares the activation energy required to create defects and destabilize the Na^+ ordering observed in $Na_2V_2(PO_4)_2F_3$ and $NaV_2(PO_4)_2F_3$, the latter being significantly greater (281 meV versus 199 meV). Dacek *et al.* (2016) proposed that any factor that could disrupt this Na^+ ordering, such as ionic substitution, could lower the activation energy for the defect formation and reduce the voltage for the Na^+ extraction. For instance, Xiong *et al.* (2018) showed that the operating voltage of $NaLi_2V_2(PO_4)_2F_3$, with foreign ions (Li^+) in the diffusion channels, was slightly lower than that of $Na_3V_2(PO_4)_2F_3$, and the voltage profiles of these two materials are quite different.

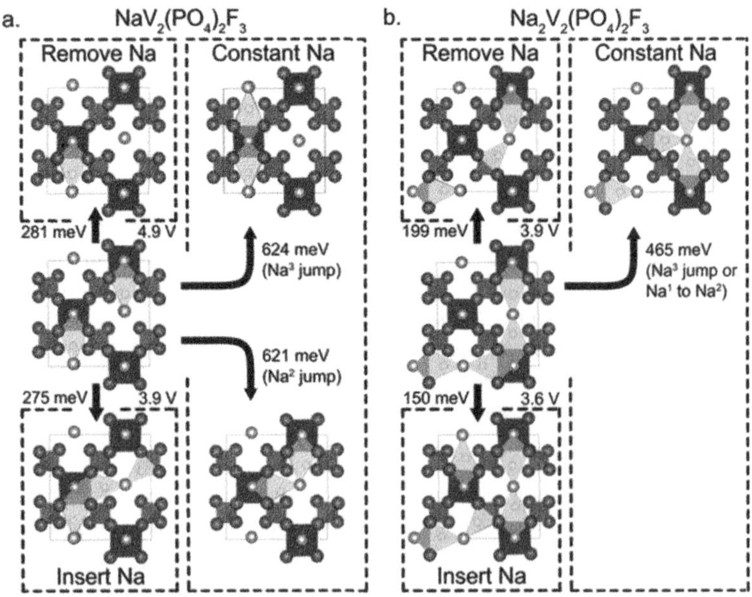

Figure 2.21. *Comparison of the defect formation energies in $NaV_2(PO_4)_2F_3$ and $Na_2V_2(PO_4)_2F_3$ for the most stable Na^+ ions distributions that allow for bulk diffusion. Reprinted with permission from Dacek S.T., Richards W.D., Kitchaev D.A., Ceder G. (2016).Copyright (2020) American Chemical Society. For a color version of this figure, see www.iste.co.uk/monconduit/batteries.zip*

In order to understand the role of cationic substitution on the redox potential of $Na_3V_2(PO_4)_2F_3$ material, Bianchini *et al.* (2017) synthesized a series of Al-substituted materials $Na_3V_{2-z}Al_z(PO_4)_2F_3$ ($0 \leq z \leq 0.3$). The redox potential of the V^{4+}/V^{3+} couple was unlikely to be affected by the Al-substitution; furthermore, the substitution was shown to be detrimental to the electrochemical performance as the number of Na^+ extracted was shown to be limited to (2-*z*). The synthesis of $Na_3VGa(PO_4)_2F_3$ was also reported; however, there was no clear evident for the activation of the V^{5+}/V^{4+} redox couple even in the high voltage region (Matts *et al.* 2015).

The crystal structure and the electrochemical properties of $Na_3(VO)Fe(PO_4)_2F_2$ have also been recently reported (Nguyen *et al.* 2019b). The material shows very low polarization which could be due to an enhancement in the Na^+ mobility and a disruption of Na^+-Na^+ charge orderings upon cycling. Nevertheless, there is no impact of the substitution on the potential value of the $\{V^{5+}{=}O\}^{3+}/\{V^{4+}{=}O\}^{2+}$ redox couple. A recent study on Al-substituted $Na_3(VO)_{2-z}Al_z(PO_4)_2F_{1+z}$ ($0 \leq z \leq 2$) phases also shows that the presence of Al^{3+} foreign ions does not have any influence on the potential value of the $\{V^{5+}{=}O\}^{3+}/\{V^{4+}{=}O\}^{2+}$ redox couple (Olchowka *et al.* 2019). The doping or substitution of cations such as Mn^{2+}, Ti^{3+} and Y^{3+} for vanadium in $Na_3V_2(PO_4)_2F_3$ by solid-state or sol-gel synthesis was also reported and the obtained results showed an improvement in the electrochemical performance of the materials due to a decrease in their charge transfer resistance (Liu *et al.* 2017a; Yi *et al.* 2018; Zhang *et al.* 2018).

In all the substituted phases, there is no extra diffraction peak corresponding to the V^{3+}—F—M charge ordering in the structure. The interaction between two metal sites in the same bi-octahedron seems to be weak, and the V^{3+} and M^{n+} ions tend to distribute randomly over all the bi-octahedral units in the structure. This random distribution of the metal ions could be the origin of the absence of Na^+ orderings during cycling.

In $Na_3V_2(PO_4)_2F_3$, each V^{3+} is coordinated by two fluorine atoms on the axial positions of the octahedron, and these fluorine anions can induce the so-called "inductive effect". If the terminal fluorine atom is replaced by other anions such as Cl^- and Br^-, with different electronegativity, it is expected to modify the electronic structure of the material, i.e. the energy of the t_{2g} and e_g orbitals of V^{3+}, and thus directly influence the potential of the redox couple V^{4+}/V^{3+}. As shown in Figure 2.22, theoretical calculations performed by Xu *et al.* have proposed that chlorine substitution for fluorine in $Na_3V_2(PO_4)_2F_3$ should lead to a significant decrease in the potential required for the extraction of the first and the third Na^+ ions, and thus the de-intercalation of all these three Na^+ ions would occur in the electrolytes' electrochemical window. From these results, it appears that the most attractive phase would be the vanadyl-rich material $Na_3(VO)_2(PO_4)_2F_{0.5}Br_{0.5}$ obtained from a combination of oxygen and bromine substitutions on fluorine positions (Figure 2.22). To the best of our knowledge, this phase and all the halogenide-substituted phases have never been obtained experimentally, as may be due to the great mismatch between the ionic radii of F^- (R_{F^-} = 1.33 Å) and Cl^-/Br^- (R_{Cl^-} = 1.81 Å and R_{Br^-} = 1.96 Å).

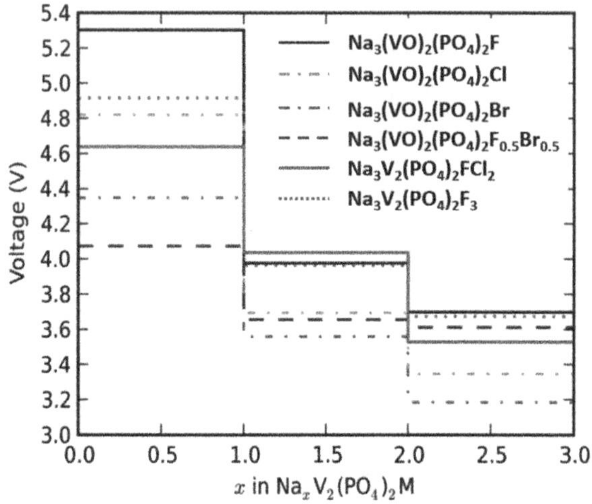

Figure 2.22. *Calculated voltage profiles of $Na_3V_2(PO_4)_2F_3$ with different anionic substitutions. Reprinted with permission from Xu M., Xiao P., Stauffer S., Stauffer S., Song J., Henkelman G., Goodenough J.B. (2014). Copyright (2020) American Chemical Society. For a color version of this figure, see www.iste.co.uk/monconduit/batteries.zip*

2.4. Conclusion and perspectives

Overall, $Na_3V_2(PO_4)_3$ and $Na_3V_2(PO_4)_2F_3$ are two of the most promising polyanionic materials that can be used at the positive electrode in NIBs and they indeed belong to two different families of materials. In $Na_3V_2(PO_4)_3$, the two vanadium octahedral sites belonging to a lantern unit are separated from each other by phosphate groups, which prevents them from having any electronic interaction: the redox activity of vanadium is thus fully independent from the nature of the second metal ion in the same lantern. In $Na_3V_2(PO_4)_2F_3$, the two vanadium ions occupy bi-octahedral units, the two octahedra being bridged through a common fluorine atom and the terminal atoms being F or O for compositions showing oxygen defects. Clearly, the nature of the terminal bonds V-X (X = F, O) in the bi-octahedral units tailors the redox properties of vanadium.

These aforementioned compounds contain three Na^+ ions per formula unit, but only two of them can be electrochemically extracted and re-inserted reversibly into the structure. A partial extraction of the third Na^+ was observed in substituted $Na_{3+x}V_{2-x}M_x(PO_4)_3$ and in $Na_3V_2(PO_4)_2F_3$; however, this process was shown to be associated with an irreversible structural modification, and not to be a long-term benefit upon cycling. A significant modification of the crystal field by partial substitution for vanadium or by altering the nature of the ligands could be key tracks to prospect further in order to achieve a reversible extraction/re-insertion of three Na^+ ions in these materials.

2.5. References

Abakumov, A. M., Zakharkin, M. V., Drozhzhin, O. A., Antipov, E. V., Chernyshov, D., Tereshchenko, I. V., and Stevenson, K. J. (2018). Enhancing Na^+ extraction limit through high voltage activation of the NASICON-type $Na_4MnV(PO_4)_3$ cathode. *ACS Applied Energy Materials*, 1(11), 5842–5846.

Aono, H., Sugimoto, E., Sadaoka, Y., Imanaka, N., and Adachi, G.-ya (1991). Electrical property and sinterability of $LiTi_2(PO_4)_3$ mixed with lithium salt (Li_3PO_4 or Li_3BO_3). *Solid State Ionics*, 47(3–4), 257–264.

Aragón, M. J., Lavela, P., Ortiz, G. F., and Tirado, J. L. (2015). Effect of iron substitution in the electrochemical performance of $Na_3V_2(PO_4)_3$ as cathode for Na-ion batteries. *Journal of the Electrochemical Society*, 162(2), A3077–A3083.

Barim, G., Cottingham, P., Zhou, S., Melot, B. C., and Brutchey, R. L. (2017). Investigating the mechanism of reversible lithium insertion into anti-NASICON $Fe_2(WO_4)_3$. *ACS Applied Materials and Interfaces*, 9(12), 10813–10819.

Barker, J., Gover, R. K. B., Burns, P., and Bryan, A. J. (2006) Hybrid-ion: A lithium-ion cell based on a sodium insertion materials. *Electrochemical and Solid-State Letters*, 9(4), A190.

Barker, J., Gover, R. K. B., Burns, P., and Bryan, A. J. (2007). $Li_{4/3}Ti_{5/3}O_4 \| Na_3V_2(PO_4)_2F_3$: An example of a hybrid-ion cell using a non-graphitic anode. *Journal of the Electrochemical Society*, 154(9), 882–887.

Bianchini, M., Brisset, N., Fauth, F., Weill, F., Elkaim, E., Suard, E., Masquelier, C., and Croguennec, L. (2014). $Na_3V_2(PO_4)_2F_3$ revisited: A high-resolution diffraction study. *Chemistry of Materials*, 26(14), 4238–4247.

Bianchini, M., Fauth, F., Brisset, N., Weill, F., Suard, E., Masquelier, C., and Croguen (2015). Comprehensive investigation of the $Na_3V_2(PO_4)_2F_3$–$NaV_2(PO_4)_2F_3$ system by operando high resolution synchrotron X-ray diffraction. *Chemistry of Materials*, 27, 3009–3020.

Bianchini, M., Xiao, P., Wang, Y., and Ceder, G. (2017). Additional sodium insertion into polyanionic cathodes for higher-energy Na-ion batteries. *Advanced Energy Materials*, 7(18), 1700514.

Bianchini, M., Lalère, F., Nguyen, H. B. L., Fauth, F., David, R., Suard, E., Croguennec, L., and Masquelier, C. (2018). $Ag_3V_2(PO_4)_2F_3$, a new compound obtained by Ag^+/Na^+ ion exchange into the $Na_3V_2(PO_4)_2F_3$ framework. *Journal of Materials Chemistry A*, 6(22), 10340–10347.

Bohnke, O., Ronchetti, S., and Mazza, D. (1999). Conductivity measurements on NASICON and NASICON-modified materials. *Solid State Ionics*, 122(1–4), 127–136.

Boilot J., Colomban, P., and Collin, G. (1986). NASICON: Amorphous to crystalline compounds. *Solid State Ionics*, 18–19, 974–980.

Boilot, J. P., Collin, G., and Colomban, P. (1987). Crystal structure of the true NASION: $Na_3Zr_2Si_2PO_{12}$. *Materials Research Bulletin*, 22(5), 669–676.

Boilot, J. P., Collin, G., and Colomban, P. (1988). Relation structure-fast ion conduction in the NASICON solid solution. *Journal of Solid State Chemistry*, 73(1), 160–171.

Boivin, E., Chotard, J.-N., Bamine, T., Carlier, D., Serras, P., Palomares, V., Rojo, T., Iadecola, A., Dupont, L., Bourgeois, L., Fauth, F., Masquelier, C. and Croguennec, L. (2017). Vanadyl-type defects in Tavorite-like NaVPO4F: from the average long range structure to local environments. *J. Mater. Chem. A*, 2017, 5, 25044–25055.

Broux, T., Bamine, T., Fauth, F., Simonelli, L., Olszewski, W., Marini, C., Ménétrier, M., Carlier, D., Masquelier, C., and Croguennec, L. (2016). Strong impact of the oxygen content in $Na_3V_2(PO_4)_2F_{3-y}O_y$ ($0 \leq y \leq 0.5$) on its structural and electrochemical properties. *Chemistry of Materials*, 28(21), 7683–7692.

Broux, T., Bamine, T., Simonelli, L., Stievano, L., Fauth, F., Ménétrier, M., Carlier, D., Masquelier, C., and Croguennec, L. (2017). V^{IV} disproportionation upon sodium extraction from $Na_3V_2(PO_4)_2F_3$ observed by operando X-ray absorption spectroscopy and solid-state NMR. *The Journal of Physical Chemistry C*, 121(8), 4103–4111.

Broux, T., Fleutot, B., David, R., Brüll, A., Veber, P., Fauth, F., Courty, M., Croguennec, L., and Masquelier, C. (2018). Temperature dependence of structural and transport properties for $Na_3V_2(PO_4)_2F_3$ and $Na_3V_2(PO_4)_2F_{2.5}O_{0.5}$. *Chemistry of Materials*, 30(2), 358–365.

Broux, T., Fauth, F., Hall, N., Chatillon, Y., Bianchini, M., Bamine, T., Leriche, J., Suard, E., Carlier, D., Reynier, Y., Simonin, L., Masquelier, C., and Croguennec, L. (2019). High rate performance for carbon-coated $Na_3V_2(PO_4)_2F_3$ in Na-ion batteries. *Small Methods*, 3(4), 1800215.

Bruce, P. G. and Miln, G. (1990). Sodium intercalation into the defect garnets $Fe_2(MoO_4)_3$ and $Fe_2(WO_4)_3$. *Journal of Solid State Chemistry*, 89(1), 162–166.

Burova, D., Shakhova, I., Morozova, P., Iarchuk, A., Drozhzhin, O. A., Rozova, M. G., Praneetha, S., Murugan, V., Tarascon, J.-M., and Abakumov, A. M. (2019). The rapid microwave-assisted hydrothermal synthesis of NASICON-structured $Na_3V_2O_{2x}(PO_4)_2F_{3-2x}$ ($0 \leq x \leq 1$) cathode materials for Na-ion batteries. *RSC Advances*. Royal Society of Chemistry, 9(34), 19429–19440.

Catti, M., Stramare, S., and Ibberson, R. (1999). Lithium location in NASICON-type Li^+ conductors by neutron diffraction. I. Triclinic α'-$LiZr_2(PO_4)_3$. *Solid State Ionics*, 123(1), 173–180.

Cao, X., Pan, A., Liu, S., Zhou, J., Li, S., Cao, G., Liu, J. and Liang, S. (2017). Chemical Synthesis of 3D Graphene-Like Cages for Sodium-Ion Batteries Applications. *Adv. Energy Mater.*, 7(20), 1700797.

Chen, F., Kovrugin, V. M., David, R., Mentré, O., Fauth, F., Chotard, J., and Masquelier, C. (2018). A NASICON-type positive electrode for Na batteries with high energy density: $Na_4MnV(PO_4)_3$. *Small Methods*, 3(4), 1800218.

Choi, J. W. and Aurbach, D. (2016). Promise and reality of post-lithium-ion batteries with high energy densities. *Nature Reviews Materials*, 16013, 1.

Chotard, J.-N., Rousse, G., David, R., Mentré, O., Courty, M., and Masquelier, C. (2015). Discovery of a sodium-ordered form of $Na_3V_2(PO_4)_3$ below ambient temperature. *Chemistry of Materials*, 27(17), 5982–5987.

Chung, S. C., Ming, J., Lander, L., Lu, J., and Yamada, A. (2018). Rhombohedral NASICON-type $Na_xFe_2(SO_4)_3$ for sodium ion batteries: Comparison with phosphate and alluaudite phases. *Journal of Materials Chemistry A*, 6(9), 3919–3925.

Clark, J. M., Barpanda, P., Yamada, A., and Islam, M. S. (2014). Sodium-ion battery cathodes $Na_2FeP_2O_7$ and $Na_2MnP_2O_7$: Diffusion behaviour for high rate performance. *Journal of Materials Chemistry A*, 2(30), 11807–11812.

Coors, W.G., Boxley, C., Robins, M., and Eccleston, A. (2015). Low temperature molten sodium secondary cell with sodium ion conductive electrolyte membrane. International Patent WO2012061823A2.

Cretin, M., Khireddine, H., and Fabry, P. (1997). NASICON structure for alkaline ion recognition. *Sensors and Actuators, B: Chemical*, 43(1–3), 224–229.

Dacek, S. T., Richards, W. D., Kitchaev, D. A., and Ceder, G. (2016). Structure and dynamics of fluorophosphate Na-ion battery cathodes. *Chemistry of Materials*, 28(15), 5450–5460.

Delmas, C., Olazcuaga, R., Cherkaoui, F., Brochu, R., and Le Flem, G. (1978). A new family of phosphates with the formula $Na_3M_2(PO_4)_3$ (M = Ti, V, Cr, Fe). *C. R. Seances Acad. Sci., Ser. C*, 287, 169–174.

Delmas, C., Braconnier, J.-J., Fouassier, C., and Hagenmuller, P. (1981). Electrochemical intercalation of sodium in Na_xCO_2 bronzes. *Solid State Ionics*, 3–4, 165–169.

Delmas, C., Cherkaoui, F., Nadiri, A., and Hagenmuller, P. (1987). A NASICON-type phase as intercalation electrode: $NaTi_2(PO_4)_3$. *Materials Research Bulletin*, 22(5), 631–639.

Delmas, C. and Nadiri, A. (1988). The chemical short circuit method. An improvement in the intercalation-deintercalation techniques. *Materials Research Bulletin*, 23(1), 65–72.

Delmas, C., Nadiri, A., and Soubeyroux, J. L. (1988). The NASICON-type titanium phosphates $ATi_2(PO_4)_3$ (A = Li, Na) as electrode materials. *Solid State Ionics*, 28–30, 419–423.

Deng, L., Sun, G., Goh, K., Zheng, L. L., Yu, F. Da, Sui, X. L., Zhao, L., and Wang, Z. B. (2019). Facile one-step carbothermal reduction synthesis of $Na_3V_2(PO_4)_2F_3$/C serving as cathode for sodium ion batteries. *Electrochimica Acta*, 298, 459–467.

Deng, Y., Eames, C., Nguyen, L. H. B., Pecher, O., Griffith, K. J., Courty, M., Fleutot, B., Chotard, J.-N., Grey, C. P., Islam, M. S., and Masquelier, C. (2018). Crystal structures, local atomic environments, and ion diffusion mechanisms of scandium-substituted sodium superionic conductor (NASICON) solid electrolytes. *Chemistry of Materials*, 30(8), 2618–2630.

Duan, W., Zhu, Z., Li, H., Hu, Z., Zhang, K., Cheng, F., and Chen, J. (2014). $Na_3V_2(PO_4)_3$@C core-shell nanocomposites for rechargeable sodium-ion batteries. *Journal of Materials Chemistry A*, 2(23), 8668–8675.

Fernández-Ropero, A. J., Zarrabeitia, M., Reynaud, M., Rojo, T., and Casas-Cabanas, M. (2018). Toward safe and sustainable batteries: $Na_4Fe_3(PO_4)_2P_2O_7$ as a low-cost cathode for rechargeable aqueous Na-ion batteries. *Journal of Physical Chemistry C*, 122(1), 133–142.

Gao, H. and Goodenough, J. B. (2016). An aqueous symmetric sodium-ion battery with NASICON-structured $Na_3MnTi(PO_4)_3$. *Angewandte Chemie–International Edition*, 55(41), 12768–12772.

Gao, H., Seymour, I. D., Xin, S., Xue, L., Henkelman, G., and Goodenough, J. B. (2018). $Na_3MnZr(PO_4)_3$: A high-voltage cathode for sodium batteries. *Journal of the American Chemical Society*, 140(51), 18192–18199.

Goodenough, J. B., Hong, H. Y. P., and Kafalas, J. A. (1976). Fast Na^+-ion transport in skeleton structures. *Materials Research Bulletin*, 11(2), 203–220.

Gopalakrishnan, J. and Kasthuri Rangan, K. (1992). $V_2(PO_4)_3$: A novel NASICON-type vanadium phosphate synthesized by oxidative deintercalation of sodium from $Na_3V_2(PO_4)_3$. *Chemistry of Materials*, 4(M), 747–749.

Gover, R. K. B., Bryan, A., Burns, P., and Barker, J. (2006). The electrochemical insertion properties of sodium vanadium fluorophosphate, $Na_3V_2(PO_4)_2F_3$. *Solid State Ionics*, 177(17–18), 1495–1500.

Grey, C. P. and Tarascon, J. M. (2016). Sustainability and in situ monitoring in battery development. *Nature Materials*, 16(1), 45–56.

Guin, M. and Tietz, F. (2015). Survey of the transport properties of sodium superionic conductor materials for use in sodium batteries', *Journal of Power Sources*, 273, 1056–1064.

Ha, K. H., Woo, S. H., Mok, D., Choi, N. S., Park, Y., Oh, S. M., Kim, Y., Kim, J., Lee, J., Nazar, L. F., and Lee, K. T. (2013). $Na_{4-\alpha}M_{2+\alpha/2}(P_2O_7)_2$ ($2/3 \leq \alpha \leq 7/8$, M = Fe, $Fe_{0.5}Mn_{0.5}$, Mn): A promising sodium ion cathode for Na-ion batteries. *Advanced Energy Materials*, 3(6), 770–776.

Hong, H. Y. P. (1976). Crystal structures and crystal chemistry in the system $Na_{1+x}Zr_2Si_xP_{3-x}O_{12}$. *Materials Research Bulletin*, 11(2), 173–182.

Hwang, J.-Y., Myung, S.-T., and Sun, Y.-K. (2017). Sodium-ion batteries: Present and future. *Chem. Soc. Rev.* Royal Society of Chemistry, 46(12), 3529–3614.

Inoishi, A., Omuta, T., Kobayashi, E., Kitajou, A. and Okada, S. (2017). A single-phase, all-solid-state sodium battery using $Na_{3-x}V_{2-x}Zr_x(PO_4)_3$ as the cathode, anode, and electrolyte. *Advanced Materials Interfaces*, 4(5), 3–7.

Jalem, R., Natsume, R., Nakayama, M., and Kasuga, T. (2016). First-principles investigation of the Na+ion transport property in oxyfluorinated titanium(IV) phosphate $Na_3Ti_2P_2O_{10}F$. *Journal of Physical Chemistry C*, 120(3), 1438–1445.

Jian, Z., Zhao, L., Pan, H., Hu, Y. S., Li, H., Chen, W., and Chen, L. (2012). Carbon coated $Na_3V_2(PO_4)_3$ as novel electrode material for sodium ion batteries. *Electrochemistry Communications*, 14(1), 86–89.

Jian, Z., Han, W., Lu, X., Yang, H., Hu, Y. S., Zhou, J., Zhou, Z., Li, J., Chen, W., Chen, D., and Chen, L. (2013). Superior electrochemical performance and storage mechanism of $Na_3V_2(PO_4)_3$ cathode for room-temperature sodium-ion batteries. *Advanced Energy Materials*, 3(2), 156–160.

Kabbour, H., Coillot, D., Colmont, M., Masquelier, C., and Mentré, O. (2011). α-$Na_3M_2(PO_4)_3$ (M = Ti, Fe): Absolute cationic ordering in NASICON-type phases. *Journal of the American Chemical Society*, 133, 11900–11903.

Kaus, M., Guin, M., Yavuz, M., Knapp, M., Tietz, F., Guillon, O., Ehrenberg, H., and Indris, S. (2017). Fast Na^+ ion conduction in NASICON-type $Na_{3.4}Sc_2(SiO_4)_{0.4}(PO_4)_{2.6}$ observed by ^{23}Na NMR relaxometry. *The Journal of Physical Chemistry C*, 121(3), 1449–1454.

Kawai, K., Zhao, W., Nishimura, S., and Yamada, A. (2018). High-voltage Cr^{4+}/Cr^{3+} redox couple in polyanion compounds. *ACS Applied Energy Materials*, 1(3), 928–931.

Kida, T., Miyachi, Y., Shimanoe, K., and Yamazoe, N. (2001). NASICON thick film-based CO_2 sensor prepared by a sol-gel method. *Sensors and Actuators, B: Chemical*, 80(1), 28–32.

Kim, J.-K., Lim, Y. J., Kim, H., Cho, G.-B., and Kim, Y. (2015). A hybrid solid electrolyte for flexible solid-state sodium batteries. *Energy Environ. Sci.*, 8(12), 3589–3596.

Klee, R., Aragón, M. J., Lavela, P., Alcántara, R., and Tirado, J. L. (2016). $Na_3V_2(PO_4)_3/C$ nanorods with improved electrode-electrolyte interface as cathode material for sodium-ion batteries. *ACS Applied Materials and Interfaces*, 8(35), 23151–23159.

Kobayashi, E., Plashnitsa, L. S., Doi, T., Okada, S., and Yamaki, J. I. (2010). Electrochemical properties of Li symmetric solid-state cell with NASICON-type solid electrolyte and electrodes. *Electrochemistry Communications*, 12(7), 894–896.

Kohler, H. and Schulz, H. (1985). NASICON solid electrolytes. Part I: The Na^+-diffusion path and its relation to the structure. *Materials Research Bulletin*, 20, 1461–1471.

Kosova, N. and Rezepova, D. (2017). $Na_{1+y}VPO_4F_{1+y}$ $(0 \leq y \leq 0.5)$ as cathode materials for hybrid Na/Li batteries. *Inorganics*, 5(2), 19.

Kosova, N. V. and Rezepova, D. O. (2018). Mixed sodium-lithium vanadium fluorophosphates $Na_{3-x}Li_xV_2(PO_4)_2F_3$: The origin of the excellent high-rate performance. *Journal of Power Sources*, 408, 120–127.

Kosova, N. V., Rezepova, D. O., Petrov, S. A., and Slobodyuk, A. B. (2017). Electrochemical and chemical Na^+/Li^+ ion exchange in Na-based cathode materials: $Na_{1.56}Fe_{1.22}P_2O_7$ and $Na_3V_2(PO_4)_2F_3$. *Journal of The Electrochemical Society*, 164(1), A6192–A6200.

Kovrugin, V. M., Chotard, J. N., Fauth, F., Jamali, A., David, R., and Masquelier, C. (2017). Structural and electrochemical studies of novel $Na_7V_3Al(P_2O_7)_4(PO_4)$ and $Na_7V_2Al_2(P_2O_7)_4(PO_4)$ high-voltage cathode materials for Na-ion batteries. *Journal of Materials Chemistry A*, 5(27), 14365–14376.

Kumar, P. R., Jung, Y. H., Lim, C. H., and Kim, D. K. (2015). $Na_3V_2O_{2x}(PO_4)_2F_{3-2x}$: A stable and high-voltage cathode material for aqueous sodium-ion batteries with high energy density. *Journal of Materials Chemistry A*, 3, 6271–6275.

Kumar, P. R., Jung, Y. H., and Kim, D. K. (2017). Influence of carbon polymorphism towards improved sodium storage properties of $Na_3V_2O_{2x}(PO_4)_2F_{3-2x}$. *Journal of Solid State Electrochemistry*, 21(1), 223–232.

Lalere, F., Leriche, J. B., Courty, M., Boulineau, S., Viallet, V., Masquelier, C., and Seznec, V. (2014). An all-solid state NASICON sodium battery operating at 200°C. *Journal of Power Sources*, 247, 975–980.

Lalère, F., Seznec, V., Courty, M., David, R., Chotard, J. N., and Masquelier, C. (2015). Improving the energy density of $Na_3V_2(PO_4)_3$-based positive electrodes through V/Al substitution. *Journal of Materials Chemistry A*, 3(31), 16198–16205.

Lalère, F., Seznec, V., Courty, M., Chotard, J. N., and Masquelier, C. (2018). Coupled X-ray diffraction and electrochemical studies of the mixed Ti/V-containing NASICON: $Na_2TiV(PO_4)_3$. *Journal of Materials Chemistry A*, 6(15), 6654–6659.

Larcher, D. and Tarascon, J.-M. (2015). Towards greener and more sustainable batteries for electrical energy storage. *Nature Chemistry*, 7(1), 19–29.

Li, C., Shen, M., Lou, X., and Hu, B. (2018). Unraveling the redox couples of V^{III}/V^{IV} mixed-valent $Na_3V_2(PO_4)_2O_{1.6}F_{1.4}$ cathode by parallel-mode EPR and in situ/ex situ NMR. *The Journal of Physical Chemistry C*, 122, 27224–27232.

Li, L., Peng, S., Lee, J. K. Y., Ji, D., Srinivasan, M., and Ramakrishna, S. (2017a). Electrospun hollow nanofibers for advanced secondary batteries. *Nano Energy*, 39, 111–139.

Li, S., Dong, Y., Xu, L., Xu, X., He, L., and Mai, L. (2014). Effect of carbon matrix dimensions on the electrochemical properties of $Na_3V_2(PO_4)_3$ nanograins for high-performance symmetric sodium-ion batteries. *Advanced Materials*, 26(21), 3545–3553.

Li, S., Ge, P., Zhang, C., Sun, W., Hou, H., and Ji, X. (2017b). The electrochemical exploration of double carbon-wrapped $Na_3V_2(PO_4)_3$: Towards long-time cycling and superior rate sodium-ion battery cathode. *Journal of Power Sources*, 366, 249–258.

Liang, X., Ou, X., Zheng, F., Pan, Q., Xiong, X., Hu, R., Yang, C., and Liu, M. (2017). Surface modification of $Na_3V_2(PO_4)_3$ by nitrogen and sulfur dual-doped carbon layer with advanced sodium storage property. *ACS Applied Materials and Interfaces*, 9(15), 13151–13162.

Lin, X., Huang, J., Tan, H., Huang, J., and Zhang, B. (2019). $K_3V_2(PO_4)_2F_3$ as a robust cathode for potassium-ion batteries. *Energy Storage Materials*, 16, 97–101.

Ling, R., Cai, S., Shen, S., Hu, X., Xie, D., Zhang, F., Sun, X., Yu, N., and Wang, F. (2017). Synthesis of carbon coated Na_2FePO_4F as cathode materials for high-performance sodium ion batteries. *Journal of Alloys and Compounds*, 704, 631–640.

Liu, Q., Wang, D., Yang, X., Chen, N., Wang, C., Bie, X., Wei, Y., Chen, G., and Du, F. (2015). Carbon-coated $Na_3V_2(PO_4)_2F_3$ nanoparticles embedded in a mesoporous carbon matrix as a potential cathode material for sodium-ion batteries with superior rate capability and long-term cycle life. *Journal of Materials Chemistry A*, 3(43), 21478–21485.

Liu, Q., Meng, X., Wei, Z., Wang, D., Gao, Y., Wei, Y., Du, F., and Chen, G. (2016). Core/double-shell structured $Na_3V_2(PO_4)_2F_3$@C nanocomposite as the high power and long lifespan cathode for sodium-ion batteries. *ACS Applied Materials & Interfaces*, 8(46), 31709–31715.

Liu, S., Wang, L., Liu, J., Zhou, M., Nian, Q., Feng, Y., Tao, Z., and Shao, L. (2019). $Na_3V_2(PO_4)_2F_3$–SWCNT: A high voltage cathode for non-aqueous and aqueous sodium-ion batteries. *Journal of Materials Chemistry A*, 7(1), 248–256.

Liu, W., Yi, H., Zheng, Q., Li, X., and Zhang, H. (2017a). Y-Doped $Na_3V_2(PO_4)_2F_3$ compounds for sodium ion battery cathodes: Electrochemical performance and analysis of kinetic properties. *J. Mater. Chem. A*, 5(22), 10928–10935.

Liu, Y., Zhou, Y., Zhang, J., Xia, Y., Chen, T., and Zhang, S. (2017b). Monoclinic phase $Na_3Fe_2(PO_4)_3$: Synthesis, structure, and electrochemical performance as cathode material in sodium-ion batteries. *ACS Sustainable Chemistry and Engineering*, 5(2), 1306–1314.

Liu, Z., Hu, Y., Dunstan, M. T., Huo, H., Hao, X., Zou, H., Zhong, G., Yang, Y., and Grey, C. P. (2014). Local structure and dynamics in the Na ion battery positive electrode material $Na_3V_2(PO_4)_2F_3$. *Chemistry of Materials*, 26(8), 2513–2521.

Manthiram, A. and Goodenough, J. B. (1987). Lithium insertion into $Fe_2(MO_4)_3$ frameworks: Comparison of M = W with M = Mo. *Journal of Solid State Chemistry*, 71(2), 349–360.

Manthiram, A. and Goodenough, J. B. (1989). Lithium Insertion into $Fe_2(SO_4)_3$ Frameworks. *Journal of Power Sources*, 26, 403–408.

Masquelier, C., Padhi, A. K., Nanjundaswamy, K. S., and Goodenough, J. B. (1998). New cathode materials for rechargeable lithium batteries: The 3-D framework structures $Li_3Fe_2(XO_4)_3$ (X = P, As). *Journal of Solid State Chemistry*, 234(135), 228–234.

Masquelier, C., Wurm, C., Rodríguez-Carvajal, J., Gaubicher, J., and Nazar, L. (2000). A powder neutron diffraction investigation of the two rhombohedral NASICON analogues: γ-$Na_3Fe_2(PO_4)_3$ and $Li_3Fe_2(PO_4)_3$. *Chemistry of Materials*, 12(2), 525–532.

Masquelier, C. and Croguennec, L. (2013). Polyanionic (phosphates, silicates, sulfates) frameworks as electrode materials for rechargeable Li (or Na) batteries. *Chemical Reviews*, 113(8), 6552–6591.

Massa, W., Yakubovich, O. V., and Dimitrova, O. V. (2002). Crystal structure of a new sodium vanadyl(IV) fluoride phosphate $Na_3\{V_2O_2F[PO_4]_2\}$. *Solid State Sciences*, 4(4), 495–501.

Matts, I. L., Dacek, S., Pietrzak, T. K., Malik, R., and Ceder, G. (2015). Explaining performance-limiting mechanisms in fluorophosphate Na-ion battery cathodes through inactive transition-metal mixing and first-principles mobility calculations. *Chemistry of Materials*, 27(17), 6008–6015.

Mba, J.-M. A., Croguennec, L., Basir, N. I., Barker, J., and Masquelier, C. (2012). Lithium insertion or extraction from/into tavorite-type $LiVPO_4F$: An in situ X-ray diffraction study. *Journal of the Electrochemical Society*, 159(8), A1171–A1175.

Le Meins, J.-M., Crosnier-Lopez, M.-P., Hemon-Ribaud, A., and Courbion, G. (1999). Phase transitions in the $Na_3M_2(PO_4)_2F_3$ family (M = Al^{3+}, V^{3+}, Cr^{3+}, Fe^{3+}, Ga^{3+}): Synthesis, thermal, structural, and magnetic studies. *Journal of Solid State Chemistry*, 148, 260–277.

Morimoto, H., Awano, H., Terashima, J., Shindo, Y., Nakanishi, S., Ito, N., Ishikawa, K., and Tobishima, S. I. (2013). Preparation of lithium ion conducting solid electrolyte of NASICON-type $Li_{1+x}Al_xTi_{2-x}(PO_4)_3$ (x = 0.3) obtained by using the mechanochemical method and its application as surface modification materials of $LiCoO_2$ cathode for lithium cell. *Journal of Power Sources*, 240, 636–643.

Nanjundaswamy, K. S., Padhi, A. K., Goodenough, J. B., Okada, S., Ohtsuka, H., Arai, H., and Yamaki, J. (1996). Synthesis, redox potential evaluation and electrochemical characteristics of NASICON-related-3D framework compounds. *Solid State Ionics*, 92(1–2), 1–10.

Nguyen, L. H. B., Broux, T., Camacho, P. S., Denux, D., Bourgeois, L., Belin, S., Iadecola, A., Fauth, F., Carlier, D., Olchowka, J., Masquelier, C., and Croguennec, L. (2019a). Stability in water and electrochemical properties of the $Na_3V_2(PO_4)_2F_3$–$Na_3(VO)_2(PO_4)_2F$ solid solution. *Energy Storage Materials*, 20, 324–334.

Nguyen, L. H. B., Olchowka, J., Belin, S., Sanz Camacho, P., Duttine, M., Iadecola, A., Fauth, F., Carlier, D., Masquelier, C., and Croguennec, L. (2019b). Monitoring the crystal structure and the electrochemical properties of $Na_3(VO)_2(PO_4)_2F$ through Fe^{3+} substitution. *ACS Applied Materials & Interfaces*, 11(42), 38808–38818.

Nguyen, L. H. B., Sanz Camacho, P., Broux, T., Olchowka, J., Masquelier, C., Croguennec, L., and Carlier, D. (2019c). Density functional theory-assisted ^{31}P and ^{23}Na magic-angle spinning nuclear magnetic resonance study of the $Na_3V_2(PO_4)_2F_3$–$Na_3V_2(PO_4)_2FO_2$ solid solution: Unraveling its local and electronic structures. *Chemistry of Materials*, 31(23), 9759–9768.

Nisar, U., Shakoor, R. A., Essehli, R., Amin, R., Orayech, B., Ahmad, Z., Kumar, P. R., Kahraman, R., Al-Qaradawi, S., and Soliman, A. (2018). Sodium intercalation/de-intercalation mechanism in $Na_4MnV(PO_4)_3$ cathode materials. *Electrochimica Acta*, 292, 98–106.

Noguchi, Y., Kobayashi, E., Plashnitsa, L. S., Okada, S., and Yamaki, J. I. (2013). Fabrication and performances of all solid-state symmetric sodium battery based on NASICON-related compounds. *Electrochimica Acta*, 101, 59–65.

Oh, S. M., Myung, S. T., Hassoun, J., Scrosati, B., and Sun, Y. K. (2012). Reversible $NaFePO_4$ electrode for sodium secondary batteries. *Electrochemistry Communications*, 22(1), 149–152.

Olchowka, J., Nguyen, L. H. B., Broux, T., Sanz Camacho, P., Petit, E., Fauth, F., Carlier, D., Masquelier, C., and Croguennec, L. (2019). Aluminum substitution for vanadium in the $Na_3V_2(PO_4)_2F_3$ and $Na_3V_2(PO_4)_2FO_2$ type materials. *Chemical Communications*, 55(78), 11719–11722.

Padhi, A. K. (1997). Effect of structure on the Fe^{3+}/Fe^{2+} redox couple in iron phosphates. *Journal of The Electrochemical Society*, 144(5), 1609.

Padhi, A. K., Manivannan, V., and Goodenough, J. B. (1998). Tuning the position of the redox couples in materials with NASICON structure by anionic substitution. *Journal of the Electrochemical Society*, 145(5), 1518–1520.

Panin, R. V., Drozhzhin, O. A., Fedotov, S. S., Khasanova, N. R., and Antipov, E. V. (2018). NASICON-type $NaMo_2(PO_4)_3$: Electrochemical activity of the Mo^{+4} polyanion compound in Na-cell. *Electrochimica Acta*, 289, 168–174.

Park, Y. U., Seo, D. H., Kim, B., Hong, K. P., Kim, H., Lee, S., Shakoor, R. A., Miyasaka, K., Tarascon, J. M., and Kang, K. (2012). Tailoring a fluorophosphate as a novel 4 V cathode for lithium-ion batteries. *Scientific Reports*, 2, 1–7.

Park, Y., Seo, D., Kwon, H., Kim, B., Kim, J., Kim, H., Kim, I., Yoo, H., and Kang, K. (2013). A new high-energy cathode for a Na-ion battery with ultrahigh stability. *Journal of the American Chemical Society*, 135, 13870–13878.

Park, Y. U., Seo, D. H., Kim, H., Kim, J., Lee, S., Kim, B., and Kang, K. (2014). A family of high-performance cathode materials for Na-ion batteries, $Na_3(VO_{1-x}PO_4)_2F_{1+2x}$ $(0 \leq x \leq 1)$: Combined first-principles and experimental study. *Advanced Functional Materials*, 24(29), 4603–4614.

Park, Y. U., Bai, J., Wang, L., Yoon, G., Zhang, W., Kim, H., Lee, S., Kim, S. W., Looney, J. P., Kang, K., and Wang, F. (2017). In situ tracking kinetic pathways of Li^+/Na^+ substitution during ion-exchange synthesis of $Li_xNa_{1.5-x}VOPO_4F_{0.5}$. *Journal of the American Chemical Society*, 139(36), 12504–12516.

Pasierb, P., Komornicki, S., Gajerski, R., Koziński, S., Tomczyk, P., and Rękas, M. (2002). Electrochemical gas sensor materials studied by impedance spectroscopy part I: Nasicon as a solid electrolyte. *Journal of Electroceramics*, 8(1), 49–55.

Patoux, S., Rousse, G., Leriche, J. B., and Masquelier, C. (2003). Structural and electrochemical studies of rhombohedral $Na_2TiM(PO_4)_3$ and $Li_{1.6}Na_{0.4}TiM(PO_4)_3$ (M = Fe, Cr) phosphates. *Chemistry of Materials*, 15(10), 2084–2093.

Patoux, S., Rousse, G., Leriche, J. B., and Masquelier, C. (2004). Crystal structure and lithium insertion properties of orthorhombic $Li_2TiFe(PO_4)_3$ and $Li_2TiCr(PO_4)_3$. *Solid State Sciences*, 6(10), 1113–1120.

Peng, M., Li, B., Yan, H., Zhang, D., Wang, X., Xia, D., and Guo, G. (2015). Ruthenium-oxide-coated sodium vanadium fluorophosphate nanowires as high-power cathode materials for sodium-ion batteries. *Angewandte Chemie–International Edition*, 54(22), 6452–6456.

Peng, M., Zhang, D., Wang, X., Xia, D., Sun, Y., and Guo, G. (2019). Three-electron reversible redox for a high-energy fluorophosphate cathode: $Na_3V_2O_2(PO_4)_2F$. *Chemical Communications*, 55(27), 3979–3982.

Petit, D., Colomban, P., Collin, G., and Boilot, J. P. (1986). Fast ion transport in $LiZr_2(PO_4)_3$: Structure and conductivity. *Materials Research Bulletin*, 21(3), 365–371.

Ponrouch, A., Dedryvere, R., Monti, D., Demet, A. E., Ateba Mba, J. M., Croguennec, L., Masquelier, C., Johansson, P., and Palacin, M. R. (2013). Towards high energy density sodium ion batteries through electrolyte optimization. *Energy & Environmental Science*, 6(8), 2361–2369.

Qi, Y., Mu, L., Zhao, J., Hu, Y.-S., Liu, H., and Dai, S. (2015). Superior Na-storage performance of low-temperature-synthesized $Na_3(VO_{1-x}PO_4)_2F_{1+2x}$ ($0 \leq x \leq 1$) nanoparticles for Na-ion batteries. *Angewandte Chemie International Edition*, 54(34), 9911–9916.

Qi, Y., Zhao, J., Yang, C., Liu, H., and Hu, Y.-S. (2019). Comprehensive studies on the hydrothermal strategy for the synthesis of $Na_3(VO_{1-x}PO_4)_2F_{1+2x}$ ($0 \leq x \leq 1$) and their Na-storage performance. *Small Methods*, 3(4), 1800111.

Reiff, W. M., Zhang, J. H., and Torardi, C. C. (1986). Topochemical lithium insertion into $Fe_2(MoO_4)_3$: Structure and magnetism of $Li_2Fe_2(MoO_4)_3$. *Journal of Solid State Chemistry*, 62(2), 231–240.

Ren, W., Zheng, Z., Xu, C., Niu, C., Wei, Q., An, Q., Zhao, K., Yan, M., Qin, M., and Mai, L. (2016). Self-sacrificed synthesis of three-dimensional $Na_3V_2(PO_4)_3$ nanofiber network for high-rate sodium-ion full batteries. *Nano Energy*. Elsevier, 25, 145–153.

Rettenwander, D., Redhammer, G. J., Guin, M., Benisek, A., Krüger, H., Guillon, O., Wilkening, M., Tietz, F., and Fleig, J. (2018). Arrhenius behavior of the bulk Na-ion conductivity in $Na_3Sc_2(PO_4)_3$ single crystals observed by microcontact impedance spectroscopy. *Chemistry of Materials*, 30(5), 1776–1781.

Saravanan, K., Mason, C. W., Rudola, A., Wong, K. H., and Balaya, P. (2013). The first report on excellent cycling stability and superior rate capability of $Na_3V_2(PO_4)_3$ for sodium ion batteries. *Advanced Energy Materials*, 3(4), 444–450.

Sauvage, F., Quarez, E., Tarascon, J. M., and Baudrin, E. (2006). Crystal structure and electrochemical properties vs. Na^+ of the sodium fluorophosphate $Na_{1.5}VOPO_4F_{0.5}$. *Solid State Sciences*, 8(10), 1215–1221.

Senguttuvan, P., Rousse, G., Arroyo Y De Dompablo, M. E., Vezin, H., Tarascon, J. M., and Palacín, M. R. (2013). Low-potential sodium insertion in a NASICON-type structure through the Ti(III)/Ti(II) redox couple. *Journal of the American Chemical Society*, 135(10), 3897–3903.

Senthilkumar, B., Murugesan, C., Sharma, L., Lochab, S., and Barpanda, P. (2018). An overview of mixed polyanionic cathode materials for sodium-ion batteries. *Small Methods*, 3, 1800253.

Serras, P., Palomares, V., Goñi, A., Gil de Muro, I., Kubiak, P., Lezama, L., and Rojo, T. (2012). High voltage cathode materials for Na-ion batteries of general formula $Na_3V_2O_{2x}(PO_4)_2F_{3-2x}$. *Journal of Materials Chemistry*, 22(41), 22301.

Serras, P., Palomares, V., Kubiak, P., Lezama, L., and Rojo, T. (2013). Enhanced electrochemical performance of vanadyl (IV) $Na_3(VO)_2(PO_4)_2F$ by ex-situ carbon coating. *Electrochemistry Communications*, 34, 344–347.

Shakoor, R. A., Seo, D. H., Kim, H., Park, Y. U., Kim, J., Kim, S. W., Gwon, H., Lee, S., and Kang, K. (2012). A combined first principles and experimental study on $Na_3V_2(PO_4)_2F_3$ for rechargeable Na batteries. *Journal of Materials Chemistry*, 22(38), 20535–20541.

Shannon, R. D. and Prewitt, C. T. (1969). Effective ionic radii in oxides and fluorides. *Acta Crystallographica Section B Structural Crystallography and Crystal Chemistry*, 25(5), 925–946.

Sharma, L., Nakamoto, K., Sakamoto, R., Okada, S., and Barpanda, P. (2019). Na_2FePO_4F fluorophosphate as positive insertion material for aqueous sodium-ion batteries. *ChemElectroChem*, 6(2), 444–449.

Sharma, N., Serras, P., Palomares, V., Brand, H. E. A., Alonso, J., Kubiak, P., Fdez-gubieda, M. L., and Rojo, T. (2014). Sodium distribution and reaction mechanisms of a $Na_3V_2O_2(PO_4)_2F$ electrode during use in a sodium-ion battery. *Chemistry of Materials*, 26, 3391–3402.

Shen, B., You, Y., Niu, Y., Li, Y., Dai, C., Hu, L., Guo, B., Jiang, J., Bao, S., and Xu, M. (2018a). Improving the performance of hard carbon//$Na_3V_2O_2(PO_4)_2F$ sodium-ion full cells by utilizing the adsorption process of hard carbon. *ACS Applied Materials and Interfaces*, 10(19), 16581–16587.

Shen, C., Long, H., Wang, G., Lu, W., Shao, L., and Xie, K. (2018b). $Na_3V_2(PO_4)_2F_3$@C dispersed within carbon nanotube frameworks as a high tap density cathode for high-performance sodium-ion batteries. *Journal of Materials Chemistry A*, 6(14), 6007–6014.

Shiva, K., Singh, P., Zhou, W., and Goodenough, J. B. (2016). $NaFe_2PO_4(SO_4)_2$: A potential cathode for a Na-ion battery. *Energy and Environmental Science*, 9(10), 3103–3106.

Song, W., Ji, X., Zhu, Y., Zhu, H., Li, F., Chen, J., Lu, F., Yao, Y., and Banks, C. E. (2014). Aqueous sodium-ion battery using a $Na_3V_2(PO_4)_3$ electrode. *ChemElectroChem*, 1(5), 871–876.

Susman, S., Delbecq, C. J., Brun, T. O., and Prince, E. (1983). Fast ion transport in the NASICON analog $Na_3Sc_2(PO_4)_3$: Structure and conductivity. *Solid State Ionics*, 9–10, 839–844.

Takada, K., Tansho, M., Yanase, I., Inada, T., Kajiyama, A., Kouguchi, M., Kondo, S., and Watanabe, M. (2001). Lithium ion conduction in $LiTi_2(PO_4)_3$. *Solid State Ionics*, 139(3–4), 241–247.

Tillement, O., Angenault, J., Couturier, J. C., and Quarton, M. (1991a). Mixed conductivity of the NASICON phase $Na_{2+x+y}Zr_{1-y}Fe^{II}{}_xFe^{III}{}_{1-x+y}(PO_4)_3$. *Solid State Ionics*, 44(3–4), 299–303.

Tillement, O., Couturier, J. C., Angenault, J., and Quarton, M. (1991b). Crystal chemistry and electrical study of $Na_xNbTi(PO_4)_3$. *Solid State Ionics*, 48(3–4), 249–255.

Torardi, C. C. and Prince, E. (1986). Structure of the lithium insertion compound $Li_2Fe_2(MoO_4)_3$ from neutron powder diffraction data. *Materials Research Bulletin*, 21(6), 719–726.

Trad, K., Carlier, D., Croguennec, L., Wattiaux, A., Ben Amara, M. and Delmas, C. (2010). $Na_3Fe_3(PO_4)_3$ Alluaudite phase: Synthesis, structure, and electrochemical properties as positive electrode in lithium and sodium batteries. *Chemistry of Materials*, 22(19), 5554–5562.

Trad K., Carlier D., Croguennec L., Wattiaux A., Lajmi B., Ben Amara M., Delmas C. A. (2010). Layered Iron(III) Phosphate Phase, $Na_3Fe_3(PO_4)_4$: Synthesis, Structure, and Electrochemical Properties as Positive Electrode in Sodium Batteries. *J. Phys. Chem. C.*, 114, 10034–10044.

Tsirlin, A. A., Nath, R., Abakumov, A. M., Furukawa, Y., Johnston, D. C., Hemmida, M., Krug Von Nidda, H. A., Loidl, A., Geibel, C., and Rosner, H. (2011). Phase separation and frustrated square lattice magnetism of $Na_{1.5}VOPO_4F_{0.5}$. *Physical Review B–Condensed Matter and Materials Physics*, 84(1), 2–13.

Wang, D., Wei, Z., Lin, Y., Chen, N., Gao, Y., Chen, G., Song, L., and Du, F. (2019). Intercalation pseudocapacitance in a NASICON-structured $Na_2CrTi(PO_4)_3$@carbon nanocomposite: towards high-rate and long-lifespan sodium-ion-based energy storage. *Journal of Materials Chemistry A*, 7(36), 20604–20613.

Wang, H., Zhang, T., Chen, C., Ling, M., Lin, Z., Zhang, S., Pan, F., and Liang, C. (2018). High-performance aqueous symmetric sodium-ion battery using NASICON-structured $Na_2VTi(PO_4)_3$. *Nano Research*, 11(1), 490–498.

Xiong, H., Liu, Y., Shao, H., and Yang, Y. (2018). Understanding the electrochemical mechanism of high sodium selective material $Na_3V_2(PO_4)_2F_3$ in Li^+/Na^+ dual-ion batteries. *Electrochimica Acta*, 292, 234–246.

Xu, M., Xiao, P., Stauffer, S., Song, J., Henkelman, G., and Goodenough, J. B. (2014). Theoretical and experimental study of vanadium-based fluorophosphate cathodes for rechargeable batteries. *Chemistry of Materials*, 26(10), 3089–3097.

Xu, X., Wen, Z., Wu, X., Yang, X., and Gu, Z. (2007). Lithium ion-conducting glass-ceramics of $Li_{1.5}Al_{0.5}Ge_{1.5}(PO_4)_3 - xLi_2O$ (x = 0.0 - 0.20) with good electrical and electrochemical properties. *Journal of the American Ceramic Society*, 90(9), 2802–2806.

Yahia, B.H., Essehli, R., Amin, R., Boulahya, K., Okumura, T., and Belharouak, I. (2018). Sodium intercalation in the phosphosulfate cathode $NaFe_2(PO_4)(SO_4)_2$. *Journal of Power Sources*, 382, 144–151.

Yan, G., Mariyappan, S., Rousse, G., Jacquet, Q., Deschamps, M., David, R., Mirvaux, B., Freeland, J. W., and Tarascon, J. (2019). Higher energy and safer sodium ion batteries via an electrochemically made disordered $Na_3V_2(PO_4)_2F_3$ material. *Nature Communications*, 10(1), 585.

Yi, H., Ling, M., Xu, W., Li, X., Zheng, Q., and Zhang, H. (2018). VSC-doping and VSU-doping of $Na_3V_{2-x}Ti_x(PO_4)_2F_3$ compounds for sodium ion battery cathodes: Analysis of electrochemical performance and kinetic properties. *Nano Energy*, 47, 340–352.

Yin, Y., Xiong, F., Pei, C., Xu, Y., An, Q., Tan, S., Zhuang, Z., Sheng, J., Li, Q., and Mai, L. (2017). Robust three-dimensional graphene skeleton encapsulated $Na_3V_2O_2(PO_4)_2F$ nanoparticles as a high-rate and long-life cathode of sodium-ion batteries. *Nano Energy*, 41, 452–459.

Zhang, B., Dugas, R., Rousse, G., Rozier, P., Abakumov, A. M., and Tarascon, J.-M. (2016a). Insertion compounds and composites made by ball milling for advanced sodium-ion batteries. *Nature Communications*, 7(1), 10308.

Zhang, J., Zhao, X., Song, Y., Li, Q., Liu, Y., Chen, J., and Xing, X. (2019). Understanding the superior sodium-ion storage in a novel $Na_{3.5}Mn_{0.5}V_{1.5}(PO_4)_3$ cathode. *Energy Storage Materials*, 5, 1–10.

Zhang, L., Huang, T., and Yu, A. (2015). Carbon-coated $Na_3V_2(PO_4)_3$ nanocomposite as a novel high rate cathode material for aqueous sodium ion batteries. *Journal of Alloys and Compounds*, 646, 522–527.

Zhang, Q., Liao, C., Zhai, T., and Li, H. (2016b). A high rate 1.2 V aqueous sodium-ion battery based on all NASICON structured $NaTi_2(PO_4)_3$ and $Na_3V_2(PO_4)_3$. *Electrochimica Acta*, 196, 470–478.

Zhang, W., Zhang, Z., Li, H., Wang, D., Wang, T., Sun, X., Zheng, J., and Lai, Y. (2019). Engineering 3D well-interconnected $Na_4MnV(PO_4)_3$ facilitates ultrafast and ultrastable sodium storage. *ACS Applied Materials & Interfaces*, 11, 35746–35754.

Zhang, Y., Zhao, H., and Du, Y. (2016c). Symmetric full cells assembled by using self-supporting $Na_3V_2(PO_4)_3$ bipolar electrodes for superior sodium energy storage. *Journal of Materials Chemistry A*, 4(19), 7155–7159.

Zhang, Y., Guo, S., and Xu, H. (2018). Synthesis of uniform hierarchical $Na_3V_{1.95}Mn_{0.05}(PO_4)_2F_3$@C hollow microspheres as a cathode material for sodium-ion batteries. *Journal of Materials Chemistry A*, 6(10), 4525–4534.

Zhou, M. and Ahmad, A. (2007). Synthesis, processing and characterization of nasicon solid electrolytes for CO_2 sensing applications. *Sensors and Actuators, B: Chemical*, 122(2), 419–426.

Zhou, S., Barim, G., Morgan, B. J., Melot, B. C., and Brutchey, R. L. (2016a). Influence of rotational distortions on Li^+- and Na^+-intercalation in anti-NASICON $Fe_2(MoO_4)_3$. *Chemistry of Materials*, 28(12), 4492–4500.

Zhou, W., Xue, L., Lü, X., Gao, H., Li, Y., Xin, S., Fu, G., Cui, Z., Zhu, Y., and Goodenough, J. B. (2016b). $Na_xMV(PO_4)_3$ (M = Mn, Fe, Ni) structure and properties for sodium extraction. *Nano Letters*, 16(12), 7836–7841.

Zhu, C., Wu, C., Chen, C., Kopold, P., van Aken, P. A., Maier, J., and Yu, Y. (2017). A high power–high energy $Na_3V_2(PO_4)_2F_3$ sodium cathode: Investigation of transport parameters, rational design and realization. *Chemistry of Materials*, 29(12), 5207–5215.

Zhu, Z., Cheng, F., Hu, Z., Niu, Z., and Chen, J. (2015). Highly stable and ultrafast electrode reaction of graphite for sodium ion batteries. *Journal of Power Sources*, 293, 626–634.

3

Hard Carbon for Na-ion Batteries: From Synthesis to Performance and Storage Mechanism

Carolina DEL MAR SAAVEDRA RIOS[1], Adrian BEDA[2,3]
Loic SIMONIN[1] and Camélia MATEI GHIMBEU[2,4]

[1] Université Grenoble Alpes, CEA, LITEN, DEHT, LM, Grenoble, France

[2] Université de Haute-Alsace, CNRS, Institut de Science des Matériaux de Mulhouse (IS2M) UMR 7361, Mulhouse, France

[3] University of Strasbourg, Strasbourg, France

[4] Réseau sur le Stockage Electrochimique de l'Energie (RS2E), FR CNRS 3459, Amiens, France

3.1. Introduction

The development of Na-ion batteries (NIBs) is currently regarded as a sustainable, cheaper and long-lasting technology (Table 3.1) compared to that of Li-ion batteries (LIBs). The design of suitable anode materials that exhibit high performance is still a key challenge in order to commercialize NIBs. Carbonaceous materials, titanium-based compounds, metal-based alloys, chalcogen materials and organic compounds (Wang *et al.* 2017) were explored until now as possible anodes. Unfortunately, there are still several limitations for each of these materials, mainly concerning a low initial coulombic efficiency (ICE), a limited cycle life, poor rate performance and a large volume expansion. In addition, an important lack of

Na-ion Batteries,
coordinated by Laure MONCONDUIT and Laurence CROGUENNEC.

comprehension of the sodium storage mechanisms into these materials limits their practical utilization (Wang *et al.* 2017).

Although LIBs and NIBs present similar components and similar electrochemical working principle, graphite, the most common anode material used in commercial LIBs, proves to exhibit poor electrochemical performance when used in NIBs (capacity ~30 mAh·g^{-1}) (Stevens and Dahn 2001). This is related to several factors such as the weaker binding energy of Na with the graphite making the reversible intercalation difficult, the larger ionic radius of Na (1.02 Å) compared to that of Li (0.76 Å), as well as the tendency of Na to coordinate in octahedral or prismatic sites leading thus to the impossibility of Na ions to form stable intercalation compounds in carbonate-based electrolytes (Liu *et al.* 2016). The higher standard redox potential of Na (–2.71 V vs. SHE) than Li (–3.04 V versus SHE) and the larger atomic weight of sodium represent other issues when considering NIBs as they trigger a lower energy density and larger electrode weight.

Element	Abundance (ppm)	Carbonate cost (USD/t)	Ionic radius (Å)	Current collector	Voltage versus SHE (V)
Li	20	6,000	0.76	Cu	–3.04
Na	23,000	200	1.02	Al	–2.71

Table 3.1. *Comparison between sodium and lithium characteristics*

Among the different carbon-based anode possibilities, hard carbons are especially interesting owing to their high precursor availability, low price, reasonable capacity (can reach ~300 mAh·g^{-1}) and low operating voltage (~ 0 V vs. Na^+/Na), which make them safer. Despite the positive aspects listed above, NIBs using hard carbon materials are still far from large-scale commercialization due to the relatively low energy density and high initial irreversible capacity, but are likely to be part of first-generation sodium cells (CNRS News 2015).

This chapter aims to provide a general overview of the hard carbons explored so far in NIBs. In the first part of the chapter, we will focus on the description of the main synthesis routes used to prepare hard carbons with a focus on the impact of the precursor type and synthesis conditions on the hard carbon characteristics. Mainly, the structural, textural and surface chemistry aspects will be emphasized through several selected examples. The performance in half-cell and full-cell will be discussed in relation with the hard carbon properties in the second part of the chapter. The last part deals with the sodium storage mechanisms, which is critical, and yet still under debate.

3.2. What is a hard carbon?

"Hard" carbon designates a type of carbon material, which is not able to graphitize (transform into graphite) even if the heating temperature exceeds 2,500°C (Oberlin and Terriere 1975). The term "non-graphitizable", "disordered" and "non-organized" carbon is widely used as synonyms. However, terms like "amorphous" and "non-graphitic" carbon are employed as well and should be avoided as recommended elsewhere (Dou *et al.* 2019) since hard carbon does not contain localized π-electrons as amorphous structures, while the "non-graphitic" term designate rather a state of carbon before graphitization process. The reason why hard carbon does not graphitize and does not have a long-range ordering along the c-axis is linked to its microstructure given by strongly cross-linked precursors. The early model used to describe such structure was proposed by Warren (1934, 1941), which considered it as randomly oriented graphitic nanodomains (Figure 3.1(a)). Later on, Franklin (1951) proposed a similar model, but in this case, the graphitic nanodomains are bonded by amorphous carbon (sp^3) domains, which prevent graphitization and lead to micropore formation (Figure 3.1(b)). While these two models were proposed based on X-ray diffraction studies, more in-depth understanding of the hard carbons structure at nanometer scale was achieved by Harris and Tsang (1997) by using TEM analyses. In this model, the graphene curvature observed in TEM images is taken into consideration and related to fragmented fullerenes-like structures, that align in various degrees, composing the micropore walls (Figure 3.1(c)). This curvature stabilized by non-hexagonal structures is believed to prevent the graphitization. Other types of defects such as vacancies, interstitial heteroatoms and empty sites may deform the regularity of graphene sheets (Dou *et al.* 2019). The hard carbon structure is considered rather complex due to the presence of the different types of porosities as well. In this regard, the widely accepted model today describing the hard carbon structure is the "house of cards" or "falling cards" model (Figure 3.1(d)) proposed by (Dahn *et al.* 1997 and Buiel *et al.* 1999). This considers hard carbon as a mixture of pseudo-graphitic micro-crystallites (sp^2 hybridized, turbostratic staked graphene layers) and amorphous regions (sp^3 hybridized) along with nanovoids (nanopores). The temperature increase generally favors the nanopores coalescence and closure above 1,000°C.

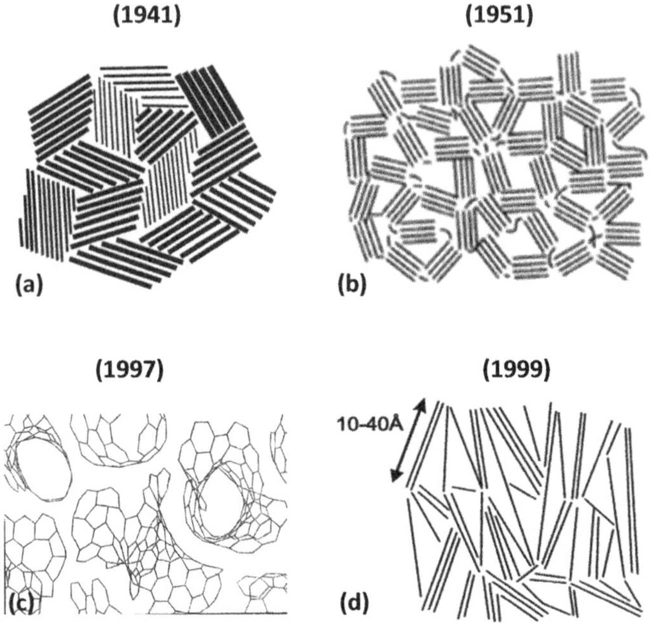

Figure 3.1. *Hard carbon models as proposed by (a) Warren (1941), (b) Franklin (1950), (c) Harris and Tsang (1997) and (d) Dahn* et al. *(1999).Reprinted with permission from Warren* et al. *(1941 Copyright (1941) by the American Physical Society. Reprinted with the permission of Haris and Tsang 1997. Copyright (1997) Taylor & Francis Ltd. Reprinted from Buiel* et al. *(1999). Copyright (1999), with permission from Elsevier*

During the synthesis of hard carbon by thermal annealing (pyrolysis) under inert atmosphere of a precursor, several events occur, highly dependent on the annealing temperature and precursor composition (Figure 3.2). Usually, water is seen below 100°C due to the dehydration/depolymerization reactions while a great amount of gases (H_2O, CO_x, CH_4) is released at around 400°C as a result of precursor thermal decomposition, with the formation of a solid residue. Due to the released gas, microporous carbonaceous material is formed called char, which is not able to graphitize contrary to fused residues that are able to undergo graphitization (Kipling *et al.* 1964). At higher temperatures (500-1000°C), the release of oxygen- and nitrogen-based functional groups and the removal of hydrogen from the structure leads to the formation of hard carbon (interlayer space, d_{002} comprise between 3.7 and 4.0 Å) (Zhang *et al.* 2016; Beda *et al.* 2018). The hydrogen amount observed in the precursor, could be linked in some extent to its ability to undergo fusion state and to be graphitized, i.e. a richer hydrocarbon precursor leading to graphitized carbon. Higher temperatures (>2,000°C) lead to glassy carbon formation ($3.4 \leq d_{002} \leq 3.6$ Å). The annealing conditions (temperature, type of gas, pressure and heating rate), the type of precursor (chemical composition and structure) and the synthesis conditions will impact the characteristics of hard carbons (porosity, structure, surface chemistry and defects) and *in-fine* their performance in NIBs. All these aspects will be developed in the following part.

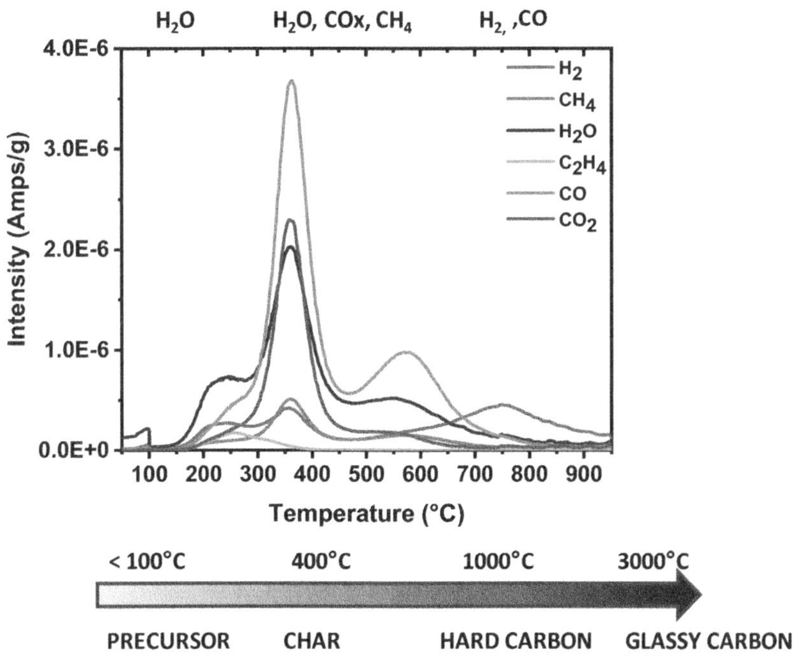

Figure 3.2. *Temperature programmed desorption coupled with mass spectrometry showing the gas release profiles during the heating of a phenolic resin precursor. Reprinted from Beda* et al. *(2018). Copyright (2018), with permission from Elsevier. For a color version of this figure, see www.iste.co.uk/monconduit/batteries.zip*

3.3. Hard carbon synthesis and microstructure

Hard carbon is the first material that has achieved performance in Na-ion half-cells close to those of graphite in LIBs. Since the early 2000s, hard carbons have been reported from a wide range of precursors such as phenolic resins, sucrose, glucose, cellulose, cotton, pitch, lignin, wood, polyaniline (PANI), polyacrylonitrile (PAN) and various biomasses (banana/pomelo/apple peels, okara, leaves, algae, etc.). To easily differentiate between them, the precursors used to develop hard carbon materials which can be grouped into three main categories: synthetic polymers, bio-polymers and raw biomass (Figure 3.3(a)). With the aim of finding the best precursor for the preparation of hard carbon, all three categories were approached owing to their own advantages.

The raw biomass is a convenient cheap precursor widely available as a renewable source of energy; this is the reason why it is widely used for hard carbon preparation (Figure 3.3(a)). However, biomass is not uniformly distributed over the world, it varies with the seasons and requires a lot of space to be stored (biomass plants), which may limit the availability. Moreover, the majority of the biomass present a significant amount of inorganic impurities in their composition, which may negatively affect the electrochemical performance, requiring a supplementary washing step generally with acids. Another crucial issue is the carbon yield after pyrolysis, which is lower (~20 %) compared to other precursors, as seen in Figure 3.3(b). This low yield has the main disadvantages of a higher production cost and the release of a greater amount of undesirable CO_x gases during their pyrolysis.

Bio-polymers (cellulose, chitosan, lignin etc.), the second category often explored to produce hard carbons, are commercially available, biodegradable and sustainable with renewable sources of raw materials and thus a low cost. As in the case of biomass, the carbon yield is generally low (~20 % for cellulose, sucrose, glucose) (Narasimman and Prabhakaran 2013; Matei Ghimbeu *et al.* 2018) but for some precursors such as chitosan and lignin (Conder *et al.* 2019; Matei Ghimbeu *et al.* 2019), a higher yield (~40–50%) can be reached.

The last category is the synthetic polymers that are largely soluble in different solvents and possess a rich chemistry offering, as the main advantage, the possibility of adapting the synthesis procedure via chemical parameters in order to control carbon characteristics. Besides, the carbon yield obtained is usually high (~40–50% for phenolic resin and PAN). The yield of carbon depends on both precursor chemical composition and structure (Irisarri *et al.* 2018). The main disadvantage exposed by this class of precursors is their toxicity. For example, phenolic resins, widely used to obtain hard carbons, are made up of phenol and formaldehyde that are carcinogenic. The high price compared to the other precursors could be also a limitation for synthetic polymers, as seen in Figure 3.3(c).

In what follows, these three categories of precursors will be discussed in detail. The main synthesis procedures reported so far to obtain hard carbon will be presented in order to highlight the main changes observed in their characteristics (structure, texture, morphology, surface chemistry, defects, etc.).

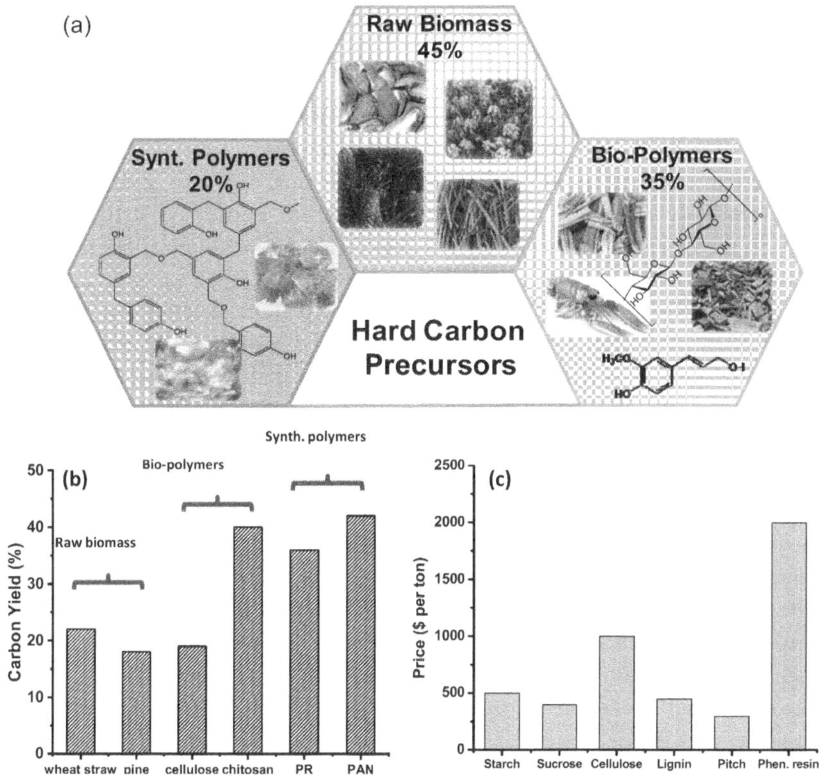

Figure 3.3. *(a) Representation of the main categories of precursors used to prepare hard carbon and their percentage of use for hard carbon preparation determined based on recent publications; (b) carbon yield exhibited by different precursors belonging to the three main categories: raw biomass (wheat straw, pine; Saavedra Rios* et al. *2018), bio-polymers (cellulose; Matei Ghimbeu* et al. *2018), chitosan (Conder* et al. *2019), synthetic polymers (phenolic resin; Beda* et al. *2018) and PAN (Zhang* et al. *2016) when thermally treated at 900°C; (c) price per ton of different precursors reported in literature (Li* et al. *2015). For a color version of this figure, see www.iste.co.uk/monconduit/batteries.zip*

3.3.1. *Synthetic precursors-based hard carbon synthesis*

Due to the high yield and synthesis flexibility they offer, synthetic precursors are often employed for hard carbon preparation. The most commonly used are phenolic resins, PAN and PANI. Phenolic resins are a family of aromatic compounds directly bonded with hydroxyl groups (Ku *et al.* 2010). They are especially interesting due to the possibility of controlling the microstructure and morphology of the hard carbon (Wang *et al.* 2017).

Generally, the simplest way to obtain hard carbon is the direct pyrolysis of the precursors at temperatures above 1,000°C. In other cases, precursors pretreatments (hydrothermal carbonization [HTC], electrospinning, sol-gel and washing) or posttreatments (washing, activation and doping) are performed with the aim of modifying the morphology, structure and chemical functionalities of hard carbons.

One of the first reports on hard carbons derived from phenolic resins for NIBs was published by Wang *et al.* (2015). The phenolic resin was obtained by mixing phenol and formaldehyde followed by the addition of sodium hydroxide solution and formaldehyde in excess. Hard carbon microspheres were obtained by hydrothermal treatment (500°C) followed by carbonization at high temperature (800–1,500°C). In a different study, Li *et al.* (2016c) used commercial phenolic resins in combination with pitch (different ratios) to obtain hard carbon materials after carbonization under Ar flow at temperatures between 1,200 and 1,600°C. Another work implying commercial phenolic resins was published by Irisarri *et al.* (2018), where the authors studied the influence of the precursor type (cellulose and lignin) and the impact of pyrolysis temperature on the material microstructure and surface chemistry. Due to the high carbon yield (68% at 500°C), high tap density (0.7 $g{\cdot}cm^{-3}$) and reversible capacity, the phenolic resin proved to be the most suitable precursor for large-scale production. Hasegawa *et al.* (2015) showed promising electrochemical performance for NIBs, as well, when testing hard carbons made of macroporous phenolic precursors. The sol-gel process was used to prepare the phenolic gel by mixing resorcinol and formaldehyde in acidic ethanol solution. The carbonization temperature was varied from 800 to 3,000°C. Zhang *et al.* (2017b) reported hard carbon prepared from a mixture of phenolic resin and sucrose in different ratios heat-treated at temperatures between 1,000 and 1,600°C. As highlighted in the above-mentioned studies, most phenolic resins used to obtain hard carbons are based on phenol/formaldehyde, which is highly toxic for both humans and the environment. Only limited attempts to find non-toxic alternatives have been made. Hard carbons from "green" phenolic resins based on plant extracted precursors such as phloroglucinol and glyoxylic acid was reported (Beda *et al.* 2018). Simple dissolution of the precursors in water-based solvents, followed by aging and thermal treatment at different temperatures (1,100–1,700°C), leads to hard carbons with various textures and structures.

Huang and his team published two articles in which lignin (Li *et al.* 2015) and phenolic resin (Li *et al.* 2016c) were mixed with pitch to increase the graphitization level of hard carbon-derived materials (Qin *et al.* 2012). By controlling the ratio between pitch and lignin/phenolic resin, as well as the pyrolysis temperature, the authors obtained materials with distinct properties (structure and texture), which were further successfully used as anode for sodium storage.

PAN is widely used as a precursor for the production of carbon fibers (Zhao *et al.* 2017; Gao *et al.* 2018). Due to the low cost and high yield, possibility of developing hard carbons for sodium storage from such precursors was explored since early 2000's when Thomas and Billaud (2002) successfully used commercial PAN-based carbon fibers heat treated at 1,650°C and 2,100°C to reversibly intercalate Na-ions. However, the higher temperature treatment resulted in a lower d-spacing (0.344 Å), which proved to be limiting for sodium storage. Zhang *et al.* (2016) reported PAN-derived hard carbons produced by electrospinning. A solution of PAN and DMF (dimethylformamide) is stretched into fibers by applying a high voltage (18 kV). The result is a fiber film that is first stabilized at 250°C to avoid PAN melting followed by thermal treatment in the range of 650–2,800°C resulting in hard carbons with distinct properties for a better understanding of Na ion storage mechanisms. Such synthesis method presents the advantage that the hard carbon film can be used as obtained in a half-cell sodium ion batteries (no other additives). A similar approach is reported by Jin *et al.* (2014) who prepared lignin–PAN based binder-free electrodes by electrospinning. The ratio lignin-PAN was studied and the results showed that increased lignin content in the system (up to 1:1 wt.%) leads to hard carbon properties more compatible with sodium storage application (larger d-spacing, higher disorder degree and lower specific surface area SSA). However, as viscosity increases with lignin addition, electrospinning process was affected at high concentration.

In order to enhance the electrochemical performance of carbon materials, doping with heteroatoms such as nitrogen (N), boron (B), sulfur (S) or phosphorus (P) is a common strategy used. Electrospinning method seems to be again preferred for N-doped carbon preparation for NIBs. Zhen *et al.* (2014) report a complex synthesis method of N-doped hollow carbon nanofibers obtained by pyrolysis of polypyrrole nanofibers (700°C under N_2) obtained using electrospun polycaprolactone as a template. In a different study, Wang *et al.* (2016b) fabricate flexible N-doped carbon films using polyamic acid as a precursor. The free-standing anodes were obtained by carbonization at 650°C for 3 h under Ar atmosphere. As S radius (102 ppm) is larger than that of N (75 ppm), larger interlayer distance of carbon material was obtained, facilitating the sodiation-desodiation process. Remarkable electrochemical performance was reported by Ye *et al.* (2016) when using co-doped (N and S) hollow carbon spheres (HCSs) obtained from phenolic resins. The precursor is obtained by dissolving resorcinol/urea/thiourea and formaldehyde in water followed by addition of polymethyl methacrylate (PMMA) that acts as a template. A thermal treatment at 600°C under nitrogen gas leads to the hollow carbon nanospheres.

Although carbon doping is often used to boost a material's performance, the thermal treatment is usually done at low temperatures (600–900°C) to preserve the heteroatoms in the structure; however, it is too low to obtain an optimal hard carbon structure.

3.3.2. *Bio-polymers derived hard carbon synthesis*

Polysaccharides are complex carbohydrate derivative structures that contain chains of linked monosaccharide subunits intermediate, recognized as an important class of bio-polymers (along with polynucleotides and polypeptides). Naturally, they can be found in plants (Liu *et al.* 2018), as well as in microbes and animal sources (Yu *et al.* 2018b). Plants, due to their availability and sustainability, are considered the most important source for production of different types of polysaccharides, i.e. cellulose (the most abundant naturally occurring polysaccharide), cotton, starch, pectin, sucrose and glucose. On the other hand, animal and microbial sources could be used as well to obtain bio-polymers: glycogen (animal and fungal cells), chitosan and chitin (arthropods), etc. Besides the wide availability and viability, polysaccharides present the advantages of low cost and along with their functional properties such as non-toxicity, biodegradability and biocompatibility, they are commonly used as precursors to prepare hard carbons for NIBs. One of the first studies on sodium storage using polysaccharides-derived carbons was published by Stevens and Dahn in the 2000s. They reported a glucose-derived hard carbon that was able to deliver very promising capacity when treated at 1,100°C (~300 $mAh{\cdot}g^{-1}$) proving that such carbons can successfully store sodium ions. A few years later (2002), Thomas and Billaud published similar results using cellulose fibers heat treated at 1,000°C. However, in the following years, very little attention has been given to sodium technology and after a long break, in 2013, Ponrouch *et al.* published a new report on hard carbon derived from sugar. The anode material was obtained by simple pyrolysis at 1,100°C for 6 h under Ar flow and employed for electrolyte optimization for NIBs (Ponrouch *et al.* 2013). In the next years, polysaccharides were widely used to prepare hard carbons with the aim of getting materials that deliver high capacity and ICE, as well as long-term stability. Bommier *et al.* (2014) tried to predict the capacity of hard carbon in sodium batteries based on materials porosity by using sucrose-derived hard carbon. The materials were obtained in two steps: sucrose dehydration at 180°C for 24 h followed by pyrolysis at temperatures between 800°C and 1,100°C. CO_2 activation was performed as well in the attempt to enhance the materials capacity. Based on their results which suggest that sodium storage is more favorable for low porosity materials, they further doped sucrose with graphene oxide (acts as dehydration agent) to lower hard carbon specific surface area down to 5.4 $m^2{\cdot}g^{-1}$ and thus improve their performance

(Luo *et al.* 2015). As result, ICE% improved from 74% to 83% while maintaining 95% capacity retention after 200 cycles. Using the same precursor but a different approach, i.e., pyrolysis at 1,100°C for 6 h followed by physical vapor deposition to obtain a carbon coating of about 2 nm thick, Ponrouch and Palacin (2015) succeeded in enhancing the performance of the hard carbon due to the decrease in SSA. The synthesis procedures involving carbohydrates are indeed simple but two practical issues limit such precursors for large-scale applications, i.e., they expand during the thermal treatment while the carbon yield is low in some cases (see Figure 3.1(b)). To overcome these problems, new approaches had to be explored and HTC prior to the pyrolysis step is one of the most used strategies. Väli *et al.* (2016, 2017a, 2017b) prepared hard carbons (for Na^+, K^+ and Li^+ storage) from D-glucose via HTC. An aqueous solution of glucose is first thermally treated in an autoclave at 200°C for 24 h, then washed, dried and pyrolyzed at 1,100°C for 2 h under Ar flow. Additionally, with the purpose to reduce surface functional groups, an additional treatment at 800°C under H_2 is performed. In a more recent study, Yamamoto *et al.* (2018), explored several polysaccharides precursors: glucose, sucrose, maltose, cellulose, glycogen and amylopectin, which were preheated at 180, 275, 300 and 350°C for 12 h followed by thermal annealing at 1,300°C for 1 h under Ar flow. The authors found that cellulose-derived hard carbon preheated at 275°C delivers the highest reversible capacity. Moreover, it was pointed out that cellulose pretreatment at 275°C induces a high cross-linking between polymer chains, which favors micropore formation that will contribute to capacity increase. Other similar reports where cellulose is stabilized prior to the thermal treatment could be found in addition, in the literature (Luo *et al.* 2013; Zhu *et al.* 2017a). Hard carbon derived from cotton precursors have been reported as well (Li *et al.* 2016b). A simple heat treatment (1,000–1,600°C) leads to hard carbon materials with interesting features for sodium storage (low surface area and high disorder degree).

Shen *et al.* propose a different approach. Instead of developing powder hard carbons, a self-standing electrode was obtained from cellulose wood fibers (Shen *et al.* 2015). The synthesis procedure consists of filtering a dispersion of softwood pulp and water leading to a cellulose fiber paper, which is first dried at 80°C and then carbonized at 1,000°C, under Ar. Because the specific surface area was high (590 $m^2 \cdot g^{-1}$), a pretreatment with 2,2,6,6-tetramethylpiperidine-1-oxyl (TEMPO) was done to lower it (130 $m^2 \cdot g^{-1}$). The advantage of developing self-sustained electrodes relies on the absence of polymeric binder, solvent, conductive carbon and current collector, and thus the price and the total weight can be reduced while the performance delivered are related just to the active material.

It can be pointed out that all these studies are limited to laboratory scale and only small amounts of materials are prepared. Palacin *et al.* (Irisarri *et al.* 2018) tried to optimize hard carbon synthesis and their performance for large-scale applications by tuning syntheses protocols and set-ups. Excellent performance was reported for pyrolyzed cellulose derived hard carbons while using a high loading of 500 g per heat treatment tube.

3.3.3. *Biomass-based hard carbon synthesis*

Recently, there is an increasing interest in the research community for biomass-based hard carbon electrodes, as can be seen from Figure 3.3(a). This interest is mainly related to the valorization of biowastes to obtain high-quality materials from lower cost precursors compared to other sources (Górka *et al.* 2016). Additionally, the processing of specific biomass precursors, such as lignocellulosic biomass, is a carbon neutral cycle, which implies that the production of hard carbon actually leads to the sequestration of CO_2.

Biomass can be technically defined as a complex, organic solid product of non-fossil nature obtained from natural or anthropogenic processes (Vassilev *et al.* 2012). It comprises the fraction of the biodegradable matter coming from products, wastes and residues from the biological origin (vegetal or animal), and the biodegradable fraction of municipal and industrial wastes (EUR-Lex n.d.). Vassilev *et al.* (2012) classify biomass into six main groups, according to the biological diversity, source and origin of the solid, as follows: (1) wood and woody products, (2) herbaceous and agricultural products, (3) aquatic products, (4) animal and human wastes, (5) industrial biomass wastes and (6) mixture of the one to five categories. Vegetal origin biomass, such as types 1 and 2 in the previous classification, is called lignocellulosic biomass and it is the main type of biomass used in research to produce hard carbon, given its chemical composition, availability and low cost (Dou *et al.* 2019). Lignocellulosic biomass is mainly composed of three structural biopolymers found in the secondary cell wall of plants: cellulose, hemicellulose and lignin (Meents *et al.* 2018), together with other minor non-structural compounds such as extractives, inorganic compounds and moisture.

The main inorganic elements found in lignocellulosic biomass are nitrogen, chlorine, calcium, potassium, magnesium, sulfur, silicon, manganese, sodium, phosphor, iron and heavy metals. These elements are present in the organic matrix as water-soluble compounds, salts, minerals or mineraloids or organically bounded to the carbon, hydrogen and oxygen of cellulose, hemicellulose and lignin. Inorganic matter in lignocellulosic biomass is mostly non-volatile and thus gathered as the remaining fraction after combustion called ash. Ashes composition may include phosphates, carbonates, silicates, chlorides, sulfates, oxides, hydroxides and nitrates which may impact the electrochemical performance of the materials. Figure 3.4 provides a representation of the organization of biomass components at different scales. It is challenging to present a typical lignocellulosic biomass composition given its high variability; so as a general overview, one can consider that cellulose content can be up to 50% in dry mass percentage, while hemicellulose, lignin and extractives can take values up to 30%, 30% and 10%, respectively. Ashes content may be as high as 60% for some biomasses (Vassilev *et al.* 2012).

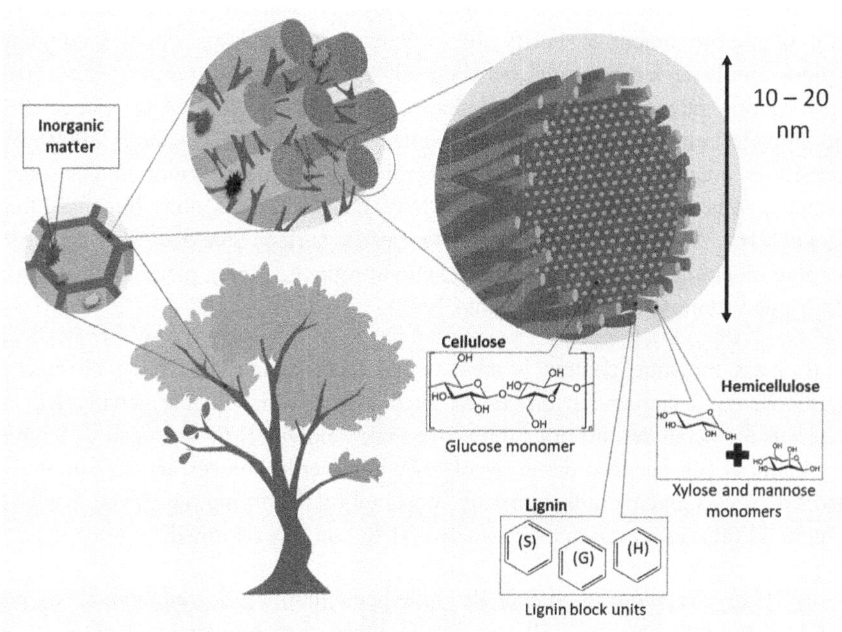

Figure 3.4. *Schematic representation of lignocellulosic biomass structure (credit to Juana Molina). For a color version of this figure, see www.iste.co.uk/monconduit/batteries.zip*

The most used biomass precursors for hard carbon reported in the literature are a wide variety of fruits wastes (Wu *et al.* 2016; Wang *et al.* 2017a; Izanzar *et al.* 2018; Zhu *et al.* 2018), fruit peels (Lotfabad *et al.* 2014; Xiang *et al.* 2017; Li *et al.* 2018; Talekar *et al.* 2018), shells of different varieties of nuts (Lv *et al.* 2015; Wang *et al.* 2016a; Dahbi *et al.* 2017; Dou *et al.* 2017; Kim *et al.* 2017; Xu *et al.* 2018; Zhang *et al.* 2018b) or agro-industrial residues (Jiang *et al.* 2016; Cao *et al.* 2017; Dou *et al.* 2017; Saavedra Rios *et al.* 2018; Wang *et al.* 2018a)). Other types of vegetal biomass such as lotus stem (Zhang *et al.* 2018a), pinecones (Zhang *et al.* 2017c), pine pollen grains (Zhang *et al.* 2018c), old loofah (Yu *et al.* 2018a), sorghum stalk (Zhu *et al.* 2017b), ramie fiber (Jiang *et al.* 2016), switchgrass (Zhang *et al.* 2017a) and pine wood and beechwood residues (Saavedra Rios *et al.* 2018). However, less common precursors of non-lignocellulosic nature, such as humic acid (Zhu *et al.* 2017c), chitosan/chitin (Conder *et al.* 2019) and algae (Meng *et al.* 2015), have been reported with competitive performance of NIBs.

Hard carbons are mainly synthesized by pyrolysis under inert atmosphere, with prior or post-treatment to purify the carbonaceous materials. Studies commonly explore pyrolysis temperatures between 600°C and 1,600°C (Górka *et al.* 2016). Slow heating rates ($<10°C\cdot min^{-1}$) enhance hard carbon yield (Anca-Couce 2016) and allow to better retain the original biomass morphology (Haas *et al.* 2009). New porosity is created in the solid upon pyrolysis, and a maximum in the specific surface area is commonly seen between 600 and 800°C for woody biomass (Anca-Couce 2016). At higher temperatures, the specific surface area decreases because of the pore blocking by tar's secondary reaction products, giving place to closed pores (Rodríguez-Reinoso and Molina-Sabio 1992).

Biomass moisture content is an essential factor when classifying biomass for thermochemical conversion processes, such as pyrolysis. In this perspective, raw materials are separate into dry (<60% moisture) and wet (≥60% moisture) biomass. Dry biomass can undergo direct pyrolysis while wet biomasses are not suitable for direct pyrolysis given a high energetic consumption for drying the elevated moisture content. In this case, pretreatments such as HTC can be performed.

In HTC, the biomass precursor undergoes thermal degradation in a water reaction medium, between 180 and 250°C, under the self-generated pressure of a closed system (5–10 MPa) (Kumar *et al.* 2018; Wang *et al.* 2018b ; Rodriguez Correa *et al.* 2019). The main products are a solid residue called hydrochar, and a liquid phase containing aqueous and organic phases. A small quantity of gas is also generated and mainly composed of CO_2 (Wang *et al.* 2018b). The hydrochar obtained has different physicochemical properties than those of chars obtained from direct pyrolysis, given that the degradation of biomass is done by different mechanisms. Hydrochars normally retain less ash content in the carbonaceous structure given that the inorganic elements dissolve into the liquid phase. Because of the continuous re-condensation reactions, hydrochars have a lower surface area than chars, resulting in the constant blocking of the pores.

Strong acid/base treatment is another pretreatment of the biomass precursor prior to pyrolysis. It is used to remove inorganic elements from raw biomass (Cao *et al.* 2017; Zhang *et al.* 2017c, 2018a, 2018c). Zhang *et al.* (2017c) and Dahbi *et al.* (2017) compared hard carbons obtained from chemically treated (with KOH and HCl solutions) and untreated pinecone and argan shell precursors. Authors found that potassium was the most abundant metal and in general, washed samples exhibit better retention capacities and rate capabilities than the unwashed ones. Washed samples may contain more active sites for sodium storage given the removal of the metals. Acid treatment seems to induce differences in the structural parameters such as higher graphitization degree and less presence of defects while enhancing the hard carbon yield. However, the punctual effect of the inorganic constituents of biomass in electrochemical performance of hard carbon still needs to be studied.

Some authors (Wu *et al.* 2016; Kim *et al.* 2017; Hasa *et al.* 2017; Xiang *et al.* 2017) employed these chemical treatments to obtain activated hard carbons (AHC) with a high specific surface area (between 200 and 800 $m^2.g^{-1}$), arguing that it is beneficial for the electrolyte percolation and sodium storage, despite the fact that coulombic efficiency is drastically reduced. Xiang *et al.* (2017) activated orange peels to make high specific surface area hard carbons. Non-AHC were prepared by direct pyrolysis at 800°C. For AHC, orange peels were first washed with 7% KOH solution, pyrolyzed at the same conditions as non-AHC samples, and then grounded and dispersed in 2 M HCl solution. The resulting AHC had a specific surface area of 638 $m^2 \cdot g^{-1}$, while non-AHC had 357 $m^2 \cdot g^{-1}$. Additionally, in AHC the higher fraction of the pore volume was made up of micropores of 0.7 nm, while non-AHC presented mostly mesoporous with a larger pore volume fraction centered at 8 nm. Concerning electrochemical behavior, authors report a high first insertion capacity of 497 $mAh \cdot g^{-1}$ for AHC and 350 $mAh \cdot g^{-1}$ for non-AHC at 0.5 $mA \cdot g^{-1}$. However, ICE in both cases is very low, around 40%. AHC presents better charge/discharge performance, better rate capability and stability through 1,000 cycles when compared to non-AHC.

Lignocellulosic biomass exhibits natural occurring morphology and porosity that may also enhance electrolyte percolation, ionic and electronic conductivity. Zhang *et al.* (2018a) have studied this feature using as precursor lotus stems from China. Lotus stem has an intrinsic hierarchical structure of channels in the longitudinal axis, which leads to the formation of a controlled three-dimensional structure after pyrolysis. The resulting hard carbon pyrolyzed at 1,400°C exhibits the first reversible capacity of 351 $mAh \cdot g^{-1}$ with an ICE of 70% at a current density of 40 $mA \cdot g^{-1}$, for a low surface area of 24.3 $m^2 \cdot g^{-1}$. In parallel, pollen grains exhibit a unique morphology of hollow spheres. Li *et al.* (2017a) synthesized hard carbon from pollen grains to expose the hollow carbon reticulate matrix in order to enhance diffusion of ions and conductivity of electrons. Hydrothermal treatment was used to get through the reticulated structure without damaging it and pyrolysis aimed to preserve this structure. Low pyrolysis temperature of 600°C produced the samples with the best performance in both reversible capacity (135 $mAh·g^{-1}$ at 100 $mA·g^{-1}$) and rate capability (50 $mAh·g^{-1}$ at 2 $A·g^{-1}$). Only 10% of the capacity was lost after 1,000 cycles at 100 $mA·g^{-1}$.

3.4. Hard carbon characteristics

Given the complexity of hard carbon structure, several analytical techniques are used to elucidate it, and subsequently to understand its relation with the electrochemical performance. The most common analysis of hard carbon microstructures comprises structural techniques such as X-ray diffraction (XRD), small angle X-ray scattering (SAXS), transmission electron microscopy (TEM), and Raman spectroscopy; porosity and surface area measurements using gas adsorption-desorption isotherms, surface chemistry analysis via X-ray photoelectron spectrometry (XPS), energy dispersive X-ray spectroscopy (EDX) and temperature programmed desorption coupled with mass spectrometry (TPD-MS). The final hard carbon structure and properties are strongly dependent on several factors, i.e., the initial precursor nature, the synthesis conditions and the final pyrolysis temperature being one of the critical parameters. The following section will briefly describe these techniques applied for hard carbon analysis along with the main important parameters that can be tuned in order to achieve the optimal hard carbon characteristics for Na-ion storage.

3.4.1. *Hard carbon structure*

The XRD technique uses an incident X-ray source that scatters against the atomic lattice of a given material to obtain information about its structure at the local scale (Bommier *et al.* 2018; Dou *et al.* 2019). It is mainly used for characterizing crystalline structures such as graphite, however, disordered carbons also present domains of randomly stacked graphene layers, allowing them to obtain structural information from this technique when applied to hard carbons. As a result, a typical diffraction pattern of hard carbon structure exhibits very broad diffraction peaks for *00l* (002) and asymmetrical *hk0* (10), (11) lines, related to the reflection of the basal graphene planes and perpendicular to them, respectively (Inagaki and Kang 2016), as presented in Figure 3.5(a). The (002) broad feature in hard carbon diffraction pattern represents the average distance between each graphene layer in the domain and is used to estimate the thickness of the domain (L_c) along the c-axis. The (10) signal is used to estimate the width of the domain (L_a) along the a-axis (Bommier *et al.* 2018) by using the Scherrer equation: $L_{a,c} = \frac{k\lambda}{B\cos\theta}$, where $k = 0.9$ for L_c and 1.84 for L_a; B is the full width at half-maximum intensity for each feature.

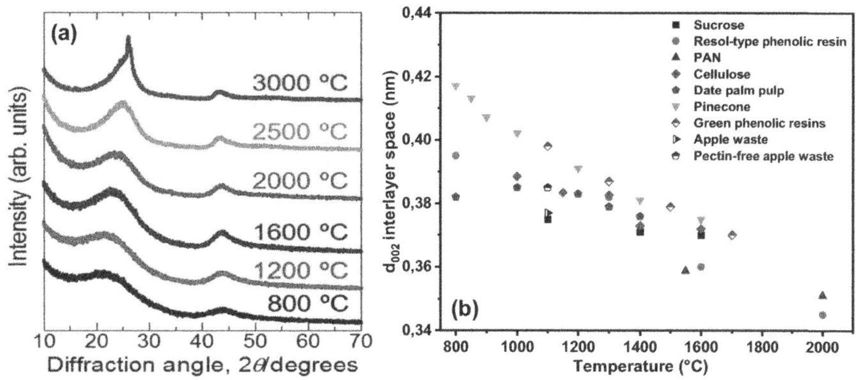

Figure 3.5. *(a) XRD pattern for hard carbon derived from phenolic resin, synthesized at different temperatures (adapted from Hasegawa* et al. *2015). (b) Evolution of the interlayer space (d_{002}) with heat treatment of various precursors : sucrose (Bommier* et al. *2015), phenolic resin (Zhang* et al. *2011), PAN (Zhang* et al. *2016), Cellulose (Simone* et al. *2016) (Matei Ghimbeu* et al. *2018), Date palm pulp (Izanzar* et al. *2018), Pinecone (Zhang* et al. *2017c), Green phenolic resins (Beda* et al. *2018), apple waste (Wu* et al. *2016), pectin-free apple waste (Dou* et al. *2018). For a color version of this figure, see www.iste.co.uk/monconduit/batteries.zip*

With the increase in the thermal treatment temperature, the graphene layers in hard carbon structure tend to get closer, as it is evidenced by the thinner diffraction features and a shift of the (002) diffraction peak toward higher 2 theta angles (Figure 3.5(a)), indicating a decrease in the distance between layers. In parallel, a growth in the domain is also observed, in both width (L_a) and thickness (L_c). Figure 3.5(b) shows literature-reported data for the diminution of the calculated d_{002} distance for different hard carbon precursors upon heat treatment. Notice also, that for a given temperature, the structure varies from one precursor to another. In parallel, for the same precursor, the chemical synthesis conditions influence the final hard carbon structure, as can be seen from the green phenolic resins obtained by varying the solvent and the drying conditions (Beda *et al.* 2018). Dou *et al.* (2018) and Wu *et al.* (2016) produced hard carbons from pectin-free pomace and apple waste, respectively, founding that the absence of pectin in the precursor and the acid pretreatment also had an influence on the final structure of the hard carbon.

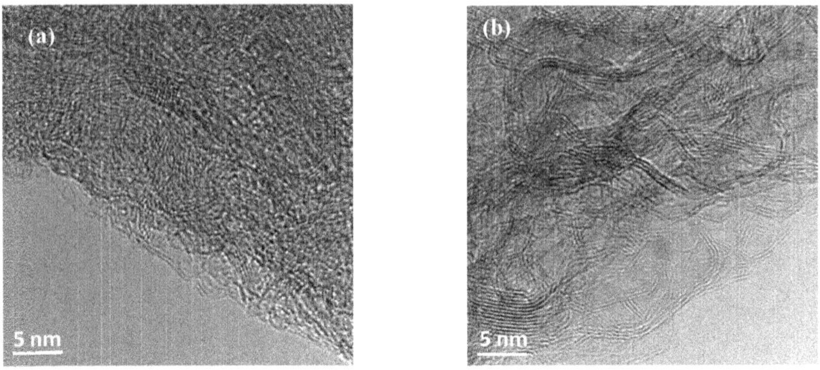

Figure 3.6. *TEM micrographs of PAN-derived hard carbon at (a) 1,550°C and (b) 2,200°C (Zhang* et al. *2016)*

Using the TEM technique, the local nanostructure of hard carbons can be observed and it is possible to evidence the presence of the distorted graphitic domains in the materials. As for XRD, TEM micrographs give evidence of an increment in the graphitization degree at higher temperatures, showing what is thought to be the alignment of more and larger graphene layers with temperature, as can be seen from Figure 3.6. In addition, it is possible to identify empty spaces between the domains, corresponding to the closed porosity in the structure. The high-resolution TEM (HR-TEM) technique allows further analysis of disordered carbons. Bommier *et al.* (2018) enlisted that recent reports deals with the estimation of the lattice fringe length, tortuosity and average curvature of graphene sheets, applied for hard carbon structure elucidation.

3.4.2. *Hard carbon porosity*

Surface area, open porosity and closed porosity are key parameters to investigate for hard carbon materials. It is well known that the greater the surface area, the highest the irreversibility, given the SEI formation (Dou *et al.* 2019). However, surface area allows better electrolyte percolation, and therefore, it is needed in some extent for a high power capability (Saurel *et al.* 2018). Although less studied and difficult to quantify, the closed porosity is thought to be an important factor contributing to sodium ions storage by bulk diffusion mechanisms (Saurel *et al.* 2018; Dou *et al.* 2019). To measure the open porosity and surface area, the most employed technique is N_2 adsorption–desorption. Normally, non-pretreated hard carbon exhibits low surface area derived from the N_2 adsorption measurement for temperatures above 1,000°C (Figure 3.7(a)). However, Matei Ghimbeu *et al.* (2018)

performed a study with cellulose-derived hard carbons, revealing that N_2 adsorption isotherms give limited information about the microporosity of the material. They used CO_2 as analytic gas and revealed that the derived surface area was significantly higher. The smaller size of the CO_2 molecule, together with a better diffusion behavior favored by the higher analysis temperature (273 K for CO_2 vs. 77 K for N_2), allows CO_2 molecule accessing smaller pores called ultramicropores (pore size < 0.7 nm) (Matei Ghimbeu *et al.* 2018). Although the role of such small pores on Na storage is not yet understood, they certainly impact molecule adsorption (water, electrolyte), material density, defects, etc. Figure 3.7(a) gathers the surface area determined from the N_2 isotherms for hard carbons synthesized at different temperatures and from different precursors. The CO_2-derived surface areas are also presented for some of the hard carbons, since only few works provided such information (Figure 3.7(b)). Worth to note that CO_2 measurements lead to higher surface area values; however, above a given temperature, the surface area tend to reach the low values similar to those calculated from N_2 isotherms due to the pore closure (Dou *et al.* 2019).

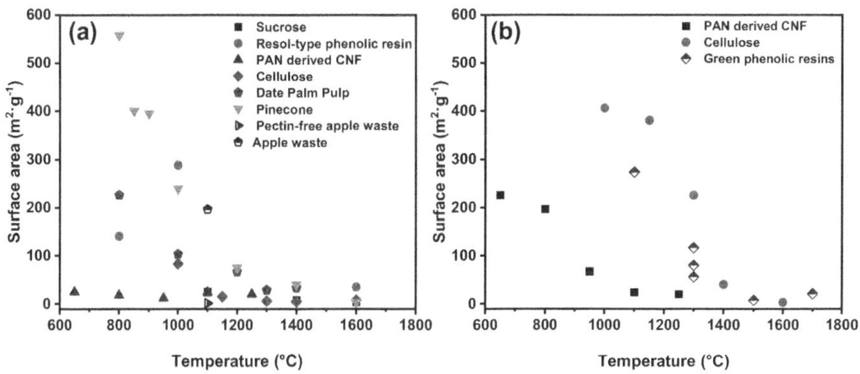

Figure 3.7. *Surface area values determined from (a) N_2 and (b) CO_2 adsorption measurements for different precursors and different synthesis conditions, reported in various studies: sucrose (Bommier* et al. *2015), phenolic resin (Zhang* et al. *2011), PAN (Zhang* et al. *2016), Cellulose (Simone* et al. *2016) (Matei Ghimbeu* et al. *2018), Date palm pulp (Izanzar* et al. *2018), Pinecone (Zhang* et al. *2017c), Green phenolic resins (Beda* et al. *2018), apple waste (Wu* et al. *2016), pectin-free apple waste (Dou* et al. *2018). For a color version of this figure, see www.iste.co.uk/monconduit/batteries.zip*

As mentioned before, the augmentation of the annealing temperature reduces the microporosity and induces the coalescence of the micropores to larger pores and the reorganization of the graphene layers into closer layers (Matei Ghimbeu *et al.* 2018). As for XRD analysis, notice that the precursor and the synthesis conditions drastically affect the surface area of the final hard carbons. Green phenolic resins precursors present very dispersed values of surface area, by N_2 and CO_2 measurements. The solvent in which the resin is prepared has a major impact on the final hard carbon porosity. The composition of the precursor, as well as the application of an acid pretreatment leads to different porosity in the material. It is confirmed by the pectin-free pomace (Dou *et al.* 2018) and the conventional apple waste derived hard carbons (Wu *et al.* 2016), in which the former has a much lower surface area compared to the latter. Worth to note that the chemical composition of the precursor is important also since oxygen-free precursors such as PAN exhibit much lower specific surface areas than the other precursors containing oxygen in their structure.

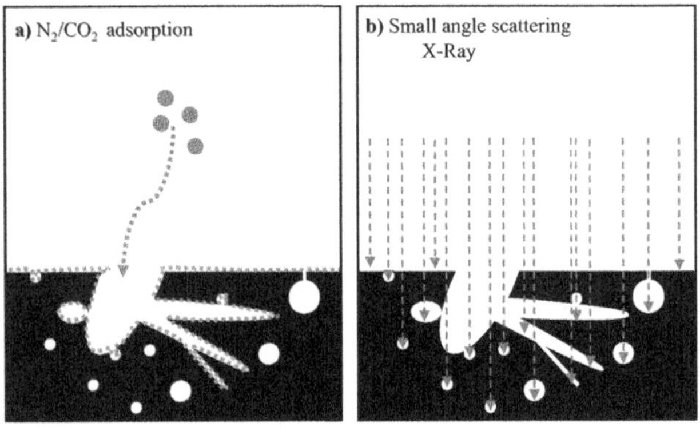

Figure 3.8. *Comparison between (a) gas adsorption (quantification of open pores including partially closed ink-bottle pores) and (b) small angle scattering techniques (analysis of all porosity, including closed pores). Adapted with permission from Saurel* et al. *(2018). Copyrights (2016) to John Wiley and Sons. For a color version of this figure, see www.iste.co.uk/monconduit/batteries.zip*

However, although gas adsorption provides useful information about the presence of ultramicropores, micropores and mesopores, in most of the cases, the isotherms and pore size distribution are not provided, which limits the understanding of hard carbon porosity. In addition, this technique does not allow quantifying the closed porosity but made possible by SAXS, which is sensitive to all the pores present in the structure by using the electronic density difference between the carbon matrix and the empty space of pores (Saurel *et al.* 2018). A schematic representation of open and closed porosity analyzed by gas adsorption and SAXS techniques is presented on Figure 3.8. SAXS uses an incident X-ray beam that diffuses into the sample. The diffused signal is then received by a detector and the intensity (which is proportional to the difference in electronic density) is plotted as a function of the 2θ scattering angle, or scattering vector Q (where $Q = 4\pi \sin(\theta)\cdot\lambda^{-1}$, and λ is the wavelength of Cu Kα = 1.5418 Å). Some SAXS patterns for char and hard carbons from cellulose synthesized at different temperatures are shown in Figure 3.9(a). At small Q values (under 0.05 nm^{-1}), the intensity signal decreases showing a Q^{-4} behavior characteristic of well-defined interfaces, such as meso- and macropores of bulk particles. When the signal drop shifts to a Q^{-3} behavior, the interface boundary is not well defined (Simone *et al.* 2016). Then, a plateau zone followed by a sharp decreasing curvature is characteristic of ultramicroporosity, which also comprises the closed porosity (Dou *et al.* 2019). A proper fitting of this curve allows quantifying the radius of gyration (Rg) of these pores, and therefore a pore volume. As for gas adsorption measurements, SAXS results also vary with heat treatment. For hard carbons produced from commercial cellulose by Simone *et al.* (2016). The intensity of the SAXS profile increased with the temperature treatment, the Rg increases with temperature in a nonlinear behavior and the pore number decreases, as it is shown in Figure 3.9(b). Recent works also proposed to combine N_2 adsorption measurements with He pycnometry to quantify the ratio between the closed and open pores (Dou *et al.* 2018; Zhang *et al.* 2018a).

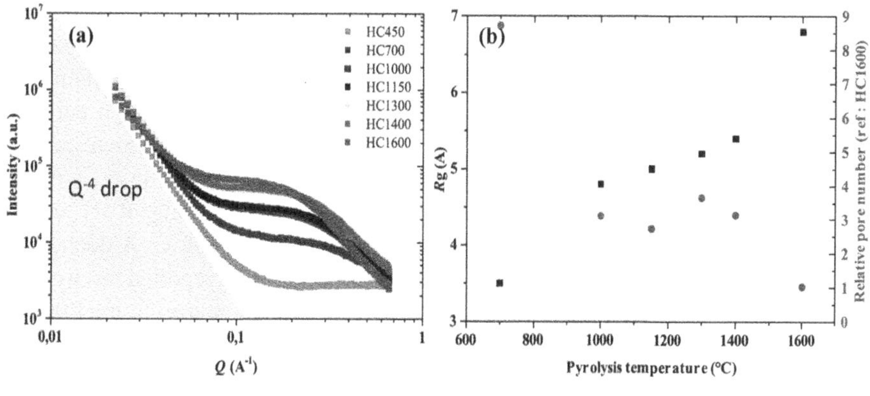

Figure 3.9. *(a) Typical SAXS patterns for hard carbon. Samples synthesized from cellulose at different temperatures. (b) Radius of gyration of hard carbon samples as a function of the temperature treatment and the relative pore number. Reprinted from Simone* et al. *(2016). Copyright (2016), with permission from Elsevier. For a color version of this figure, see www.iste.co.uk/monconduit/batteries.zip*

3.4.3. *Hard carbon surface chemistry*

Hard carbon surface chemistry plays an important role in sodium storage as the material surface will interact with the surrounding environment (electrolyte, Na^+). At a molecular level, a high chemical reactivity is exhibited by the surface atoms as their atomic and electronic structure changes. For example, following the thermal treatment, hard carbons surface develops various chemical complexes (generally oxygen functional groups), which influence the electrochemical performance. As the hard carbons are obtained by heating the carbon precursor in a wide range of temperatures, the functional groups will vary as well. Presence of heteroatoms (N, S, P, H) and physisorbed water can also play an important role in both the maximum reversible capacity and the irreversible capacity. Besides heating temperature, precursor selection may also modify the surface chemistry. The most common techniques used to characterize hard carbon surface chemistry are TPD-MS, XPS and EDX. However, compared to the hard carbon structure and porosity investigations that are reported in most papers, the surface chemistry is less studied.

TPD-MS is an analysis method that investigates changes taking place on the bulk material when the temperature is varied in a controlled manner. TPD-MS is mainly used to identify the oxygen functional groups present on the hard carbon structure. During hard carbon heating, the functional groups are decomposed releasing CO, CO_2, H_2O and H_2 at temperatures corresponding to their thermal stability. Of high interest when analyzing hard carbons is also the amount of functional groups. Typically, CO can be derived from groups such as phenol, ether, carbonyl and quinone while CO_2 is released as a result of acidic groups (carboxyl and/or anhydride groups) decomposition (Figure 3.10(a)). Only a few studies implying this characterization technique have been reported for hard carbons, although currently employed for activated porous carbons. As the annealing temperature increases, the amount of functional groups decreases, as shown in Figure 3.10. Zhang *et al.* (2016) showed such dependence of CO_x groups by thermally treating PAN carbon nanofibers at various temperatures ranged between 650 and 2,800°C. A decrease in functional groups, amount from 0.6 to 0.01 $mmol \cdot g^{-1}$, was reported as well for cellulose-derived hard carbons when increasing the temperature from 1,000 to 1,600°C (Matei Ghimbeu *et al.* 2018).

Similar results were reported by Irisarri *et al.* when preparing hard carbons from phenolic resins, i.e., the oxygen functional groups decrease from 0.86 $mmol \cdot g^{-1}$ at 1,200°C to 0.64 $mmol \cdot g^{-1}$ at 1,500°C. Besides temperature dependence, the oxygenated groups significantly vary with the precursor used as exposed in Figure 3.10(b). More insights on the surface chemical composition are provided by XPS (maximum 10 nm depth), a popular technique used to analyze hard carbons for providing valuable quantitative and chemical state information. The hybridization state in which carbon atoms are found (usually a mix of sp^2 and sp^3) and the percentage of different oxygen-based functional groups (COOR, C-OR, C=O, etc.) can be elucidated, likewise. Elements (S, N, metals) bounded to carbon can be also identified. As the quantitative information is accurate, XPS results are often used to determine the C to O ratio. Figures 3.10(c) and (d) shows the O and C content evolution with annealing temperature of few selected hard carbon materials, derived from different precursors. It is well illustrated the temperature dependence of both O and C content: while the O amount decreases with the increase in pyrolysis temperature, the C content increases, as expected. Such tendency is observed for most of the hard carbon materials, however, for a given temperature both O and C content is highly dependent on precursor type, especially at low temperatures. For example, at 900°C, the hard carbon made out of peat moss has ~10 wt% O content whereas PAN has only 2.5 wt%.

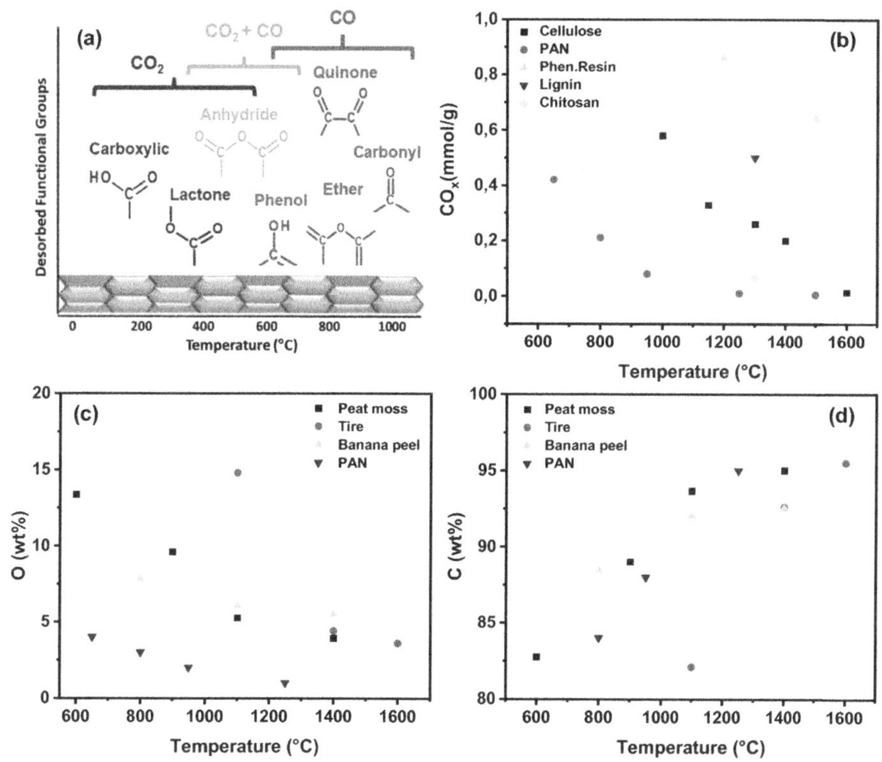

Figure 3.10. *Schematic representation of oxygen-based functional groups evolving from carbon surface during thermal annealing (a). Evolution of total oxygen–based functional groups amount (CO_X) versus heating temperature of different precursors (PAN - Zhang* et al. *2016; Phenolic resin - Irisarri* et al. *2018; Cellulose - Matei Ghimbeu* et al. *2018; Lignin - Matei Ghimbeu* et al. *2019; Chitosan - Conder* et al. *2019). (c) Oxygen and (d) carbon content of hard carbon materials derived from different precursors as revealed by XPS (Peat moss - Ding* et al. *2013; Banana peel - Lotfabad* et al. *2014; Tire - Li* et al. *2016d; PAN - Zhang* et al. *2016). For a color version of this figure, see www.iste.co.uk/monconduit/batteries.zip*

Elemental analysis often used to get additional information on the bulk composition usually reveals similar trends as the XPS technique. However, when the materials contain impurities, their amount cannot be determined. In addition, the quantification of oxygen is not possible and is usually estimated by the difference 100% – % (CNHS). In such case, EDX may provide information about the bulk composition and detect the presence of impurities. For example, in the case of biomass waste derived carbons, EDX is an important tool for such a purpose (Zhang *et al.* 2017c). For the doped hard carbons, EDX is often used to determine if the doping with heteroatoms was uniformly achieved (Hong *et al.* 2018). However, the technique is limited, especially for light elements such as N or B that are used as dopants. Therefore, depending on hard carbon nature, a pallet of complementary techniques is required to give a complete overview about the nature and quantity of functional groups, dopants or impurities constituting the material.

3.4.4. *Hard carbon structural defects*

As mentioned before, hard carbon structure consists of a mixture stacked graphene clusters randomly oriented and micropores. Depending on the annealing temperature, the materials present heteroatoms and/or impurities, therefore, the final hard carbon structure presents defects. Since it is impossible to prepare defect-free hard carbons, understanding their influence on the electrochemical performance is fundamental. There are several types of defects that could be accounted in the hard carbon structure: point defects (lattice mono- or di-vacancies), cluster defects (Stone–Wales defects) and boundaries or edges. Tsai *et al.* (2015) studied the effect of point and cluster defects on Na intercalation into disordered carbons by computational method. They pointed out that, the first Na^+ ion intercalated is stored close to the defect sites, as result of the strong binding energy ion-defect. In addition, it was also found that structural defects can enhance sodium intercalation, especially through mono- and di-vacancies while, in the case of Stone–Wales defects, the improvement is less significant.

Raman spectroscopy is widely used for characterizing the degree of order in carbon allotropes, and thus it is a powerful complement to XRD structural analysis. It consists of using an incident laser beam to generate light scattering from the studied sample (Inagaki and Kang 2016). Generally, visible light wavelengths are used, which only resonates with the π electrons of sp^2-hybridized carbons. Consequently, the obtained Raman spectrum does not provide direct information regarding sp^3-hybridized carbons related to amorphous carbon allotropes (Ferreri and Robertson 2000; Dou *et al.* 2019). The Raman spectrum of hard carbon presents two main relative well-defined features, the G and the D band. The G band located around 1,580 cm^{-1} is related to the in-plane vibrations of sp^2 carbons, not only carbons in sixfold rings. The D band placed around 1,355 cm^{-1} is associated with the ring breathing modes that only become active in the presence of disorder, therefore, it is directly related to the presence of sixfold aromatic rings, and has no intensity for defects-free materials (Ferreri, Robertson 2000; Inagaki and Kang 2016; Bommier *et al.* 2018). As hard carbon structure combines disordered and graphitic domains, D-band has a high intensity and is proportional to the defect's concentration. The graphitic domains (crystalline graphite) are identified by the G-band. By determining the ratio between the area/intensities of D and G bands (I_D/I_G or I_G/I_D), the amount of defects (disorder degree) can be estimated. Figure 3.11(a) shows trends for the structural defects present in the hard carbon structure determined by both I_D/I_G and I_G/I_D ratios.

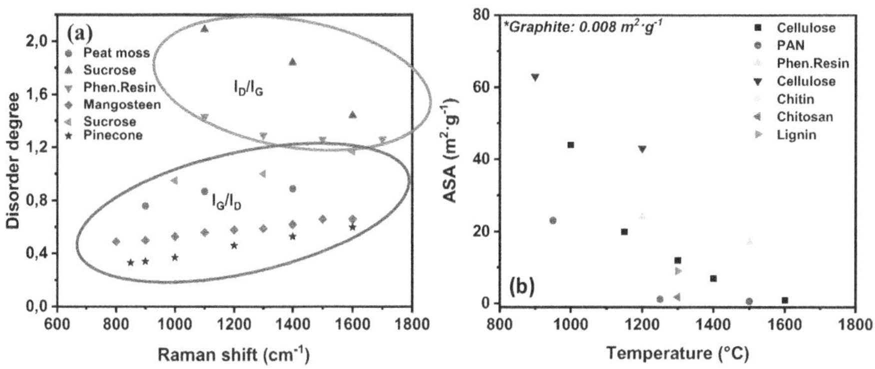

Figure 3.11. *(a) Disorder degree determined by either the I_D/I_G or I_G/I_D ratio based on Raman spectroscopy. Evolution according to the pyrolysis temperature of different precursors reported in various studies (Peat moss – Ding* et al. *2013; Sucrose - Bommier* et al. *2014; Sucrose - Li* et al. *2017b; Mangosteen - Wang* et al. *2017; Pinecone - Zhang* et al. *2017c; Phenolic resin - Beda* et al. *2018). (b) Active surface area (carbon edges defects) of different hard carbon materials determined by TPD-MS measurements (Graphite - Spahr* et al. *2006; PAN - Zhang* et al. *2016; Phenolic resin, Cellulose - Irisarri* et al. *2018; Cellulose - Matei Ghimbeu* et al. *2018; Lignin - Matei Ghimbeu* et al. *2019; Chitin, Chitosan - Conder* et al. *2019). For a color version of this figure, see www.iste.co.uk/monconduit/batteries.zip*

For hard carbon materials obtained from different precursor sources (raw biomass and polysaccharides), a decrease in the I_D/I_G ratio is observed with temperature increase indicating less structural defects and increase ordering. This phenomenon can be explained by the fact that at lower temperatures, all the graphene sheets are independent and thus only sp^2 carbons are detected. However, when temperature increases, the graphene sheets get closer to each other and begin to align, but punctual defects are still created at the junction points. Consequently, sp^3 carbons are detected and their intensity increases with temperature. When I_G/I_D value is taken into account, the trend is opposite, as expected. As for most of hard carbon characteristics, the disorder degree does not rely just on the heating temperature, precursor selection playing an important role (Figure 3.11(a)).

However, it has been pointed out that Raman does not take into consideration the defects caused by micropores (<2 nm) and moreover it is not quantitative. Therefore, TPD-MS can be used to determine the amount of defects including contribution of micropores. The active surface area (ASA) is correlated to all different types of defects present on the carbon edges such as stacking faults, single and multiple vacancies, and dislocations. Presence of such active sites is important since they can interact with other species (i.e. Na^+). ASA is determined by O_2 chemisorption on a vacuum-cleaned hard carbon and quantification of surface oxygenated complexes subsequently formed. The assessment of such parameter requires specific set-up which is not very common in literature and only a few studies have reported such values. The few works found (Figure 3.11(b)) show that the ASA values decrease with the increase in carbonization temperature for all precursors used, cellulose, PAN or phenolic resin, in line with the Raman results. At higher temperatures (>1.700°C) a very low amount of defects is usually seen, no matter the precursor used. Moreover, the dependence on the precursor used is clear, e.g., 18 $m^2 \cdot g^{-1}$ for phenolic resin vs. ~0.5 for PAN/cellulose at 1500°C. Although most of the studies limit the temperature to ~1,600°C (to avoid a high degree of graphitization which hinders sodium storage), Zhang paper shows clearly that PAN derived hard carbons have very low ASA values in the temperature range of 1,500–2,800°C. These values are comparable with those found for graphite (Spahr *et al.* 2006; Novák *et al.* 2007).

3.5. Electrochemical performance

3.5.1. *Materials performance*

Since the work of Stevens and Dahn (2000) on sodium insertion into hard carbon in 2000s, many groups had reported electrochemical performance of these materials in half coin cells configuration. Generally, specific capacities vary from 200 to 360 $mAh \cdot g^{-1}$, depending on the hard carbon properties, such as their morphology, composition, porosity/specific surface area, microstructure, defects concentration, etc., which are deeply related to the synthesis parameters: precursor type, solvents, pre-treatments, pyrolysis temperature, etc. The electrochemical performance also depends strongly on test conditions such as electrode mass, C-rate, cut-off voltage or the presence/absence of a potentiostatic step. Table 3.2 summarizes some electrochemical performance of different hard carbons at different cycling conditions.

Precursor	Temperature (°C)	ICE (%)	Irrev. Cap. ($mAh{\cdot}g^{-1}$)	Rev. Cap. ($mAh{\cdot}g^{-1}$)	C rate ($mA{\cdot}g^{-1}$)	Capacity retention	Reference
Pitch/phenolic resin (1:1)	1400	86	301	259	30 (0.1 C)	94% in 100 cycles	(Li *et al.* 2016c)
Pitch/phenolic resin (1:1)	1600	84	250	210	30 (0.1 C)	90% in 100 cycles	(Li *et al.* 2016c)
Phenol/ formaldehyde resin	1250	60	517	311	20	>80% in 100 cycles	(Wang *et al.* 2015)
Resorcinol/ formaldehyde resin	1600	95	368	350	20	94% in 100 cycles	(Hasegawa *et al.* 2015)
Phloroglucinol/ glyoxylic acid resin	1300	76	355	270	37.2	92% in 100 cycles	(Beda *et al.* 2018)
Glucose	1100	88	349	307	33*	N/A	(Stevens and Dahn 2000)
Sucrose (coated)	1000	83	335	279	30 (0.1 C)	80% in 100 cycles	(Li *et al.* 2014)
Sucrose (coated)	1600	84	381	320	30 (0.1 C)	93% in 100 cycles	(Li *et al.* 2014)
Cellulose	1400	83	375	310	37.2	90% in 200 cycles	(Simone *et al.* 2016)
Cellulose	1600	84	366	307	37.2	94% in 200 cycles	(Simone *et al.* 2016)
Mangosteen shell	1500	83	396	329	20 (0.1 C)	98% in 100 cycles	(Wang *et al.* 2017)
Lotus stem	1400	70	501	351	40	94% in 450 cycles (at 100 $mA{\cdot}g^{-1}$)	(Zhang *et al.* 2018a)
Apple waste	1100	61	400	245	4 (0.02 C)	100% in 80 cycles (at 20 $mA{\cdot}g^{-1}$)	(Wu *et al.* 2016)
Apple and pomace – pectin free	1100	66	450	297	20 (0.1 C)	96% in 230 cycles	(Dou *et al.* 2018)
Banana peel (activated)	1100	78	455	355	0.05	88% in 300 cycles (at 0.1 $A{\cdot}g^{-1}$)	(Lotfabad *et al.* 2014)

*$\mu A.cm^{-2}$.

Table 3.2. *Electrochemical performance of different HC reported in literature*

Regarding polysaccharides precursors, glucose, sucrose and cellulose were employed and the best results in terms of first reversible capacities, for glucose derived hard carbon were obtained by Stevens and Dahn in 2000 (Stevens and Dahn 2000), i.e., 300 $mAh{\cdot}g^{-1}$. For sucrose, 362 $mAh{\cdot}g^{-1}$ was obtained by Qiu *et al.* (2017) at 20 $mA{\cdot}g^{-1}$. The group of Palacin achieved capacities around 340 $mAh{\cdot}g^{-1}$ at 2 C rate with sucrose-derived carbon-coated materials cycled at high temperature (75°C) (Ponrouch and Palacín 2015). For cellulose, the best reversible capacities were achieved by Komaba *et al.* (2017) (349 $mAh{\cdot}g^{-1}$ at 25 $mA{\cdot}g^{-1}$), Qiu *et al.* (2017) (324 $mAh{\cdot}g^{-1}$ at 20 $mA{\cdot}g^{-1}$) and Simone *et al.* (2016) (308 $mAh{\cdot}g^{-1}$ at 37.2 $mAh{\cdot}g^{-1}$). In terms of raw biomass, 360 $mAh{\cdot}g^{-1}$ were attained using Mapple leaf (Li *et al.* 2016a), 355 $mAh{\cdot}g^{-1}$ employing banana peel (Lotfabad *et al.* 2014) and 329 $mAh{\cdot}g^{-1}$ provided by mangosteen (Wang *et al.* 2017a). Finally, fossil resources (coal or petroleum) derived polymers have been also studied. The best results were obtained by Hasegawa *et al.* (2015) for phenolic resins, with up to 353 $mAh{\cdot}g^{-1}$ specific capacity. For more exhaustive descriptions, the reader can refer to the review article of Saurel *et al.* (2018).

Hard carbon materials present very different electrochemical behavior at different synthesis conditions, especially at different pyrolysis temperatures, given the evolution of the microstructure and both open and closed porosity. Figure 3.12 shows the relation between the reversible capacity and the ICE with the annealing temperature for some HC-derived from different precursors. At low temperatures (below 1,000°C), the hard carbons exhibit a low first reversible capacity (<200 $mAh{\cdot}g^{-1}$) and low ICE (<50%). With the augmentation of the annealing temperature, the graphitization degree of the hard carbon structure improves via the removal of surface functional groups, defects concentration and porosity. In parallel, the specific surface area decreases, most pores closes and creation of closed pores occur. Between 1,300°C and 1,600°C, the reversible capacity improves significantly reaching values over 350 $mAh.g^{-1}$ while the ICE increases to around 80%, respectively, for some of the precursors. The specific mechanism behind sodium uptake by different structure is still under debate, as seen in the next part of the chapter. However, it is clear that hard carbons synthesized between 1,300 and 1,600°C present in general the best performance. Higher annealing temperatures (>1,600°C) lead to a more organized structure in which the sodium uptake is limited, as can be seen from the decrease of the first reversible capacities recorded for all the given precursors in Figure 3.12. In addition, the rate capability of hard carbon materials is of great importance, which is rather poor, compared to graphite and soft carbon capabilities (Li *et al.* 2014; Saurel *et al.* 2018), and decreases with annealing temperature, as can be seen from the values listed in Table 3.2.

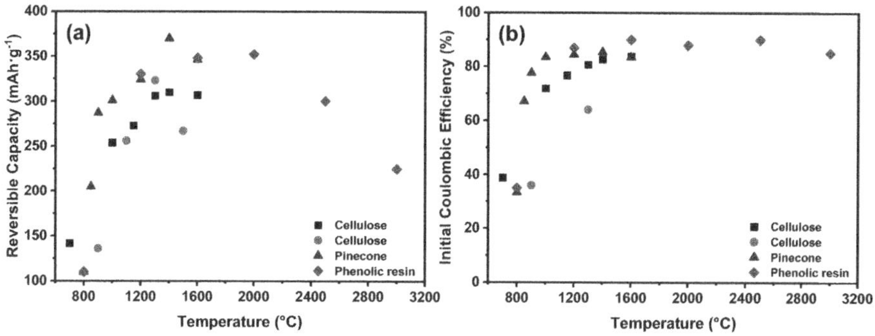

Figure 3.12. *(a) Reversible capacity and (b) ICE of different hard carbons as a function of pyrolysis temperature: cellulose (Simone* et al. *2016), cellulose (Qiu* et al. *2017), Pinecone (Zhang* et al. *2017c), phenolic resin (Hasegawa* et al. *2015). For a color version of this figure, see www.iste.co.uk/monconduit/batteries.zip*

Open porosity represents a key feature regarding the electrochemical behavior of hard carbon. Bommier *et al.* (2014) intensely studied the relationship between the open porosity and the final reversible capacity. By using sucrose derived hard carbon, an inverse proportionality was found between the pore volume and the developed surface area with the final reversible capacity.

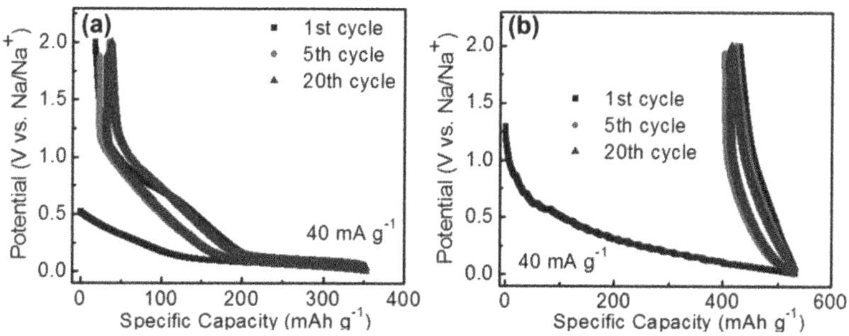

Figure 3.13. *Charge/discharge profiles for (a) low surface area hard carbon and (b) high surface area hard carbon. Reprinted from Bommier* et al. *(2014). Copyright (2014), with permission from Elsevier. For a color version of this figure, see www.iste.co.uk/monconduit/batteries.zip*

Interestingly, very high surface area hard carbons (>1,000 $m^2 \cdot g^{-1}$) present very different charge/discharge profiles than lower surface area carbons, as can be seen from Figure 3.13. Qiu *et al.* (2017) found similar results using a templated hard carbon produced from phenolic resin. Both works argue that at very high surface area values, an important capacitive contribution takes place. The electrolyte's decomposition increases significantly and the volume of carbon for storing Na ions is drastically reduced.

The morphology of the final hard carbon also influences its performance. Although most precursors result in random-like morphology, spherical particles obtained by HTC of carbohydrates or polymerization of phenolic resins as well as fiber-like materials made by electrospinning were reported. Hard carbon hollow nanospheres were obtained by Tang *et al.* (2012) using latex templates and glucose. The initial reversible capacity and ICE were 223 $mAh{\cdot}g^{-1}$ (at 40 $mA{\cdot}g^{-1}$) and 42%, respectively. However, this carbon is able to retain approximately 88% of the initial capacity after 100 cycles at 100 $mA{\cdot}g^{-1}$ cycling rate and 50 $mAh{\cdot}g^{-1}$ at 10 $A{\cdot}g^{-1}$, showing an outstanding rate capability. Hollow spheres were also obtained from pollen grains, as presented before, taking advantage of the natural occurring morphology of this precursor. Sucrose-derived microspheres were synthesized by Qiu *et al.* (2017) obtaining a significantly high reversible capacity of 362 $mAh{\cdot}g^{-1}$ and 86% of columbic efficiency. Regarding fiber-like morphologies, Bai *et al.* (2015) studied the influence of the fiber morphology on the electrochemical performance of the hard carbon. They prepared PVC nanofibers by electrospinning and further pyrolyzed them at 700°C. Results showed that the PVC nanofibers had better performance than PVC particles (reference hard carbon). A reversible capacity of 271 $mAh{\cdot}g^{-1}$ was obtained at 12 $mA{\cdot}g^{-1}$, 70% of ICE, 79% retention rate after 120 cycles and 147 $mAh{\cdot}g^{-1}$ at 240 $mA{\cdot}g^{-1}$ of cycling rate. PVC particles delivered only 206 $mAh{\cdot}g^{-1}$ of reversible capacity at 12 $mA{\cdot}g^{-1}$, 61% of ICE and 61% retention after 120 cycles. At this point, however, it is difficult to analyze the real impact of the carbon morphology on the electrochemical performances because it is a parameter difficult to change while maintaining the other characteristics as constant. More systematic studies are required in this direction.

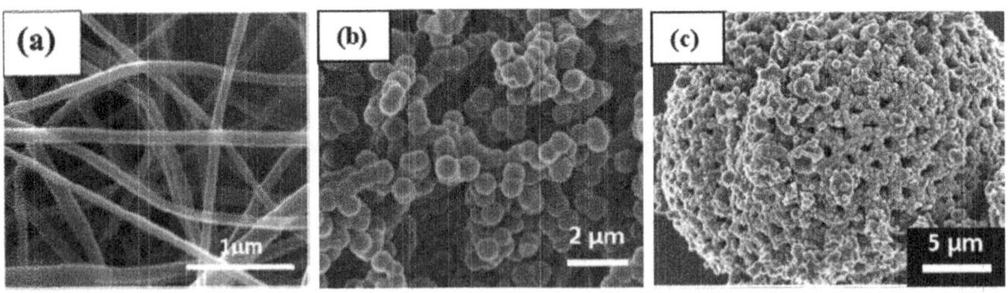

Figure 3.14. *SEM images of (a) PVC nanofibers (Reprinted with permission from Bai* et al. *2015 Copyright (2015) American Chemical Society), (b) hard carbon microspheres (Reprinted from Tang* et al. *2012, Copyright (2012) with permission from Wiley & Sons, Inc.) and (c) hollow reticulate hard carbon from pollen grains (Reprinted from Li* et al. *2017a, Copyright (2017), with permission from Elsevier)*

3.5.2. *Full Na-ion system performance*

As already mentioned, hard carbon is the most promising negative material for Na-ion application in terms of capacity, cycling behavior and power rate capability. From this point of view, almost all full Na-ion systems tests reported use hard carbons as negative electrode. However, few examples have been reported so far. Moreover, comparing performance is rather complicated as capacities/energy densities are sometimes reported as a function of positive, negative, full active mass or full system. These systems used both transition metal oxide and polyanionic positive electrodes, which complicates the comparison.

In 2011, Komaba's group showed results on coin-cell based $NaNi_{0.5}Mn_{0.5}O_2$/hard carbon systems. They reported around 200 mAh·g^{-1} (hard carbon mass) (Komaba *et al.* 2011). Later, Johnson *et al.* (2012) and Ponrouch *et al.* (2013) proposed, respectively, $NaNi_{1/3}Mn_{1/3}Co_{1/3}$/hard carbon and $Na_3V_2(PO_4)_2F_3$/hard carbon systems. This last system showed a capacity of 100 mAh·g^{-1} and very good performance in terms of cycling life and power rate capability. Zhang *et al.* (2016) prepared a full Na-ion using a PAN-derived CNF anode and $Na_2Fe_2(SO_4)_3$ as the cathode and achieved 70 mAh·g^{-1} of capacity (based on the cathode weight).

More recently, several proofs of concept have been established using "real" Na-ion cell formats such as prismatic or stacked pouch cells (Faradion n.d.) and 18,650 cylindrical cells (CNRS News 2015; Broux *et al.* 2019). Broux *et al.* (2019) showed impressive results in terms of cycling behavior (20% capacity loss after almost 4,000 cycles at 1 C and 100% depth of discharge) and rate capability (75% capacity retention at 100 C in discharge and at 10 C in charge) for the system $Na_3V_2(PO_4)_2F_3$.

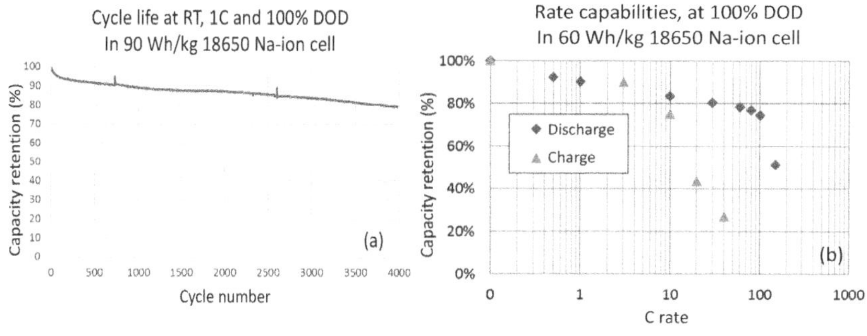

Figure 3.15. *Performance obtained in 18,650 cells: (a) long-term cyclability at a 1 C rate upon charge and discharge for a 75 Wh kg^{-1} prototypes and (b) rate capabilities for 55 Wh kg^{-1} prototypes (Reprinted with permission from Broux* et al. *2019 Copyright (2019) to John Wiley & Sons, Inc.). For a color version of this figure, see www.iste.co.uk/monconduit/batteries.zip*

Demonstrations of integration in real systems have also been presented. FARADION first demonstrated integration in an E-bike (En ligneGreen Car Congress n.d.). The French network of Electrochemical Storage (RS2E) integrated it into a remote military robot (Ministère des Armées n.d.) and RS2E-spin off company TIAMAT presented a Na-ion powered electric scooter (Industrie-techno 2018). Very recently, the Chinese Na-ion battery company HiNa demonstrated the integration in a small electric vehicle (HiNa Battery Technology Co., Ltd n.d.).

3.5.3. *Sodium insertion mechanisms in hard carbon*

Although hard carbon development is an important issue to obtain competing materials for future energy storage systems, the fundamental understanding of the sodium storage mechanisms in hard carbon is a key aspect in order to design sustainable high-performing anodes for NIBs.

The early description of a Na-based charge storage mechanism in hard carbon was reported in the 2000s by Stevens and Dahn (2000, 2001) using the "falling cards" structural model. They proposed, based on XRD and SAXS analyses, that sodium is inserted in the carbon structure via a similar mechanism as for lithium, implying a two-stage "insertion-adsorption" process. The Na is first intercalated between the "pseudo-graphitic" nanodomains at high potentials (>0.1 V), giving rise to a slope potential region while at much lower potentials (<0.1V), sodium is adsorbed in the carbon micropores, resulting in a plateau region. Since this pioneering work, several other works confirmed this mechanism (Alcántara *et al.* 2005; Gotoh *et al.* 2013; Irisarri *et al.* 2015; Simone *et al.* 2016), as schematically illustrated in Figure 3.16.

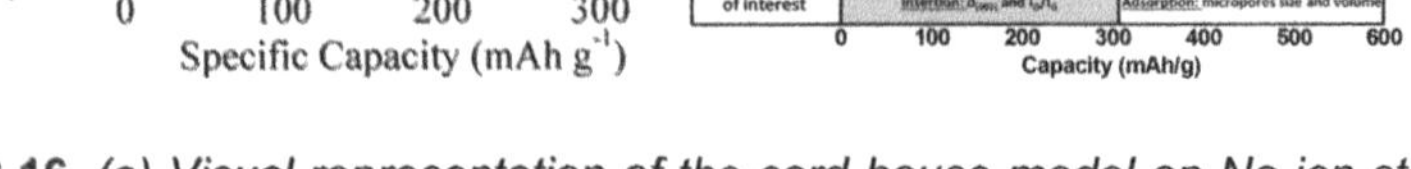

Figure 3.16. *(a) Visual representation of the card-house model on Na-ion storage in hard carbon showing the two distinct phases: intercalation between the graphitic domains and pore filling (Reprinted with permission from Bommier* et al. *2015, Copyright (2015) to American Chemical Society). (b) Potential versus capacity profile for hard carbon when tested against sodium metal counter electrodes. The different steps of the mechanism are also labeled. Reprinted from Irisarri* et al. *(2015). For a color version of this figure, see www.iste.co.uk/monconduit/batteries.zip*

However, starting to the mid-2010s, the researcher's experimental observations could no longer rely on the "insertion-adsorption" mechanism. Therefore, different propositions of mechanisms have been postulated and to date, there is still no consensus about a general sodium ion mechanism in hard carbons. This is mainly related to the complexity of hard carbon structure and to several factors that are still not well understood, as proposed by Saurel *et al.* (2018): (1) the lack of knowledge regarding the accessibility of the electrolyte in the microporosity and precise characterization of porosity (open vs. closed porosity); (2) the contradictory structural results determined by both local (HR-TEM, Raman) and averaging (XRD) techniques induced by the inhomogeneous nature of hard carbon and (3) the wide range of susceptible mechanisms and their possible overlapping along with the missing combined characterization techniques, which would allow to confirm/infirm the intercalation mechanism (XRD, Raman) and/or micropore filling mechanism (SAXS, NMR). We would add here a fourth point, that is the theoretical and modeling works, which are limited and certainly could contribute to understanding such complex mechanisms.

Taking into consideration these aspects, the more recent proposed mechanisms will be described. Cao *et al.* (2012) studied the insertion of sodium in hallow carbon nanowires and they first ascribe the low potential region (0.2–0 V) to the intercalation of sodium between the graphitic layers by similitude with the behavior observed on lithium insertion into graphite. They suggest that the sloping region (1–0.2 V) might be related to the charge transfer on the surface of graphitic microdomains. Theoretical calculations were undertaken as well and revealed that a graphitic interlayer space larger than 0.37 nm is required for sodium insertion/extraction. This value represents a balance between the attractive van der Waals interactions between the carbon layers and the repulsive interactions between the carbon layers and the Na ions. Later in 2013, Ding *et al.* (2013) prepared hard carbon at various annealing temperatures and observed that the interlayer space and the microporosity decreases while the graphitization in the materials increases. They noticed that the capacity gained in the plateau region increases with the increase in the temperature and they assigned this observation to the insertion of sodium between the graphitic layers rather than to the adsorption in the porosity. To sustain their hypothesis, the authors synthesized highly microporous carbon that revealed no potential plateau in the low voltage region, therefore excluding adsorption of Na in this region. Moreover, *ex situ* XRD studies showed a reversible shift of the (002) peak toward lower 2theta values in the low potential region. All these observations strengthened the idea that the low potential plateau is related to Na intercalation between graphitic plans and not to the sodium adsorption in the porosity. Further works sustained this mechanism based on similar XRD observations (Komaba *et al.* 2011; Lotfabad *et al.* 2014).

In 2015, Bommier *et al.* (2015) pointed out another carbon characteristic that may be involved in the Na storage, namely the carbon defects. The authors evidenced a linear decrease in carbon defects expressed as I_D/I_G report (when increasing the pyrolysis temperature) and the sloping capacity. In consequence, they assign the sloping region to the Na interaction with the carbon defect sites. Based on XRD analyses, they further confirm the insertion of the sodium in the plateau region and in minor extent to the adsorption of Na on the pore surface (Figure 3.17(a)). Qiu *et al.* (2017) performed a systematic work on the characterization of hard carbons obtained at different temperature and employed *in situ* XRD and *ex situ* NMR techniques along with computational simulations, which allowed to propose an "adsorption-intercalation" mechanism of Na in hard carbon, (Figure 3.17(b)). Other works (Zhang *et al.* 2016; Matei Ghimbeu *et al.* 2018) suggested in addition, an impact coming from the carbon surface chemistry (oxygen-containing functional groups and heteroatoms) in the slope region.

Matei Ghimbeu *et al.* (2018) attempted to discriminate the contribution of oxygen surface groups, porosity and defects (Figure 3.17(c)). They first pointed out the importance of using complementary gases such as CO_2 to assess correctly the porosity of hard carbons since carbons estimated "non-porous" based on N_2 adsorption may present large amounts of small pores, ultamicropores (<0.7 nm). The authors also quantified the structural defects as active surface sites (ASA) and they observed that the increase in the pyrolysis temperature of cellulose triggers a decrease in the functional group amount, defects and porosity, which is accompanied by the reduction of the slope region capacity. In addition, the plateau capacity was observed to increase with the interlayer distance decrease. These results allowed them to assign the slope capacity to the insertion of sodium in the porosity and to the interactions with the defects, while the plateau region to the intercalation of sodium between the graphitic layers. Therefore, all these works listed above sustain rather an "adsorption-intercalation" mechanism.

Zhang *et al.* (2016) proposed a supplementary mechanism based on Na filling in the graphitic nanopores formed on hard carbons derived from PAN fibers, heated up to very high temperatures, i.e., 2,000 to 2,800°C. This temperature range extension allowed them to observe for the first time a single plateau region with no contribution the from slope region. Depending on the annealing temperature, a three-stage insertion mechanism "adsorption-intercalation-filling" has been proposed (Figure 3.17(d)).

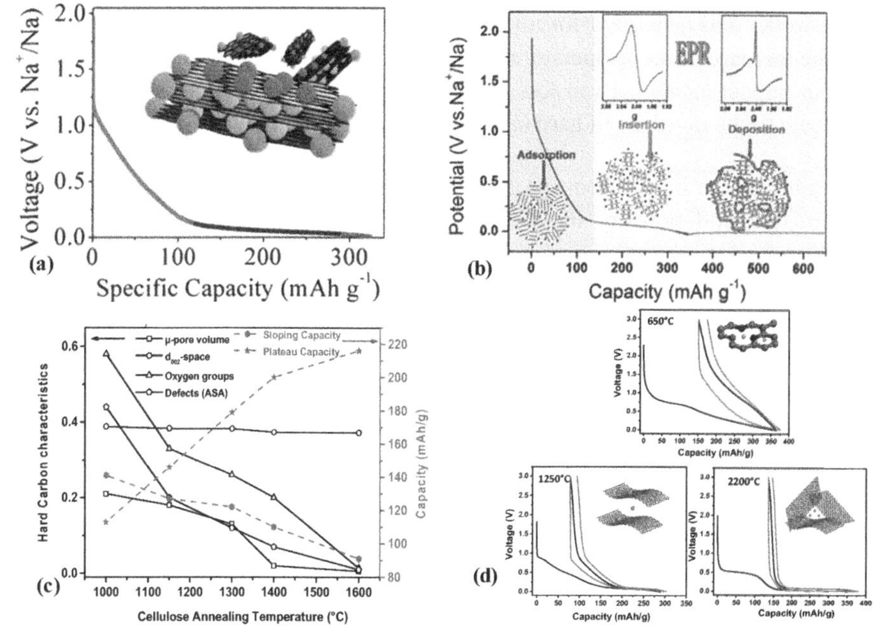

Figure 3.17. *Representation of different storage mechanisms in hard carbons. (a) Potentiogram along with simplified representation of three stage storage mechanism (slope region: defect sites; plateau region: intercalation between graphite shits and small contribution of Na adsorption on pore surface) (Reprinted with permission from Bommier* et al. *2015, Copyright (2015) to American Chemical Society); (b) "adsorption-insertion" mechanism involving adsorption on defect sites, insertion between the graphitic planes and Na deposition (Reprinted with permission from Qiu* et al. *2017, Copyright (2017) to John Wiley & Sons, Inc.); (c) correlation between the slope/plateau capacities and HC characteristics versus annealing temperature (Reprinted from Matei Ghimbeu* et al. *2018, Copyright (2018), with permission from Elsevier) and (d) three-stage mechanism involving adsorption–insertion–filling depending on the pyrolysis temperature (Reprinted with permission from Zhang* et al. *2016, Copyright (2016) to John Wiley & Sons, Inc.). For a color version of this figure, see www.iste.co.uk/monconduit/batteries.zip*

3.6. Conclusion

We have shown in this chapter that hard carbon materials derived from cheap, available and biosourced precursors can be successfully synthesized by simple pyrolysis process. Their main characteristics (porosity, structure, surface chemistry and defects) can be easily tuned by varying the precursor type and the processing conditions. Mainly, the precursors and the annealing temperature, are varied in most of the works and generally, for all precursors, the carbon characteristics decreases with the increase of the temperature, although for a given temperature the precursors have an impact as well. Even less studied, the chemical synthesis parameters also contribute to the diversity of hard carbon features.

Regarding the hard carbon properties, there are still challenges to determining the structure and porosity of such materials due to their low crystallinity and porosity. A panel of techniques must be used to correctly asses them. The use of nitrogen adsorption should be systematically combined with CO_2 adsorption and SAXS measurements to better determine the open and closed porosity. The contribution of ultra-micropores and closed pores to Na storage/deposition is not yet well understood and requires further understanding. Regarding the surface chemistry (functional groups, dopants, inorganic impurities, etc.), it was found to be less studied than the porosity and structure, as well as its impact on the performance. Moreover, defects are rarely quantified, although theoretical and experimental studies suggest an important role in the Na storage.

Coming to the electrochemical performance, interplaying characteristics (small surface area, large d-space, presence of defects and functional groups) must be carefully balanced in order to reach satisfactory performance. Most electrochemical tests are performed in half-cell and several materials reached reasonable capacity close to ~300 $mAh{\cdot}g^{-1}$ at a C/10 rate. However, there is still a need for more in-depth studies to be undertaken, to evaluate the long-term cycling behavior, capacity retention at variable C rates and the behavior at different temperature. In addition, the optimization of carbon electrode and electrolyte formulations, of the electrochemical conditions that are less studied, would certainly lead to performance improvement.

Comprehensible studies of SEI formation and the parameters leading to its control must be still approached. Understanding the storage mechanism is still required and difficulty arises due to the complex carbon structure, which is challenging to assess along with limitations of *in situ* and *ex situ* postmortem techniques. The evaluation of hard carbons in full batteries is rarely considered, and represents an important point to be addressed in order to assess complete overall performance in real systems and to favor up-scaling and commercialization.

3.7. References

Alcántara, R., Lavela, P., Ortiz, G.F., and Tirado, J.L. (2005). Carbon microspheres obtained from resorcinol-formaldehyde as high-capacity electrodes for sodium-ion batteries. *Electrochemical and Solid-State Letters*, 8(4), A222–A225.

Anca-Couce, A. (2016). Reaction mechanisms and multi-scale modelling of lignocellulosic biomass pyrolysis. *Progress in Energy and Combustion Science*, 53, 41–79.

Bai, Y., Wang, Z., Wu, C., Xu, R., Wu, F., Liu, Y., Li, H., Li, Y., Lu, J., and Amine, K. (2015). Hard carbon originated from polyvinyl chloride nanofibers as high-performance anode material for Na-ion battery. *ACS Applied Materials & Interfaces*, 7(9), 5598–5604.

Beda, A., Taberna, P.-L., Simon, P., and Matei Ghimbeu, C. (2018). Hard carbons derived from green phenolic resins for Na-ion batteries. *Carbon*, 139, 248–257.

Bommier, C., Luo, W., Gao, W.-Y., Greaney, A., Ma, S., and Ji, X. (2014). Predicting capacity of hard carbon anodes in sodium-ion batteries using porosity measurements. *Carbon*, 76, 165–174.

Bommier, C., Mitlin, D., and Ji, X. (2018). Internal structure–Na storage mechanisms–Electrochemical performance relations in carbons. *Progress in Materials Science*, 97, 170–203.

Bommier, C., Surta, T.W., Dolgos, M., and Ji, X. (2015). New mechanistic insights on Na-ion storage in nongraphitizable carbon. *Nano Letters*, 15(9), 5888–5892.

Broux, T., Fauth, F., Hall, N., Chatillon, Y., Bianchini, M., Bamine, T., Leriche, J.-B., Suard, E., Carlier, D., Reynier, Y., Simonin, L., Masquelier, C., and Croguennec, L. (2019). High rate performance for carbon-coated $Na_3V_2(PO_4)_2F_3$ in Na-ion batteries. *Small Methods*, 3(4), 1800215.

Buiel, E.R., George, A.E., and Dahn, J.R. (1999). Model of micropore closure in hard carbon prepared from sucrose. *Carbon*, 37(9), 1399–1407.

Cao, L., Hui, W., Xu, Z., Huang, J., Zheng, P., Li, J., and Sun, Q. (2017). Rape seed shuck derived-lamellar hard carbon as anodes for sodium-ion batteries. *Journal of Alloys and Compounds*, 695(Supplement C), 632–637.

Cao, Y., Xiao, L., Sushko, M.L., Wang, W., Schwenzer, B., Xiao, J., Nie, Z., Saraf, L.V., Yang, Z., and Liu, J. (2012). Sodium ion insertion in hollow carbon nanowires for battery applications. *Nano Letters*, 12(7), 3783–3787.

CNRS News. (2015). A battery revolution in motion. [Online]. Available at: https://news.cnrs.fr/articles/a-battery-revolution-in-motion [Accessed 30 August 2019].

Conder, J.M., Vaulot, C., Marino, C., Villevieille, C., and Ghimbeu, C.M. (2019). Chitin and chitosan–Structurally-related precursors of dissimilar hard carbons for Na-ion battery. *ACS Appl. Energy Mater.*, 2, 4841–4852.

Dahbi, M., Kiso, M., Kubota, K., Horiba, T., Chafik, T., Hida, K., Matsuyama, T., and Komaba, S. (2017). Synthesis of hard carbon from argan shells for Na-ion batteries. *Journal of Materials Chemistry A*, 5(20), 9917–9928.

Ding, J., Wang, H., Li, Z., Kohandehghan, A., Cui, K., Xu, Z., Zahiri, B., Tan, X., Lotfabad, E.M., Olsen, B.C., and Mitlin, D. (2013a). Carbon nanosheet frameworks derived from peat moss as high performance sodium ion battery anodes. *ACS Nano*, 7(12), 11004–11015.

Dahn, J.R., Xing, W., Gao, Y. (1997). The "falling cards model" for the structure of microporous carbons. *Carbon*, 35(6), 825–830.

Dou, X., Geng, C., Buchholz, D., and Passerini, S. (2018). Research update: Hard carbon with closed pores from pectin-free apple pomace waste for Na-ion batteries. *APL Materials*, 6(4), 047501.

Dou, X., Hasa, I., Hekmatfar, M., Diemant, T., Behm, R.J., Buchholz, D., and Passerini, S. (2017). Pectin, hemicellulose, or lignin? Impact of the biowaste source on the performance of hard carbons for sodium-ion batteries. *ChemSusChem*, 10(12), 2668–2676.

Dou, X., Hasa, I., Saurel, D., Vaalma, C., Wu, L., Buchholz, D., Bresser, D., Komaba, S., and Passerini, S. (2019). Hard carbons for sodium-ion batteries: Structure, analysis, sustainability, and electrochemistry. *Mater. Today*. 23, 2019, 87–104.

En ligneGreen Car Congress. (n.d.). *Faradion demonstrates proof-of-concept sodium-ion electric bike*. [Online]. Available at: https://www.greencarcongress.com/2015/05/20150515-faradion.html [Accessed 1 September 2019].

EUR-Lex. (n.d.). Directive 2009/28/EC of the European Parliament and of the Council of 23 April 2009 on the promotion of the use of energy from renewable sources and amending and subsequently repealing Directives 2001/77/EC and 2003/30/EC (Text with EEA relevance). [Online]. Available at: http://eur-lex.europa.eu/eli/dir/2009/28/oj [Accessed 29 January 2018].

Faradion. (n.d.). *Strong performance*. [Online]. Available at: https://www.faradion.co.uk/technology-benefits/strong-performance/ [Accessed 1 September 2019].

Ferreri, A.C. and Robertson, J. (2000). Interpretation of Raman spectra of disordered and amorphous carbon. *Physical Review*, 61, 14095–14107.

Franklin, R. E. (1951). Crystallite Growth in Graphitizing and Non-Graphitizing Carbons. *Proc. R. Soc. A Math. Phys. Eng. Sci.* 209, 196–218.

Gao, X., An, Y., Zhang, W., Yu, M., Ci, L., and Feng, J. (2018). Self-supporting soft carbon fibers as binder-free and flexible anodes for high-performance sodium-ion batteries. *Materials Technology*, 33(12). https://www.tandfonline.com/doi/abs/10.1080/10667857.2018.1509525.

Górka, J., Vix-Guterl, C., and Matei Ghimbeu, C. (2016). Recent progress in design of biomass-derived hard carbons for sodium ion batteries. *C*, 2(4), 24.

Gotoh, K., Ishikawa, T., Shimadzu, S., Yabuuchi, N., Komaba, S., Takeda, K., Goto, A., Deguchi, K., Ohki, S., Hashi, K., Shimizu, T., and Ishida, H. (2013). NMR study for electrochemically inserted Na in hard carbon electrode of sodium ion battery. *Journal of Power Sources*, 225(Supplement C), 137–140.

Haas, T.J., Nimlos, M.R., and Donohoe, B.S. (2009). Real-time and post-reaction microscopic structural analysis of biomass undergoing pyrolysis. *Energy & Fuels*, 23(7), 3810–3817.

Harris, P.J.F. and Tsang, S.C. (1997). High-resolution electron microscopy studies of non-graphitizing carbons. *Philosophical Magazine A*, 76(3), 667–677.

Hasa, I., Hassoun, J., and Passerini, S. (2017). Nanostructured Na-ion and Li-ion anodes for battery application: A comparative overview. *Nano Research*, 10, 1–28.

Hasegawa, G., Kanamori, K., Kannari, N., Ozaki, J., Nakanishi, K., and Abe, T. (2015). Hard carbon anodes for Na-ion batteries: Toward a practical use. *ChemElectroChem*, 2(12), 1917–1920.

HiNa Battery Technology Co., Ltd. (n.d.). Na-ion battery–Sodium ion battery. [Online]. Available at: http://www.hinabattery.com/en/index.php?catid=12 [Accessed 1 September 2019].

Hong, Z., Zhen, Y., Ruan, Y., Kang, M., Zhou, K., Zhang, J.-M., Huang, Z., and Wei, M. (2018). Rational design and general synthesis of S-doped hard carbon with tunable doping sites toward excellent Na-ion storage performance. *Advanced Materials*, 30(29), 1802035.

Inagaki, M. and Kang, F. (2016). *Materials Science and Engineering of Carbon: Characterization*. Butterworth-Heinemann, Oxford, UK, and Cambridge, USA.

Industrie-techno. (2018). Trottinette électrique: la batterie sodium-ion a sa première application. [Online]. 2018. Available at: /article/trottinette-electrique-la-batterie-sodium-ion-a-sa-premiere-application.53164 [Accessed 1 September 2019].

Irisarri, E., Amini, N., Tennison, S., Ghimbeu, C.M., Gorka, J., Vix-Guterl, C., Ponrouch, A., and Palacin, M.R. (2018). Optimization of large scale produced hard carbon performance in Na-ion batteries: Effect of precursor, temperature and processing conditions. *Journal of The Electrochemical Society*, 165(16), A4058–A4066.

Irisarri, E., Ponrouch, A., and Palacin, M.R. (2015). Review—hard carbon negative electrode materials for sodium-ion batteries. *Journal of The Electrochemical Society*, 162(14), A2476–A2482.

Izanzar, I., Dahbi, M., Kiso, M., Doubaji, S., Komaba, S., and Saadoune, I. (2018). Hard carbons issued from date palm as efficient anode materials for sodium-ion batteries. *Carbon*, 137, 165–173.

Jiang, Q., Zhang, Z., Yin, S., Guo, Z., Wang, S., and Feng, C. (2016). Biomass carbon micro/nano-structures derived from ramie fibers and corncobs as anode materials for lithium-ion and sodium-ion batteries. *Applied Surface Science*, 379, 73–82.

Jin, J., Yu, B., Shi, Z., Wang, C., and Chong, C. (2014). Lignin-based electrospun carbon nanofibrous webs as free-standing and binder-free electrodes for sodium ion batteries. *Journal of Power Sources*, 272, 800–807.

Kim, D., Lee, E., Slater, M., Lu, W., Rood, S., and Johnson, C.S. (2012). Layered $Na[Ni_{1/3}Fe_{1/3}Mn_{1/3}]O_2$ cathodes for Na-ion battery application. *Electrochemistry Communications*, 18, 66–69.

Kim, K., Lim, D.G., Han, C.W., Osswald, S., Ortalan, V., Youngblood, J.P., and Pol, V.G. (2017). Tailored carbon anodes derived from biomass for sodium-ion storage. *ACS Sustainable Chemistry & Engineering*, 5(10), 8720–8728.

Kipling, J.J., Sherwood, J.N., Shooter, P.V., and Thompson, N.R. (1964). The pore structure and surface area of high-temperature polymer carbons. *Carbon*, 1(3), 321–328.

Komaba, S., Murata, W., Ishikawa, T., Yabuuchi, N., Ozeki, T., Nakayama, T., Ogata, A., Gotoh, K., and Fujiwara, K. (2011). Electrochemical Na insertion and solid electrolyte interphase for hard-carbon electrodes and application to Na-ion batteries. *Advanced Functional Materials*, 21(20), 3859–3867.

Komaba, S., Yamamoto, H., Kubota, K., Muratsubaki, S., Dahbi, M., and Manivannan, A. (2017). (Invited) Hard carbons made from cellulose for Na- and K-ion batteries. *Meeting Abstracts*, MA2017-01(1), 44.

Ku, H., Trade, M., Nixon, R., and Wong, P. (2010). Flexural properties of phenolic resin reinforced with glass powder: Preliminary results. *Journal of Applied Polymer Science*, 116(1), 347–354.

Kumar, M., Olajire Oyedun, A., and Kumar, A. (2018). A review on the current status of various hydrothermal technologies on biomass feedstock. *Renewable and Sustainable Energy Reviews*, 81, 1742–1770.

Li, H., Shen, F., Luo, W., Dai, J., Han, X., Chen, Y., Yao, Y., Zhu, H., Fu, K., Hitz, E., and Hu, L. (2016a). Carbonized-leaf membrane with anisotropic surfaces for sodium-ion battery. *ACS Applied Materials & Interfaces*, 8(3), 2204–2210.

Li, R., Huang, J., Xu, Z., Qi, H., Cao, L., Liu, Y., Li, W., and Li, J. (2018). Controlling the thickness of disordered turbostratic nanodomains in hard carbon with enhanced sodium storage performance. *Energy Technology*, 6(6), 1080–1087.

Li, W., Huang, J., Feng, L., Cao, L., Ren, Y., Li, R., Xu, Z., Li, J., and Yao, C. (2017a). Controlled synthesis of macroscopic three-dimensional hollow reticulate hard carbon as long-life anode materials for Na-ion batteries. *Journal of Alloys and Compounds*, 716, 210–219.

Li, Y., Hu, Y.-S., Li, H., Chen, L., and Huang, X. (2015). A superior low-cost amorphous carbon anode made from pitch and lignin for sodium-ion batteries. *Journal of Materials Chemistry A*, 4(1), 96–104.

Li, Y., Hu, Y.-S., Titirici, M.-M., Chen, L., and Huang, X. (2016b). Hard carbon microtubes made from renewable cotton as high-performance anode material for sodium-ion batteries. *Advanced Energy Materials*, 6(18), 1600659.

Li, Y., Mu, L., Hu, Y.-S., Li, H., Chen, L., and Huang, X. (2016c). Pitch-derived amorphous carbon as high performance anode for sodium-ion batteries. *Energy Storage Materials*, 2, 139–145.

Li, Y., Paranthaman, M.P., Akato, K., Naskar, A.K., Levine, A.M., Lee, R.J., Kim, S.-O., Zhang, J., Dai, S., and Manthiram, A. (2016d). Tire-derived carbon composite anodes for sodium-ion batteries. *Journal of Power Sources*, 316, 232–238.

Li, Y., Xu, S., Wu, X., Yu, J., Wang, Y., Hu, Y.-S., Li, H., Chen, L., and Huang, X. (2014). Amorphous monodispersed hard carbon micro-spherules derived from biomass as a high performance negative electrode material for sodium-ion batteries. *Journal of Materials Chemistry A*, 3(1), 71–77.

Li, Z., Bommier, C., Chong, Z.S., Jian, Z., Surta, T.W., Wang, X., Xing, Z., Neuefeind, J.C., Stickle, W.F., Dolgos, M., Greaney, P.A., and Ji, X. (2017). Mechanism of Na-ion storage in hard carbon anodes revealed by heteroatom doping. *Advanced Energy Materials*, 7(18), 1602894.

Liu, Y., Merinov, B.V., and Goddard, W.A. (2016). Origin of low sodium capacity in graphite and generally weak substrate binding of Na and Mg among alkali and alkaline earth metals. *Proceedings of the National Academy of Sciences*, 113(14), 3735–3739.

Liu, Y., Sun, Y., and Huang, G. (2018). Preparation and antioxidant activities of important traditional plant polysaccharides. *International Journal of Biological Macromolecules*, 111, 780–786.

Lotfabad, E.M., Ding, J., Cui, K., Kohandehghan, A., Kalisvaart, W.P., Hazelton, M., and Mitlin, D. (2014a). High-density sodium and lithium ion battery anodes from banana peels. *ACS Nano*, 8(7), 7115–7129.

Luo, W., Bommier, C., Jian, Z., Li, X., Carter, R., Vail, S., Lu, Y., Lee, J.-J., and Ji, X. (2015). Low-surface-area hard carbon anode for na-ion batteries via graphene oxide as a dehydration agent. *ACS Applied Materials & Interfaces*, 7(4), 2626–2631.

Luo, W., Schardt, J., Bommier, C., Wang, B., Razink, J., Simonsen, J., and Ji, X. (2013). Carbon nanofibers derived from cellulose nanofibers as a long-life anode material for rechargeable sodium-ion batteries. *Journal of Materials Chemistry A*, 1(36), 10662–10666.

Lv, W., Wen, F., Xiang, J., Zhao, J., Li, L., Wang, L., Liu, Z., and Tian, Y. (2015). Peanut shell derived hard carbon as ultralong cycling anodes for lithium and sodium batteries. *Electrochimica Acta*, 176, 533–541.

Matei Ghimbeu, C., Górka, J., Simone, V., Simonin, L., Martinet, S., and Vix-Guterl, C. (2018). Insights on the Na+ ion storage mechanism in hard carbon: Discrimination between the porosity, surface functional groups and defects. *Nano Energy*, 44, 327–335.

Matei Ghimbeu, C., Zhang, B., Martinez de Yuso, A., Rety, B., and Tarascon, J.-M. (2019) Valorizing low cost and renewable lignin as hard carbon for Na-ion batteries: Impact of lignin grade. *Carbon*, 153, 634–647.

Meents, M.J., Watanabe, Y., and Samuels, A.L. (2018). The cell biology of secondary cell wall biosynthesis. *Annals of Botany*, 121(6), 1107–1125.

Meng, X., Savage, P.E., and Deng, D. (2015). Trash to treasure: From harmful algal blooms to high-performance electrodes for sodium-ion batteries. *Environmental Science & Technology*, 49(20), 12543–12550.

Ministère des Armées. (2017). Des batteries innovantes au service des armées. [Online]. Available at: https://www.defense.gouv.fr/english/actualites/articles/video-des-batteries-innovantes-au-service-des-armees [Accessed 1 September 2019].

Narasimman, R. and Prabhakaran, K. (2013). Preparation of carbon foams with enhanced oxidation resistance by foaming molten sucrose using a boric acid blowing agent. *Carbon*, 55, 305–312.

Novák, P., Ufheil, J., Buqa, H., Krumeich, F., Spahr, M.E., Goers, D., Wilhelm, H., Dentzer, J., Gadiou, R., and Vix-Guterl, C. (2007). The importance of the active surface area of graphite materials in the first lithium intercalation. *Journal of Power Sources*, 174(2), 1082–1085.

Oberlin, A. and Terriere, G. (1975). Graphitization studies of anthracites by high resolution electron microscopy. *Carbon*, 13(5), 367–376.

Ponrouch, A. and Palacín, M.R. (2015). On the high and low temperature performances of Na-ion battery materials: Hard carbon as a case study. *Electrochemistry Communications*, 54, 51–54.

Ponrouch, A., Dedryvère, R., Monti, D., Demet, A.E., Mba, J.M.A., Croguennec, L., Masquelier, C., Johansson, P., and Palacín, M.R. (2013). Towards high energy density sodium ion batteries through electrolyte optimization. *Energy & Environmental Science*, 6(8), 2361–2369.

Qin, X., Lu, Y., Xiao, H., Wen, Y., and Yu, T. (2012). A comparison of the effect of graphitization on microstructures and properties of polyacrylonitrile and mesophase pitch-based carbon fibers. *Carbon*, 50(12), 4459–4469.

Qiu, S., Xiao, L., Sushko, M.L., Han, K.S., Shao, Y., Yan, M., Liang, X., Mai, L., Feng, J., Cao, Y., Ai, X., Yang, H., and Liu, J. (2017). Manipulating adsorption–insertion mechanisms in nanostructured carbon materials for high-efficiency sodium ion storage. *Advanced Energy Materials*, 7(17), 1700403.

Rodriguez Correa, C., Hehr, T., Voglhuber-Slavinsky, A., Rauscher, Y., and Kruse, A. (2019). Pyrolysis vs. hydrothermal carbonization: Understanding the effect of biomass structural components and inorganic compounds on the char properties. *Journal of Analytical and Applied Pyrolysis*, 140, 137–147.

Rodríguez-Reinoso, F. and Molina-Sabio, M. (1992). Activated carbons from lignocellulosic materials by chemical and/or physical activation: An overview. *Carbon*, 30(7), 1111–1118.

Saavedra Rios, C. del M., Simone, V., Simonin, L., Martinet, S., and Dupont, C. (2018). Biochars from various biomass types as precursors for hard carbon anodes in sodium-ion batteries. *Biomass and Bioenergy*, 117, 32–37.

Saurel, D., Orayech, B., Xiao, B., Carriazo, D., Li, X., and Rojo, T. (2018). From charge storage mechanism to performance: A roadmap toward high specific energy sodium-ion batteries through carbon anode optimization. *Advanced Energy Materials*, 8(17), 1703268.

Shen, F., Zhu, H., Luo, W., Wan, J., Zhou, L., Dai, J., Zhao, B., Han, X., Fu, K., and Hu, L. (2015). Chemically crushed wood cellulose fiber towards high-performance sodium-ion batteries. *ACS Applied Materials & Interfaces*, 7(41), 23291–23296.

Simone, V. (2016). Développement d'accumulateurs sodium-ion. PhD thesis, Université Grenoble Alpes, Grenoble.

Simone, V., Boulineau, A., de Geyer, A., Rouchon, D., Simonin, L., and Martinet, S. (2016). Hard carbon derived from cellulose as anode for sodium ion batteries: Dependence of electrochemical properties on structure. *Journal of Energy Chemistry*, 25(5), 761–768.

Spahr, M.E., Buqa, H., Würsig, A., Goers, D., Hardwick, L., Novák, P., Krumeich, F., Dentzer, J., and Vix-Guterl, C. (2006). Surface reactivity of graphite materials and their surface passivation during the first electrochemical lithium insertion. *Journal of Power Sources*, 153(2), 300–311.

Stevens, D.A. and Dahn, J.R. (2000). High capacity anode materials for rechargeable sodium-ion batteries. *Journal of the Electrochemical Society*, 147(4), 1271–1273.

Stevens, D.A. and Dahn, J.R. (2001). The mechanisms of lithium and sodium insertion in carbon materials. *Journal of the Electrochemical Society*, 148(8), A803–A811.

Talekar, S., Patti, A.F., Vijayraghavan, R., and Arora, A. (2018). Complete utilization of waste pomegranate peels to produce a hydrocolloid, punicalagin rich phenolics, and a hard carbon electrode. *ACS Sustainable Chemistry & Engineering*, 6(12), 16363–16374.

Tang, K., Fu, L., White, R.J., Yu, L., Titirici, M.-M., Antonietti, M., and Maier, J. (2012). Hollow carbon nanospheres with superior rate capability for sodium-based batteries. *Advanced Energy Materials*, 2(7), 873–877.

Thomas, P. and Billaud, D. (2002). Electrochemical insertion of sodium into hard carbons. *Electrochimica Acta*, 47(20), 3303–3307.

Tsai, P., Chung, S.-C., Lin, S., and Yamada, A. (2015). Ab initio study of sodium intercalation into disordered carbon. *Journal of Materials Chemistry A*, 3(18), 9763–9768.

Väli, R., Jänes, A., and Lust, E. (2017b). Alkali-metal insertion processes on nanospheric hard carbon electrodes: An electrochemical impedance spectroscopy study. *Journal of the Electrochemical Society*, 164(11), E3429–E3437.

Väli, R., Jänes, A., Thomberg, T., and Lust, E. (2016). D-glucose derived nanospheric hard carbon electrodes for room-temperature sodium-ion batteries. *Journal of the Electrochemical Society*, 163(8), A1619–A1626.

Väli, R., Jänes, A., Thomberg, T., and Lust, E. (2017a). Synthesis and characterization of d-glucose derived nanospheric hard carbon negative electrodes for lithium- and sodium-ion batteries. *Electrochimica Acta*, 253, 536–544.

Vassilev, S.V., Baxter, D., Andersen, L.K., Vassileva, C.G., and Morgan, T.J. (2012). An overview of the organic and inorganic phase composition of biomass. *Fuel*, 94, 1–33.

Wang, H., Shi, Z., Jin, J., Chong, C., and Wang, C. (2015). Properties and sodium insertion behavior of phenolic resin-based hard carbon microspheres obtained by a hydrothermal method. *Journal of Electroanalytical Chemistry*, 755, 87–91.

Wang, H., Yu, W., Shi, J., Mao, N., Chen, S., and Liu, W. (2016a). Biomass derived hierarchical porous carbons as high-performance anodes for sodium-ion batteries. *Electrochimica Acta*, 188(Supplement C), 103–110.

Wang, K., Jin, Y., Sun, S., Huang, Y., Peng, J., Luo, J., Zhang, Q., Qiu, Y., Fang, C., and Han, J. (2017a). Low-cost and high-performance hard carbon anode materials for sodium-ion batteries. *ACS Omega*, 2(4), 1687–1695.

Wang, Q., Zhao, C., Lu, Y., Li, Y., Zheng, Y., Qi, Y., Rong, X., Jiang, L., Qi, X., Shao, Y., Pan, D., Li, B., Hu, Y.-S., and Chen, L. (2017). Advanced nanostructured anode materials for sodium-ion batteries. *Small*, 13(42), 1701835.

Wang, Q., Zhu, X., Liu, Y., Fang, Y., Zhou, X., and Bao, J. (2018a). Rice husk-derived hard carbons as high-performance anode materials for sodium-ion batteries. *Carbon*, 127, 658–666.

Wang, S., Xia, L., Yu, L., Zhang, L., Wang, H., and Lou, X.W. (2016b). Free-standing nitrogen-doped carbon nanofiber films: Integrated electrodes for sodium-ion batteries with ultralong cycle life and superior rate capability. *Advanced Energy Materials*, 6(7), 1502217.

Wang, T., Zhai, Y., Zhu, Y., Li, C., and Zeng, G. (2018b). A review of the hydrothermal carbonization of biomass waste for hydrochar formation: Process conditions, fundamentals, and physicochemical properties. *Renewable and Sustainable Energy Reviews*, 90, 223–247.

Warren, B.E. (1934). X-ray diffraction study of carbon black. *The Journal of Chemical Physics*, 2(9), 551–555.

Warren, B.E. (1941). X-ray diffraction in random layer lattices. *Physical Review*, 59(9), 693–698.

Wu, L., Buchholz, D., Vaalma, C., Giffin, G.A., and Passerini, S. (2016). Apple-biowaste-derived hard carbon as a powerful anode material for Na-ion batteries. *ChemElectroChem*, 3(2), 292–298.

Xiang, J., Lv, W., Mu, C., Zhao, J., and Wang, B. (2017). Activated hard carbon from orange peel for lithium/sodium ion battery anode with long cycle life. *Journal of Alloys and Compounds*, 701(Supplement C), 870–874.

Xu, S.-D., Zhao, Y., Liu, S., Ren, X., Chen, L., Shi, W., Wang, X., and Zhang, D. (2018). Curly hard carbon derived from pistachio shells as high-performance anode materials for sodium-ion batteries. *Journal of Materials Science*, 53(17), 12334–12351.

Yamamoto, H., Muratsubaki, S., Kubota, K., Fukunishi, M., Watanabe, H., Kim, J., and Komaba, S. (2018). Synthesizing higher-capacity hard-carbons from cellulose for Na- and K-ion batteries. *Journal of Materials Chemistry A*, 6(35), 16844–16848.

Ye, J., Zang, J., Tian, Z., Zheng, M., and Dong, Q. (2016). Sulfur and nitrogen co-doped hollow carbon spheres for sodium-ion batteries with superior cyclic and rate performance. *Journal of Materials Chemistry A*, 4(34), 13223–13227.

Yu, C., Hou, H., Liu, X., Yao, Y., Liao, Q., Dai, Z., and Li, D. (2018a). Old-loofah-derived hard carbon for long cyclicity anode in sodium ion battery. *International Journal of Hydrogen Energy*, 43(6), 3253–3260.

Yu, Y., Shen, M., Song, Q., and Xie, J. (2018b). Biological activities and pharmaceutical applications of polysaccharide from natural resources: A review. *Carbohydrate Polymers*, 183, 91–101.

Zeng, L., Li, W., Cheng, J., Wang, J., Liu, X., and Yu, Y. (2014). N-doped porous hollow carbon nanofibers fabricated using electrospun polymer templates and their sodium storage properties. *RSC Advances*, 4(33), 16920–16927.

Zhang, B., Ghimbeu, C.M., Laberty, C., Vix-Guterl, C., and Tarascon, J.-M. (2016). Correlation between microstructure and Na storage behavior in hard carbon. *Advanced Energy Materials*, 6(1), 1501588.

Zhang, F., Yao, Y., Wan, J., Henderson, D., Zhang, X., and Hu, L. (2017a). High temperature carbonized grass as a high performance sodium ion battery anode. *ACS Applied Materials & Interfaces*, 9(1), 391–397.

Zhang, H., Ming, H., Zhang, W., Cao, G., and Yang, Y. (2017b). Coupled carbonization strategy toward advanced hard carbon for high-energy sodium-ion battery. *ACS Applied Materials & Interfaces*, 9(28), 23766–23774.

Zhang, N., Liu, Q., Chen, W., Wan, M., Li, X., Wang, L., Xue, L., and Zhang, W. (2018a). High capacity hard carbon derived from lotus stem as anode for sodium ion batteries. *Journal of Power Sources*, 378, 331–337.

Zhang, S., Li, Y., and Li, M. (2018b). Porous hard carbon derived from walnut shell as an anode material for sodium-ion batteries. *JOM*, 70(8), 1387–1391.

Zhang, T., Mao, J., Liu, X., Xuan, M., Bi, K., Zhang, X.L., Hu, J., Fan, J., Chen, S., and Shao, G. (2017c). Pinecone biomass-derived hard carbon anodes for high-performance sodium-ion batteries. *RSC Advances*, 7(66), 41504–41511.

Zhang, Y., Li, X., Dong, P., Wu, G., Xiao, J., Zeng, X., Zhang, Y, and Sun, X. (2018c). Honeycomb-like hard carbon derived from pine pollen as high-performance anode material for sodium-ion batteries. *ACS Applied Materials & Interfaces*, 10(49), 42796–42803.

Zhang, Z., Brydson, R., Aslam, Z., Reddy, S., Brown, A., Westwood, A., and Rand, B. (2011). Investigating the structure of non-graphitising carbons using electron energy loss spectroscopy in the transmission electron microscope. *Carbon*, 49(15), 5049–5063.

Zhao, P.-Y., Yu, B.-J., Sun, S., Guo, Y., Chang, Z.-Z., Li, Q., and Wang, C.-Y. (2017). High-performance anode of sodium ion battery from polyacrylonitrile/humic acid composite electrospun carbon fibers. *Electrochimica Acta*, 232, 348–356.

Zhu, H., Shen, F., Luo, W., Zhu, S., Zhao, M., Natarajan, B., Dai, J., Zhou, L., Ji, X., Yassar, R.S., Li, T., and Hu, L. (2017a). Low temperature carbonization of cellulose nanocrystals for high performance carbon anode of sodium-ion batteries. *Nano Energy*, 33, 37–44.

Zhu, X., Jiang, X., Liu, X., Xiao, L., and Cao, Y. (2017b). A green route to synthesize low-cost and high-performance hard carbon as promising sodium-ion battery anodes from sorghum stalk waste. *Green Energy & Environment*, 2(3), 310–315.

Zhu, Y., Chen, M., Li, Q., Yuan, C., and Wang, C. (2017c). High-yield humic acid-based hard carbons as promising anode materials for sodium-ion batteries. *Carbon*, 123(Supplement C), 727–734.

Zhu, Y., Chen, M., Li, Q., Yuan, C., and Wang, C. (2018). A porous biomass-derived anode for high-performance sodium-ion batteries. *Carbon*, 129, 695–701.

4

Non-Carbonaceous Negative Electrodes in Sodium Batteries

Vincent GABAUDAN[1,2], Moulay Tahar SOUGRATI[1,2,3], Lorenzo STIEVANO[1,2,3] and Laure MONCONDUIT[1,2,3]

[1] *Institut Charles Gerhardt Montpellier, University of Montpellier, CNRS, France*

[2] *Réseau Français sur le Stockage Electrochimique de l'Energie (RS2E), FR CNRS 3459, Amiens, France*

[3] *ALISTORE-European Research Institute, CNRS FR 3104, Amiens, France*

4.1. Introduction

Finding an adapted negative electrode for Na-ion batteries (NIB) is one of the critical issues for their successful development. Nevertheless, the field of anode materials for NIB is also the one where most differences are found when compared to lithium-ion batteries (LIB).

The first difference concerns Na metal, which also represents the lowest potential limit for a viable sodium-based systems: the standard potential of the Na^+/Na couple in carbonate solvents (those commonly used in commercial LIB) is about 0.3 V higher than that of the corresponding Li^+/Li couple. The theoretical specific capacity of sodium metal is 1,166 mAh g^{-1}, much lower than that of lithium, ca. –70%. Then, chemically speaking, sodium metal is much more reactive that lithium not only toward ambient atmosphere and water, but also against non-aqueous electrolytes. From the safety point of view, the relatively low melting point of sodium (98°C) makes it significantly less safe than lithium. Finally, an additional important issue of sodium metal is dendrite formation upon plating, which is even more problematic than in the case of lithium. All these issues make batteries based on Na metal negative electrodes undesirable from the points of view of long-term stability of the battery and safety, at least when using common non-aqueous electrolytes. Therefore, the use of veritable Na-ion systems working with the "rocking-chair" mechanism of LIB and employing suitable negative electrode materials is necessary.

The majority of the works performed until now in the quest for suitable anode materials for NIB have concerned carbon-based negative electrodes, as already described in detail in Chapter 3. Shortly, contrary to lithium-based systems, graphite does not intercalate sodium ions with conventional electrolytes, and the alternative carbonaceous negative electrode materials that have been proposed, such as hard carbon, are characterized by relatively low specific capacities up to 300 mAh g^{-1} and undergo substantial irreversible capacity during the first sodiation (Palomares *et al.* 2012, 2013; Irisarri *et al.* 2015, 2018; Cui *et al.* 2017)

For these reasons, the research of suitable anode materials alternative to carbon-based anodes has continued during the past years, and some of these materials as well as their average potential and capacities are summarized in Figure 4.1.

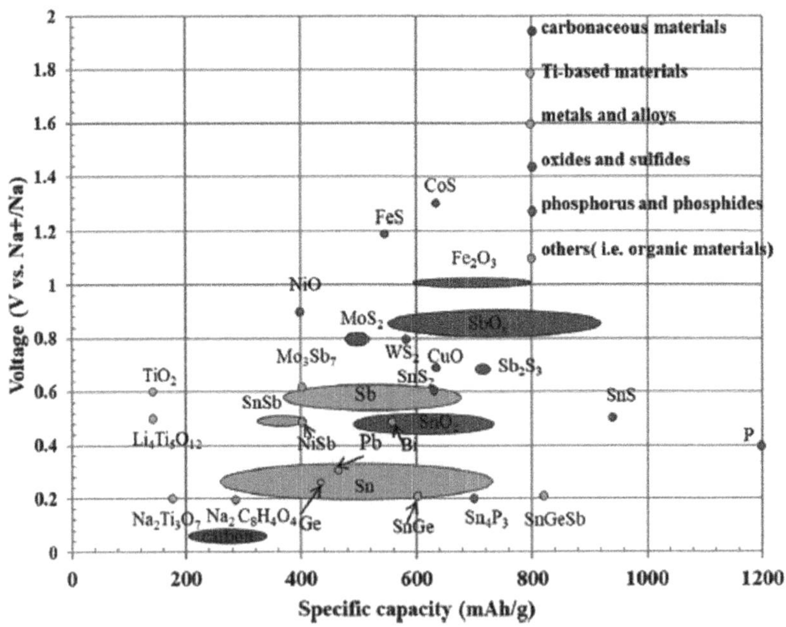

Figure 4.1. *Main families of negative electrodes for NIB (Li* et al. *2018a). For a color version of this figure, see www.iste.co.uk/monconduit/batteries.zip*

These studies can be thus classified similarly to those concerning LIB, i.e., based on their reaction mechanism. We can therefore divide them into materials undergoing insertion, alloying or conversion reaction, respectively. In general, however, both mechanism and the performance of electrode materials in NIB cannot be extrapolated from those occurring in LIB, and their electrochemical reactions have been carefully reexamined.

The first group of insertion materials concerns solids that are able to reversibly intercalate and deintercalate sodium ions, or even adsorb them through a pseudo-capacitive mechanism, without undergoing wide structural rearrangement. In this, we can find materials very similar to those already studied in the case of lithium, such as titanates and titanium oxides (Wang *et al.* 2013a; Wu *et al.* 2015), transition metal chalcogenides (MoS_2, WS_2, $SnSe_2$, etc.) (Gao *et al.* 2015; Liu *et al.* 2015b), MXenes (Yu 2016) and organic anodes (Nithya and Gopukumar 2015).

The second group, alloys materials, can provide high energy density due to the multiple electron exchange associated with the alloying/dealloying process. However, this mechanism leads to a high volume change that could represent as in LIB and even worse, a real penalty. Nevertheless, elements as phosphorus or antimony, with an optimized electrode formulation, seem to work better in the case of sodium than in the case of lithium.

In the third group of conversion materials (CM), one finds mainly metal oxides and sulfides, even though also other families such as metal selenides, carbodiimides, halides and phosphides have been explored. CM can deliver high specific and gravimetric capacities with excellent cycling stability, but they face at least one critical issue related to their low initial coulombic efficiency or high volume expansion.

All these systems will be treated in detail in the following sections.

4.2. Insertion materials

4.2.1. *Insertion anodes based on titanium oxide and titanates*

As in the case of LIB, transition metal oxides (TMO) have been also largely explored as potential electrode candidates for NIB (Kim *et al.* 2012; Slater *et al.* 2012; Pan *et al.* 2013; Xu *et al.* 2013; Dahbi *et al.* 2014; Su *et al.* 2016; Cui *et al.* 2017; Fang *et al.* 2020) both as positive and negative electrodes. In the case of positive electrodes, TMO are mostly layered rock salt phases already containing sodium in the "exchangeable" layer and work via a reversible sodium (de)intercalation mechanism, as mentioned in Chapter 1. In the case of negative electrodes, however, most simple

TMO such as Fe_2O_3, Co_3O_4, MnO, CuO and NiO react electrochemically with sodium through a conversion reaction, thus undergoing a complete structural rearrangement. The only TMO family that can thus be employed effectively in the design of insertion-based negative electrodes is that of titanium-based oxides. Among the interesting features of this family of materials are their non-toxicity, their chemical stability and their low production cost. It is thus not surprising that TiO_2 materials have been considered as practical negative electrode materials in NIB. Moreover, their working potential is relatively high compared to the Na^+/Na redox couple, making them inherently safe toward Na plating and the consequent growth of Na metal dendrites. Several different Ti-based phases have been studied until now, starting from the different TiO_2 polymorphs to already sodiated (and lithiated) phases such as ramsdellite $Na_2Ti_3O_7$, layered $NaTiO_2$ or spinel $Li_4Ti_5O_{12}$.

4.2.1.1. *Titanium oxide TiO_2*

Many different polymorphs of TiO_2 are reported in the literature, but only a few of them, namely anatase, rutile, brookite and TiO_2(B) (Figure 4.2), have been explored for application in LIB (Aravindan *et al.* 2015).

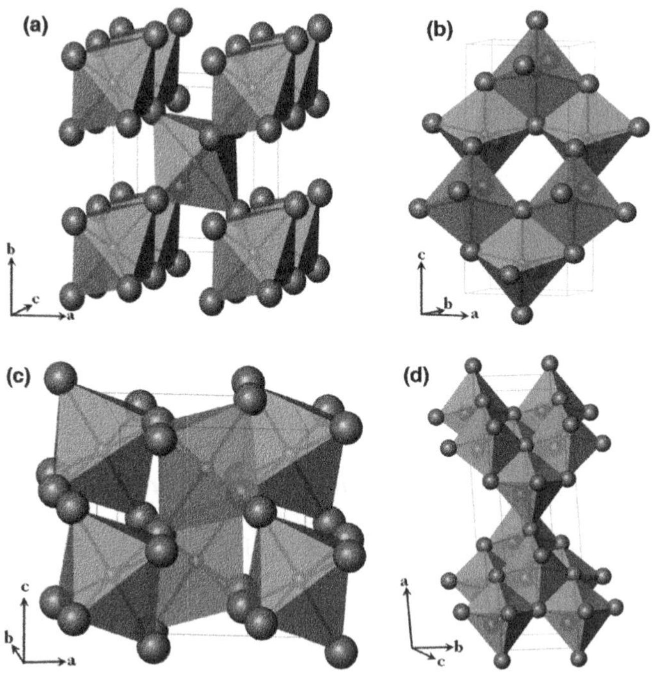

Figure 4.2. *Crystal structures of (a) rutile, (b) anatase, (c) TiO2(B) and (d) brookite. For a color version of this figure, see www.iste.co.uk/monconduit/batteries.zip*

The same phases were also investigated as anode materials in NIB, with results that are quite different from those of LIB. For instance, the best polymorph in the case of LIB is $TiO_2(B)$, which is able to react electrochemically with lithium through a mixed intercalation/pseudo-capacitive mechanism leading to capacities as high as 250 mAh g^{-1} if the morphology and the formulation are appropriately tuned (Opra *et al.* 2019). The same polymorph, however, provides much worse performance when reacting electrochemically with sodium (Huang *et al.* 2013). Improved operation could be nevertheless obtained when $TiO_2(B)$ was coupled to graphene in three-dimensional porous morphologies, enhancing the pseudo-capacitive behavior of the material (Chen *et al.* 2015). It is interesting to note that, in spite of the bad observed performance, Dawson and Robertson (2016) predicted by hybrid density functional theory (DFT) calculations that $TiO_2(B)$ should be the most favorable polymorphs for Na insertion.

A slightly different picture is obtained in the case of rutile: in fact, previous studies in LIB have shown that Li transport in the solid is strongly anisotropic since it occurs only along the c-axis channels, thus leading to capacities well below the theoretical one and in any case much lower than that of $TiO_2(B)$ (Yang *et al.* 2009). For bulk rutile, therefore, the performance is quite bad with lithium, and even worse with sodium. Recently, however, it has been shown that rutile microspheres, anchored by nanoneedle clusters, were shown to exhibit interesting energy storage behavior in NIB, reaching a stable capacity of 120 mAh g^{-1} for more than 200 cycles and especially very good rate capabilities (Zhang *et al.* 2015b).

The most interesting and studied case, however, remains that of anatase TiO_2. Anatase is stable and can be easily synthesized at relatively low temperature. In its structure, two TiO_6 octahedra share two adjacent edges with two other ones, forming planar double chains. In the case of LIB, Li diffusion occurs along a 3D reaction path connecting the octahedral interstitial sites. The Li insertion mechanism consists of three steps, which are represented electrochemically by a gradual decrease in the potential, followed by a plateau at about 1.7 V and finally a new decreasing slope: these processes correspond to a solid solution insertion, followed by a biphasic reaction expressed by the transformation of tetragonal to orthorhombic lattice structure and a final pseudo-capacitive step (Aravindan *et al.* 2015). These reactions are perfectly reversible, and the crystal structure of anatase is preserved during the electrochemical reaction (Fehse *et al.* 2014).

The mechanism of anatase with Na, however, is very different, and has been the matter of debate for quite some time. In fact, during the first discharge, which is different from the following ones, an irreversible plateau corresponding to the insertion of Na^+ in the interstitial sites of anatase goes along with drastic loss of the long-range order. Dambournet and co-workers (Li *et al.* 2017c) studied the electrochemical mechanism in anatase using a combination of high angle annular dark-field scanning transmission electron microscope (HAADF-STEM) and PDF analysis, showing that the electrochemically sodiated phase displays a disordered layered-like rhombohedral structure built from the stacking of intermixed Ti and Na slabs. This structure is completely different from all known lithiated anatase phases, and contains a high concentration of vacancies. During charge, the local structure of the anatase is recovered only with a short-range order of about 1 nm. These two disordered phases are formed alternatively during galvanostatic cycling, while no sign of the known ordered stoichiometric $NaTiO_2$ is ever obtained. Given the amorphous nature of the two phases, the electrochemical signature is characterized by sloping galvanostatic profiles.

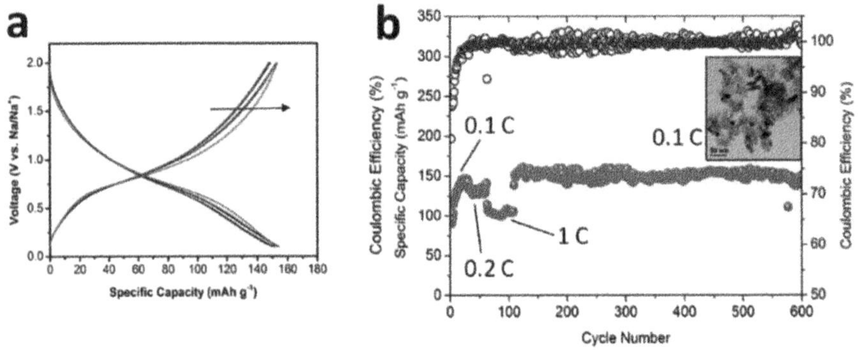

Figure 4.3. *(a) Charge/discharge potential profiles of rhombohedral anatase particles at 0.1 C (50 mA g^{-1}) after five preconditioning cycles at 1 C (500 mA g^{-1}) and (b) corresponding cycling performance between 0.1 and 2.0 V versus Na^{+}/Na (adapted from Longoni* et al. *2017). For a color version of this figure, see www.iste.co.uk/monconduit/batteries.zip*

Interestingly, the electrochemical properties of anatase seem to depend on the morphology of the cycled particles rather than on their surface area as one might expect. Longoni *et al.* showed that, by using TiO_2 particles grown along specific crystallographic directions and by employing an appropriate conductive wrapping agent (graphene), it is possible to obtain a TiO_2-based composite electrode delivering a stable capacity of 150 mAh g^{-1} for more than 600 cycles in the 1.5–0.1 V potential range (Figure 4.3) (Longoni *et al.* 2017). Su *et al.* could further improve the performance of anatase TiO_2 by appropriately tuning the morphology, increasing the capacity to above 200 mAh g^{-1} for more than 200 cycles when using hollow nanospheres (Su *et al.* 2015). In addition, they showed by DFT calculations that the energy barriers for the insertion of Na^{+} into anatase are relatively low, explaining the excellent high-rate performance of their materials. Similarly, Ling *et al.* (2017)

prepared nanoporous anatase that, adequately amorphized via a preliminary quick formation process in the initial cycles, withstand a stable capacity of about 200 mAh g^{-1} with no fading after 500 cycles, even in fast charging conditions (20 C, 6.7 A g^{-1}). Further improvements of both cycling and rate-capability performance of anatase can be obtained by appropriately tuning the electrode formulation. Ling *et al.* (2018), for instance, showed that using sodium alginate as the binder instead of the conventional polyvinylidene difluoride (PVDF), it is possible to achieve a stable capacity of 180 and 82 mAh g^{-1} at 1 and 10 C for 500 and 3,600 cycles, respectively.

4.2.1.2. *$Na_2Ti_3O_7$ sodium titanate*

$Na_2Ti_3O_7$, commonly abbreviated as NTO, has been long considered as one of the most promising anodes, as it exhibits a high theoretical capacity 177 mAh g^{-1} (corresponding to the intercalation of two additional Na^+ ions and the reduction of 2/3 of the Ti^{4+} to Ti^{3+}) at the very low average potential of 0.3 V versus Na^+/Na (Senguttuvan *et al.* 2011; Rudola *et al.* 2013; Xu *et al.* 2014). NTO crystallizes in the monoclinic space group P21/m, and consists of a layered structure of edge-sharing $(Ti_3O_7)^{2-}$ units forming zig-zag 2D sheets with Na^+ ions, which are accommodated between them (see Figure 4.4). Such a structure enables the fast diffusion of the Na^+ ions between the layers of TiO_6 octahedra.

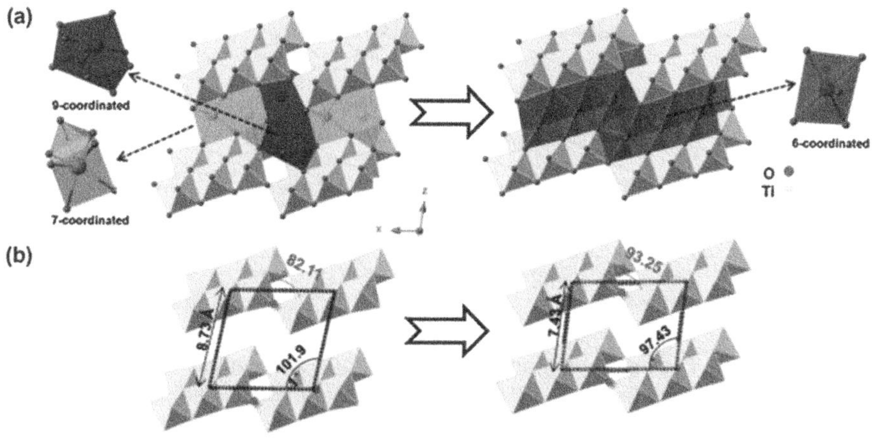

Figure 4.4. *The phase transformation (a) and the related structural modifications (b) upon Na intercalation in $Na_2Ti_3O_7$ (adapted from Xu* et al. *2014). For a color version of this figure, see www.iste.co.uk/monconduit/batteries.zip*

The intercalation of the additional Na^+ ion during sodiation also occurs within the titanium oxide layers, with a drastic modification of the coordination of the preexisting

Na^+ ions (Figure 4.4). In fact, while in pristine NTO the coordination of Na^+ varies between 7 and 9 depending upon the occupied site, in the fully sodiated $Na_4Ti_3O_7$ all Na^+ centers are hexa-coordinated (Xu *et al.* 2014). During the sodiation, the lattice structure is notably modified, and the widening of the joint angle between two neighboring Ti–O layers together with the shearing of TiO_6 slabs enables the accommodation of more Na^+ ions. Such a reversible mechanism testifies the structural flexibility of this network, which makes it unique among layered intercalation compounds.

In spite of such promising properties, the possible application of $Na_2Ti_3O_7$ is hampered by its poor performance in terms of capacity retention, coulombic efficiency and electronic conductivity. Since the latter is also limiting in terms of rate capability, several studies were centered on its improvement either by nanosizing/nanostructuring (Zhang *et al.* 2014; Dong *et al.* 2015; Sun *et al.* 2019), carbon coating (Xu *et al.* 2014; Li *et al.* 2017b) or doping (Chen *et al.* 2017). In particular, carbon-coated-NTO coaxial nanorods demonstrated excellent electrochemical performance, with a reversible specific capacity of 142 and 84 mAh g^{-1} after 100 cycles at 1 and 10 C (see Figure 4.5), respectively (Li *et al.* 2017b). The carbon coating seems to be particularly efficient, when coupled to the nanorod morphology, in increasing the electronic conductivity of the material.

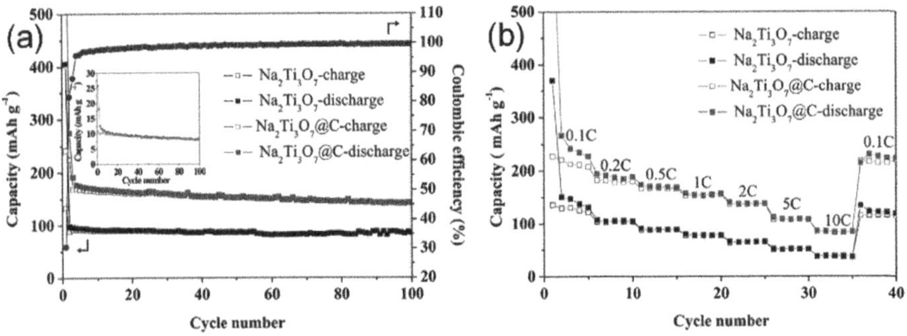

Figure 4.5. *(a) Cycling performance at 1 C and (b) rate capability of $Na_2Ti_3O_7$ and $Na_2Ti_3O_7$@C electrodes (1 C = 311 mA g^{-1}) (adapted from Li* et al. *2017b). For a color version of this figure, see www.iste.co.uk/monconduit/batteries.zip*

Interesting performance is also observed for the inversed morphology, with NTO coated over carbon nanotubes (CNT) prepared by a sol-gel method (Dong *et al.* 2015). In this case, after the first cycle with a huge irreversible capacity loss due to surface reaction of the electrolyte over the very large exposed surface area, the material stabilizes at about 100 mAh g^{-1} after 200 cycles at 170 mA g^{-1} and especially excellent capacity and stability at high cycling rates (100 mAh g^{-1} at 1,700 mA g^{-1}). A detailed kinetics analysis reveals a predominant pseudo-capacitive contribution of the Na^+ storage mechanism, which appears to be highly beneficial to fast long-term cyclability.

Finally, Nb-doping during sol-gel synthesis also brings higher reversible capacity, better cyclability and considerably enhanced rate capability, even though to a reduced extent compared to the other two other strategies (Chen *et al.* 2017). Thus, Nb-doping should be applied in combination with other strategies such as nanostructuring and/or carbon coating to provide an electrode material with acceptable properties for a possible practical application in NIB.

4.2.1.3. *$Li_4Ti_5O_{12}$ lithium titanate*

$Li_4Ti_5O_{12}$, usually abbreviated as LTO, has a cubic spinel crystal structure (Fd-3m space group) with a unit cell parameter *a* taking values between 8.35 and 8.37 Å. In this lattice, Li^+ ions occupy alone the tetrahedral 8a sites and share the octahedral 16d sites with Ti^{4+} in the ratio 1/5, while the oxide ions are set in the 32e sites forming a typical cubic-close packing arrangement. LTO has long been considered as a promising candidate for the negative electrode in LIB due to its high working potential relative to Li^+/Li, structural stability with no volume change during the biphasic Li^+ insertion and good compatibility with typical LIB electrolytes (Zaghib *et al.* 1999; Yi *et al.* 2015). In fact, the Li^+ insertion occurs with a structural rearrangement, with the existing Li^+ ions moving from the 8a sites to the empty 16c sites, which are also occupied by the newly inserted Li^+. In this way, the material transforms into rock salt $Li_7Ti_5O_{12}$ via a "zero-strain" phase transition with a global volume change of less than 0.1%, corresponding to a specific theoretical capacity of 155 mAh g^{-1}. Such structural properties offer one of the best cycle life performances among all oxide-based anodes while the high working potential allows lithium metal plating to be avoided ensuring high safety at high rates.

In spite of an identical theoretical specific capacity of 155 mAh g^{-1}, the insertion mechanism of Na^+, however, is completely different from that of Li^+. In fact, during sodiation in NIB, LTO forms two new phases via the global reaction:

$$2\,Li_4Ti_5O_{12} + 6\,Na^+ + 6\,e^- \leftrightarrows Li_7Ti_5O_{12} + Na_6LiTi_5O_{12}$$

with Na^+ occupying the 16c sites in $Na_6LiTi_5O_{12}$ and the simultaneous migration of Li^+ toward neighboring LTO resulting in the formation of $Li_7Ti_5O_{12}$ (Natarajan *et al.* 2019). The three-phase separation involves the formation of many moving phase boundaries in the materials, leading to electrode cracking and detaching from the current collector, and thus poor cycling performance.

Several strategies have been proposed to improve the overall cycling performance of LTO in NIB, such as nanostructuring, carbon coating, doping or preparation of composites with other metal oxides. The preparation of tailored grain sizes and morphologies such as spheres, nanosheets or porous structures lead to slight improvements of the low rate performance. On the other hand, carbon coating, metal doping with lanthanides, copper and boron, and especially the preparation of composites with graphene are also efficient methods to prepare more efficient LTO. Additional performance improvements can be obtained by appropriate electrode formulation, for instant using carboxymethyl-cellulose as the binder.

4.2.1.4. *Layered titanates*

Composed of only earth-abundant elements, the Ti^{3+}-based $NaTiO_2$ is a promising intercalation anode material for NIB. It crystallizes in the classical layered rock salt O3-type rock salt structure (space group R-3m) typical of $LiCoO_2$ and of many layered $NaMO_2$ metal oxides currently tested as positive electrode materials in NIB. Similar to them, also $NaTiO_2$ is able to deintercalate about half of its Na^+ ions providing an electrochemical reversible capacity of 152 mAh g^{-1} at the voltage of 0.6–1.6 V with excellent cyclability and good rate capability (Figure 4.6).

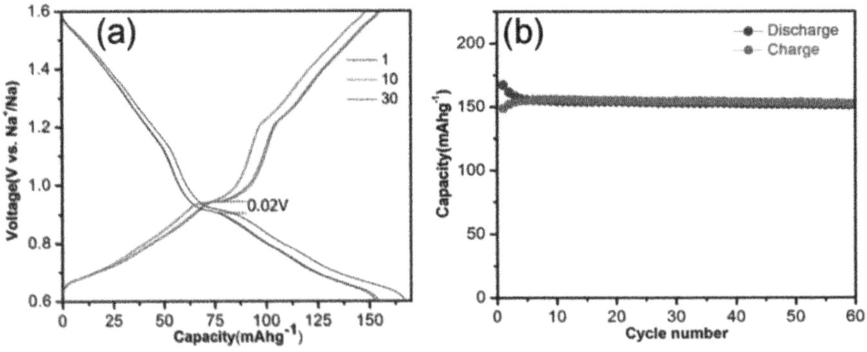

Figure 4.6. *(a) Electrochemical charge/discharge signature at C/10 and (b) cycle life of $NaTiO_2$ in the 0.6–1.6 V range versus Na^+/Na (adapted from Wu* et al. *2015). For a color version of this figure, see www.iste.co.uk/monconduit/batteries.zip*

In such material, DFT calculations reveal very low Na^+ diffusion barriers, consistent with the excellent electrochemical performance. *In situ* XRD results, on the other hand, reveal a complicated O3 to O'3 oxidation mechanism featuring superstructure evolution coupled to an unusually small lattice variation. In fact, during Na^+ deintercalation, a specific vacancy ordering is observed, yielding a constant interslab distance and slightly changing in-plane Ti–Ti distances in the newly formed O'3 phase (Wu *et al.* 2015).

In addition to $NaTiO_2$, a new Ti-based layered cathode material has been recently reported: $Na_{0.66}[Li_{0.22}Ti_{0.78}]O_2$, with a P2-type rock salt structure, shows a reversible capacity of 116 mAh g^{-1} at an average voltage of 0.75 V, sufficiently far from the sodium plating voltage to make the battery potentially safer (Wang *et al.* 2013b). Such material also exhibits excellent cycle life due to its negligible volume change (0.77%) during Na^+ (de)insertion, making it a very efficient zero-strain material. This electrochemical reaction goes through an unusual quasi-single-phase electrochemical mechanism, which is in contrast to most of the other P2-type materials, which usually show P2 to OP4 transitions upon oxidation.

4.2.2. *Insertion anodes based on transition metal chalcogenides*

Transition metal chalcogenides, with typical formula MX_2, where M is a transition metal cation and X is a chalcogenide ion, usually have a layered structure that can accommodate relatively large cations such as Na^+ in the interlayer slabs, which makes them promising candidates as negative electrode materials for NIB. They are often applied in the form of nanocomposites with conductive carbon or graphene additives with improved Na^+ ion storage performance in terms of capacity, rate capability and cycling stability. Their reaction with sodium usually features the first intercalation step, leading to the formation of Na_xMX_2, followed by a conversion reaction, which produces separated nanostructured Na_2X and M metal phases. Since the two processes usually occur at different potentials, it is possible to control the mechanism and limit the reaction to the insertion step by adjusting the working voltage window of the cell.

The most known member of this family is surely MoS_2, a typical 2D-layered metal dichalcogenide. MoS_2 is built on covalently bonded two-dimensional S-Mo-S trilayers,which are stacked by van der Waals interactions. This 2D structuration facilitates intercalation of large cations such as Na+ (Xiao *et al.* 2017). However, the MoS_2 layers tend to restack to reduce their surface energy, which compromises its electrochemical performance. Therefore, an appropriate expansion of MoS_2 is desirable to improve its intercalation properties. Moreover, the poor electronic and ionic conductivity of MoS_2 and its large volume expansion during the electrochemical processes are additional obstacles to its practical application in NIB. Many efforts have thus been made in the past years to improve its performance, and the most successful ones have concerned the preparation of nanocomposites with graphene. In these materials, the synergistic effects of graphene and MoS_2 lead to improved capacity, rate capability and cycling stability (Liang *et al.* 2018).

Complete information about the intercalation of Na in MoS_2 are obtained by several complementary studies: first of all, Wang *et al.* (2014b) used XRD and *ex situ* STEM to follow the global phase change, the lattice expansion and Na^+ coordination in MoS_2. They observed that Na^+ intercalation occurs through a series of partially staged two-phase structural transitions, with the Na^+ ions first inserted in every other interlayer of MoS_2 to form hexagonal $Na_{0.5}MoS_2$, corresponding to the upper potential plateau at 0.85 V. Before these interlayers are fully occupied, the Na ions begin to fill the other interlayers between the S-Mo-S slabs. The large strain and unfavorable energy induce a gliding of the sulfur plane along an intralayer atomic plane and the phase transition from 2H- to 1T-MoS_2, where the Mo coordination changes from trigonal to octahedral (Figure 4.7).

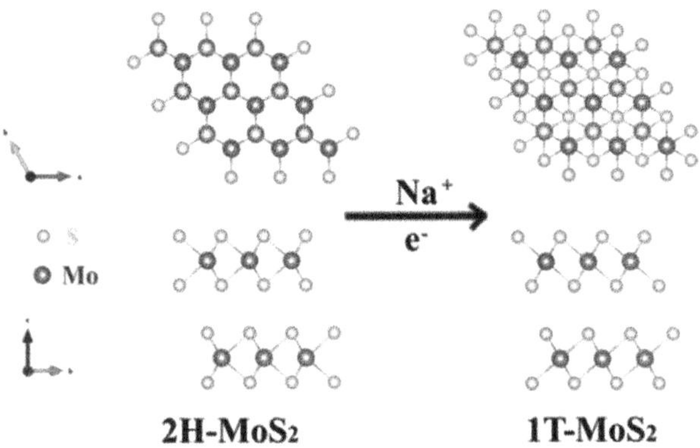

Figure 4.7. *Schematic diagram for phase transition from 2H-MoS_2 to 1T-MoS_2. The 2H-MoS_2 shows hexagonal lattice with threefold symmetry and the atomic stacking sequence (S-Mo-S) of ABA while the 1T-MoS_2 shows an atomic stacking sequence (S-Mo-S') of ABC where the bottom S' plane occupies the hollow center of 2H hexagonal lattice* (*Wang* et al. *2014b*). *For a color version of this figure, see www. iste.co.uk/monconduit/batteries.zip*

On the other hand, Gao *et al.* have recently applied *in situ* high-resolution transmission electron microscopy (HR-TEM) to follow the dynamics of the intercalation of Na^+ in MoS_2 nanosheets, confirming that Na^+ insertion occurs through a two-phase reaction mechanism initiated by the nucleation of a new phase $NaMoS_2$ and followed by a propagation of the phase boundary with a much smaller velocity than that observed for the similar lithiation. The transport of Na^+ fractures the layered structure into a series of nanodomains characterized by a high concentration of defects. Finally, additional new insights on the reaction mechanism of Na with exfoliated MoS_2 were recently provided by Gonzales *et al.* (2017) using *operando* electron paramagnetic resonance (EPR) spectroscopy.

EPR clearly suggests that the mechanism of sodiation proceed first at the Mo^{5+} structural defect sites located at the surface of MoS_2. Such reaction is however characterized by a low reversibility. When the exfoliated MoS_2 is formulated in composites together with CNT, an improved reversible behavior is observed due to the enhanced conductivity created by the synergistic combination of the few-layer chalcogenide and the CNT network. The improved electronic percolation allows a good cycling stability even at low cutoff potentials, improving the obtained specific capacity and energy density. In summary, these studies point to a complex mechanism based on a sequence of two-phase stages, with the simultaneous creation of highly reactive Mo^{5+} defects at the surface of the material as the reaction proceeds, with an increasing surface reactivity as the size of the MoS_2 particles decreases.

Similarly to MoS_2, WS_2 (as well as other sulfides) also has been the object of the interest as a negative electrode for NIB. In particular, WS_2 nanowires prepared via a simple solvothermal route followed by a heat treatment show a high capacity exceeding 600 mAh g^{-1}at 100 mA g^{-1} with a mixed intercalation-conversion mechanism, leading to poor cyclability (Liu *et al.* 2015b). However, if the lower limit of the cycling potential window is restricted to 0.3 V, the electrochemical reaction occurs through a highly reversible pure intercalation mechanism, with an excellent stable capacity of 330 mAh g^{-1} for 1,400 cycles at high rate of 1,000 mA g^{-1}. Also in this case, the intercalation of Na^+ at high potential produces a gradually expanded and exfoliated interlayer structure, facilitating both electron and Na^+ diffusion kinetics as well as creating more active surface sites for Na storage.

4.2.3. *Insertion MXene-based anodes*

MXenes (Naguib *et al.* 2014) constitute a family of 2D transition metal carbides and carbonitrides with the general formula $M_{n+1}X_n$ (where M = Sc, Ti, V, Cr, Zr, Nb, Mo, Hf, Ta; X = C, N; and typically n = 1, 2 or 3). They can be produced by chemical etching in the HF of their precursors, the so-called MAX phases (Figure 4.8), with general formula $M_{n+1}AX_n$ (n = 1, 2, 3), A is a 13 or 14 group element (often Al or Si). They were first proposed as negative electrodes for LIB, with a mixed mechanism possibly including both capacitive/pseudo-capacitive storage at the surface and Li^+ intercalation depending on composition, structure, morphology and surface chemistry; the latter in particular being influenced by the HF treatment, which might produce different surface functional groups (e.g., OH^-, F^-, O^{2-}) (Kajiyama *et al.* 2016; Li *et al.* 2019; Perssonand Rosen 2019; Bu *et al.* 2020; Liu *et al.* 2020).

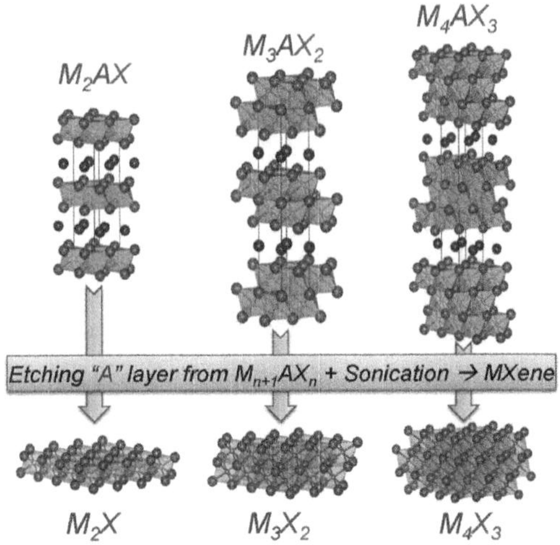

Figure 4.8. *Preparation of MXenes form the corresponding MAX phases (Naguib* et al. *2014). For a color version of this figure, see www.iste.co.uk/monconduit/batteries.zip*

A similar behavior is observed also in the case of Na^+ intercalation: as resumed in a recent review by Chen *et al.* (2018a), most of the work has been related, as in the case of lithium to $Ti_3C_2T_X$-type compounds (where T is the surface terminal group) with varied morphologies and architectures. In particular, due to their layered structure, hydrophilic terminal groups and excellent electrical conductivity hybrid composite nanomaterials with MXenes can be easily prepared and show improved reversible capacity in NIB at very high rates, reaching power performance close to that of supercapacitors, in particular when the supercapacitive behavior is exalted (Wang *et al.* 2015c). Nevertheless, simulations based on DFT calculations suggest that there is no particular advantage in using MXenes with Na instead of Li due to the larger size of the intercalating Na^+ ion (Eames and Islam 2014). On the contrary, compounds based on Mg^{2+} generally exhibit a higher (theoretical) gravimetric capacity than Li^+ intercalated compounds, as the two-electron redox chemistry offered by the divalent Mg^{2+} prevails over its heavier mass. Moreover, the modification of the functional groups at the surface as well as of the transition metal species can significantly affect the cell voltage and the gravimetric capacity of a MXene: in particular, surface termination by oxygen tends to promote the highest capacity, whereas termination by H and/or OH should be avoided if possible since they result in a lower capacity (Yu 2016). In summary, most published results on MXenes indicate them as promising negative electrode materials for NIB for power applications. In particular, when used in composites, they can provide fast electron pathways as well as helping buffering the volume variation during cycling because of their delaminated morphology.

4.2.4. *Insertion organic anodes*

An interesting approach to obtain electrode materials from renewable resources with minimum energy consumption is using them as organic electrode materials. Based on both small organic molecules and polymers, they were mainly tested in the past as positive electrode at relatively low potential for LIB, and extended also to NIB and Mg-ion batteries, as comprehensively resumed by Schon *et al.* Some advances, however, have been obtained also in the field of negative electrode organic electrodes (Schon *et al.* 2016). In fact, the redox potential of organic species can be tuned adequately by adding specific functionalities: for instance, the addition of electron-donating groups to the materials lowers the reduction potential, whereas the addition of electron-withdrawing ones increases it. The production of negative electrodes, therefore, is usually related to the preparation of molecules containing electron-donating groups.

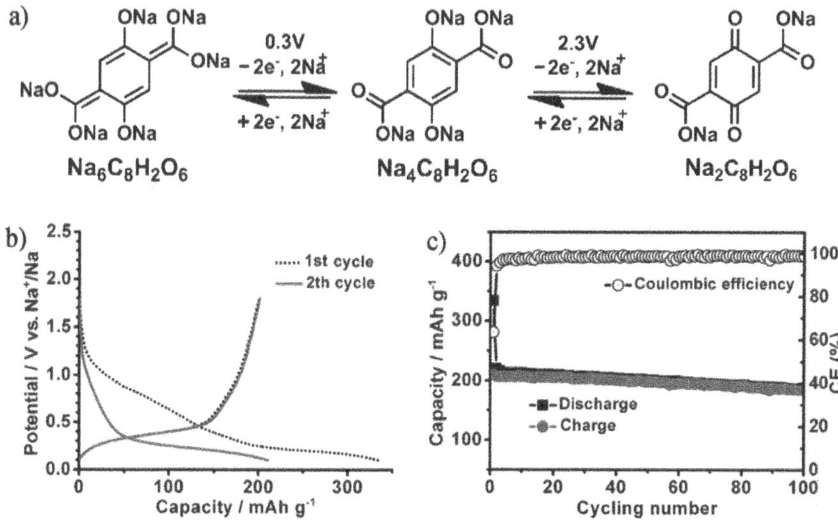

Figure 4.9. *(a) Redox reaction mechanisms of $Na_4C_8H_2O_6$ at 0.3 and 2.3 V, (b) initial discharge/charge profile of a $Na_4C_8H_2O_6$/Na half-cell at a rate of C/10 and (c) capacity retention and Coulombic efficiency at C/10 (adapted from Wang* et al. *2014a). For a color version of this figure, see www.iste.co.uk/monconduit/batteries.zip*

The most studied systems as negative electrodes are based on the redox activity of terephthalates, widely abundant and simple to handle materials with a redox activity based on the two carboxylate groups (Park *et al.* 2012). Among them, the most studied one is surely the tetra-sodium salt of 2,5-dihydroxyterephthalic acid ($Na_4C_8H_2O_6$). Such materials can be used both as positive and as negative electrodes, depending on the redox activity of the couples $Na_2C_8H_2O_6/Na_4C_8H_2O_6$ and $Na_4C_8H_2O_6/Na_6H_2O_6C_8$ at 2.3 and 0.3 V, respectively, with a reversible capacity for each reaction exceeding 180 mAh g^{-1} with excellent cycling performance (Figure 4.9) (Wang *et al.* 2014a).

Similar to this compound, many other organic systems such as croconic acid, gallic acid and benzene diacrylate, as well as some biomolecules containing the quinone group as the redox active center, such as juglone, have been proposed. Some strategies were also proposed to improve the performance of such organic materials, which are usually characterized by low densities and ionic conductivities. For instance, the introduction of extended π-conjugation in the molecule structure produces enhanced intermolecular interactions and packing, which then facilitate Na^+ diffusion in the compound. Finally, also some polymeric systems similar to such molecules were proposed as negative electrode materials for NIB with alternate success (Nithya and Gopukumar 2015; Schon *et al.* 2016).

4.3. Negative electrode materials based on electrochemical alloying with sodium

Given that *p*-bloc elements form numerous Na-rich alloys with sodium, they can provide very interesting high energy density negative electrodes for NIB. Tin, antimony, bismuth, lead, germanium and phosphorus offer high theoretical capacities (Table 4.1) resulting from the multiple electron exchange associated with the alloying process. The potentials for these reactions were predicted by DFT calculations (Chevrier and Ceder 2011). Indeed, alloys have very high energy densities and low redox potentials, two highly desirable properties for NIB applications. The main alloy systems are summarized in Table 4.1.

Similarly to LIB, the main challenge associated with alloy materials for NIB is the large volume change occurring during the sodiation/desodiation processes, causing the pulverization of the electrode and the loss of electrical contact. This drawback is in principle more serious in NIB than in LIB because of the larger ionic radius of Na^+ (1.02 Å) compared to Li^+ (0.76 Å), which is expected to significantly impact long-term cycling in NIB electrodes. Moreover, activation energy (E_a) barriers to Na^+ diffusion in solids are expected to be larger compared to Li^+, indicating a more difficult sodiation of alloy materials in comparison with lithiation.

M	Theoretical specific capacity ($mAh.g^{-1}$)	Na_xM (x max)	Volume change (%)	Voltage vs. Na^+/Na (V)
Si	954	NaSi	244	0.5
P	2596	Na_3P	390	0.4
Ge	369	NaGe	225	0.3
Sn	847	$Na_{3.75}Sn$	420	0.2
Sb	660	Na_3Sb	393	0.6
Bi	385	Na_3Bi	365	0.6
Pb	485	$Na_{3.75}Pb$	250	0.2
In	467	Na_2In	365	0.4

Table 4.1. *Highly sodiated Na_xM phases with the corresponding capacity, volume change and voltage versus Na^+/Na*

The main alloy materials are described hereafter together with the most common approaches to reduce the consequences of volume changes and to boost electronic conduction. Different nanostructuration approaches to accommodate the stress and strain without pulverizing, and to reduce the ionic and electronic transport pathways are described. Several effective strategies have been developed to solve these issues with respect to the sluggish Na diffusion and large volume expansion, including the optimization of electrolytes and binders.

4.3.1. *Silicon and germanium*

Silicon and germanium are two promising anode materials for LIB due to their high theoretical capacities. Moreover, silicon is abundant in Earth's crust and is relatively cheap. Germanium is more expensive but has a higher electric conductivity compared to Si. Unfortunately, Si exhibits sluggish sodiation performance in NIB, attributed to its small interstitial space and high stiffness (Chou *et al.* 2015). Different maximum values for alkali insertion in Si and Ge are founded in the literature (NaSi (Legrain *et al.* 2013), $Na_{0.76}Si$ (Jung *et al.* 2014), NaGe (Baggetto *et al.* 2013a) and $Na_{1.6}Ge$ (Lu *et al.* 2016)). The theoretical capacity corresponding to the formation of NaSi is about 954 mAh g^{-1} with a calculated volume deformation of about 244%. Experimentally, crystalline silicon (c-Si) has been considered to be inactive or at least partially inactive toward Na, whereas amorphous silicon (a-Si) appears to work better and thus qualifies as a potential negative electrode candidate. The limited electrochemical activity of c-Si is related to the large insertion barrier and energy difference between surface and subsurface sites for Na^+ insertion (Malyi *et al.* 2013). The limited sodiation has the advantage of eluding large volume changes, and NaSi represents indeed a promising system. Moreover, the diffusion coefficient (D_{Na}) at room temperature in a-NaSi is around 10^{-8} cm^2s^{-1}, approximately 1 order of magnitude higher than the corresponding D_{Li} in a-LiSi. These enhanced properties enable facile Na intercalation and migration. Recently, partially amorphous Si nanoparticles ($\approx$20 nm) delivered a reversible capacity of 248 mA h g^{-1} for 100 cycles (Xu *et al.* 2016).

Germanium (Ge) alloys with Na to form NaGe, delivering a theoretical capacity of 369 mAh.g^{-1} (Chevrier and Ceder 2011). If compared with Si, Ge performs reasonably well in terms of mechanical stability and ionic conductivity. Ge has a thermodynamic driving force for sodiation that is much stronger than Si. Moreover, it exhibits moderate volume expansion (if we compare with other posttransition elements) and bulk modulus upon sodiation that allows relatively fast Na^+ conductivity (Na diffusivity 1.2×10^{-7} cm^2 s^{-1}). The experimental sodiation of Ge was first studied by Baggetto *et al.* (2013a). The potential profile of Ge during sodiation exhibits a single wide plateau. The XRD reveals that pristine amorphous Ge remains amorphous after full sodiation, with no formation of crystalline phases. The reaction works well at high charge rates, 170 C, for a capacity limited to 220 mAh g^{-1}, even though the capacity declines rapidly after 15 cycles. However, the application of Ge is held back by its poor cyclability (Nithya and Gopukumar 2015).

Even nanostructured or amorphized silicon and germanium do not show interesting cycling performance in NIB. An alternative strategy proposed recently is the use of siloxene or germananes obtained from Si- and Ge-based layered Zintl phases (Loaiza *et al.* 2020a). Layered $CaSi_2$ and $CaGe_2$ contain puckered $Si_6(Ge_6)$ rings interconnected to form planes, with several stacking sequences possible for the layers, resulting in a whole set of polymorphs. After topotactic deintercalation of Ca, the Si(Ge) 2D backbone is preserved, forming the so-called siloxene ($Si_6OH_3H_3$) and germanane (GeH_n). Their electrochemical behavior was tested in NIB (Loaiza *et al.* 2019, 2020b) showing very promising results, with the layered structure serving as a buffer for the volume expansion of the associated electrochemical reaction, while enhancing electronic conductivity and ion diffusivity. These properties produce an excellent capacity retention (Si: 300 mAh g^{-1}, Ge 500 mAh g^{-1}) offering a real improvement compared to bulk Si and Ge (Figure 4.10). The electrochemical mechanism is still under study, and likely does not correspond to a simple alloying process.

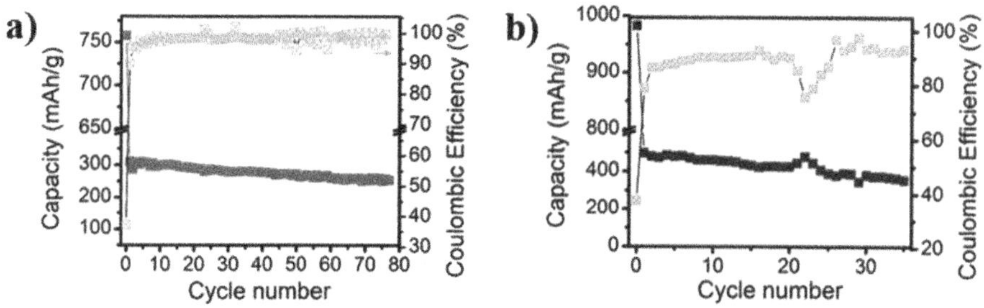

Figure 4.10. *Capacity retention versus Na^+/Na at C/20 of (a) siloxene and (b) germanane (adapted from Loaiza* et al. *2020b). For a color version of this figure, see www.iste.co.uk/monconduit/batteries.zip*

4.3.2. *Tin*

The most sodiated Na-Sn phase obtained electrochemically was shown to be $Na_{15}Sn_4$, in good agreement with the Na-Sn phase diagram (1 Sn for 3.75 Na) (Komaba *et al.* 2012). The intermediate species formed along the sodiation are still a matter of debate, since those suggested by DFT have not been clearly identified experimentally. Ellis *et al.* proposed four sodiation steps, with the successive formation of $NaSn_3$, NaSn, Na_9Sn_4 and $Na_{15}Sn_4$ (Ellis *et al.* 2012). Wang *et al.* (2012) proposed a sodiation of crystalline Sn is two steps with first the growth of an amorphous a-$Na_{0.5}Sn$ phase through a limited volumetric expansion of 60% followed by a second step with the formation of amorphous Na-rich species and finally crystalline $Na_{15}Sn_4$. The total volumetric change after full sodiation is approximately 420%. The formation of amorphous intermediates before that of $Na_{15}Sn_4$ likely decreases the stress in the electrode (Zhang *et al.* 2018a). The volume contraction upon charge leads to cracks in the electrode and rapid capacity fading. In addition, low kinetics lead to limited rate capability and power performance. To address these issues, considerable efforts in terms of nanostructuration have been achieved. Among the strategies aiming to improve the Sn electrode performance in NIB, the use of inactive or active elements (with Ge, Sb, Bi) to accommodate volume change has been proposed (Abel *et al.* 2014).

Another approach is to introduce carbon into the electrode, acting as a buffer to absorb the volume change and limit metal aggregation during cycling. Embedding Sn nanoparticles in a spherical carbon network by aerosol spray pyrolysis was efficient to avoid this aggregation (Figure 4.11) (Liu *et al.* 2015a).

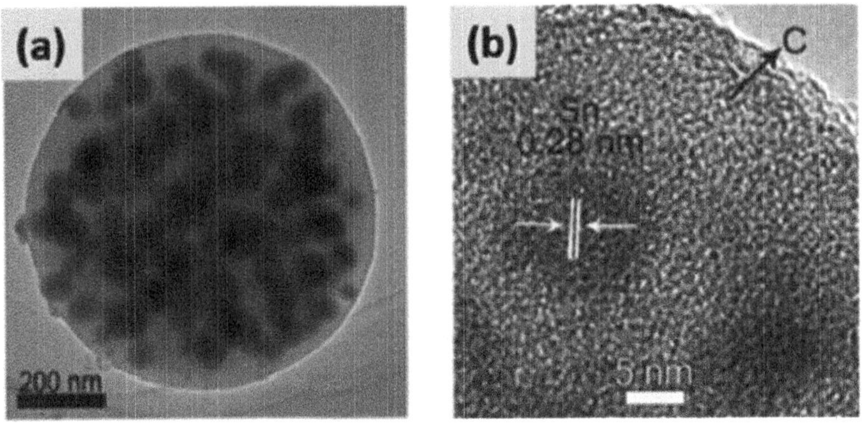

Figure 4.11. *(a) Transmission electron microscope (TEM) image and (b) high-resolution TEM image of Sn nanoparticles in a spherical carbon network (adapted from Liu* et al. *2015a)*

Sn particles smaller than 10 nm allow the strain to decrease, effectively mitigate pulverization and also limit the aggregation of Sn particles during cycling. Such nanoparticles, homogeneously embedded in a carbon matrix, deliver a stable capacity of 415 mAh g^{-1} after 500 cycles at 1 A g^{-1}. The control of the cut-off voltage window is also an efficient strategy as shown by Komaba's group (Fukunishi *et al.* 2016). The low voltage cut-off at 0.7 V increases the reversibility of Sn sodiation by limiting the formation of the solid electrolyte interphase (SEI), which is not stable at lower potentials, thus improving the cycling life. Nanostructuration and/or tailoring Sn/C composites, applied to address the main issues, i.e., volume change during cycling, low kinetics and particle aggregation, are nevertheless not sufficient to limit efficiently the rapid capacity fade during cycling. Moreover, these approaches penalize the volumetric energy density.

4.3.3. *Phosphorus*

Among the three allotropes of phosphorus (white, black and red), the red one is relatively stable and commercially available. It has been the most studied polymorph as negative electrode material for NIB. Moreover, P is the most abundant among pnictogens, ranking 11th in Earth's crust. Due to its low atomic weight (30.97 g mol^{-1}) and the possible formation of highly sodiated phases (Na_3P) at about 0.4 V versus Na^+/Na, P has the highest theoretical capacity (2,596 mA h g^{-1}) and could be promising for NIB if the following issues could be tackled: poor conductivity ($\sim 10^{-14}$ S cm^{-1} at room temperature), large volume change and unstable SEI (Ni *et al.* 2018).

Due to its low electronic conductivity, it is necessary to mix phosphorus with relatively large fractions of carbon. Kim *et al.* (2013) were first to report phosphorus as negative electrode for NIB in 2013, reaching a capacity of 1,890 mAh g^{-1}. They prepared a red P/carbon composite by ball milling (7:3) with carbon black and polyacrylic acid (PAA) in a 7:1:2 ratio, meaning a very low P loading in the final electrode. Inspired by the works on LIB (Marino *et al.* 2011), plethora of P/C composites were proposed with in some cases interesting results. Indeed, a nanostructured P/C composite containing high carbon loading (>30%) was prepared through simple sublimation of P (given the low sublimation temperature of red P) onto a porous carbon framework. Such composite exhibited high capacity and reasonable cyclability. Nanostructuration strategies, such as engineered 3D P nanostructures, were efficient to limit the volume swelling, to enhance the exchange interface between P and the electrolyte, and to provide good performance for application in microbatteries (Ni and Li 2018). Extremely long cycling (>1,000 cycles) was achieved with a ZIF-8-derived microporous carbon matrix, displaying a reversible capacity of about 600 mAh g^{-1} (of composite) at 150 mA g^{-1}, in relation with a reversible sodiation reaction limited to Na_xP (with $x < 3$) intermediates rather than to the higher sodiated unstable Na_3P (Li *et al.* 2017d).

The addition of 1D or 2D carbon nanomaterials such as CNT, graphene or reduced graphene oxide (rGO) to the formulation was shown to improve the conductivity of C/P composites, leading to outstanding capacity retention of 80% over 2,000 cycles (Zhu *et al.* 2015).

Another approach to counterbalance the low conductivity of P and improve the electron percolation was achieved when red P was chemically bonded to carboxyl group-functionalized CNTs via P–O–C bonds (Song *et al.* 2015). Such a system delivered a discharge capacity of 2,134 mAh g^{-1} (/ P) in half-cells versus Na, with a capacity retention of 92% after 100 cycles (Figure 4.12).

To address the volume swelling issue leading to the pulverization of phosphorus and the loss of electrical contact of the P electrode, amorphous and/or porous P are preferred to crystalline P. A porous P/N-doped graphene (P@GN) paper effectively buffers the volume change due to free expansion of the nanosheets during sodiation along both longitudinal and transverse directions, showing a capacity retention beyond 85% over 350 cycles (Pei *et al.* 2015; Zhang *et al.* 2016a).

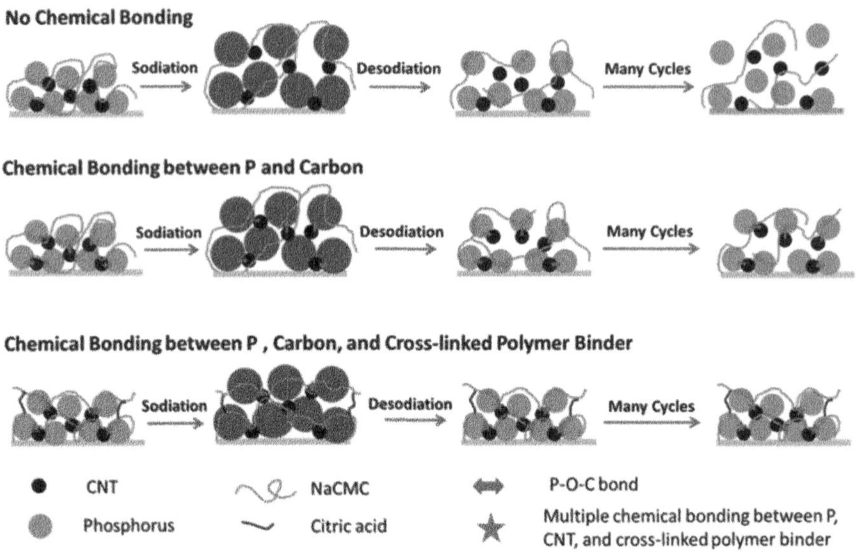

Figure 4.12. *Schematic illustration of the chemical bonding between phosphorus, carbon nanotube and cross-linked polymer binder (adapted from* Song et al. *2015). For a color version of this figure, see www.iste.co.uk/monconduit/batteries.zip*

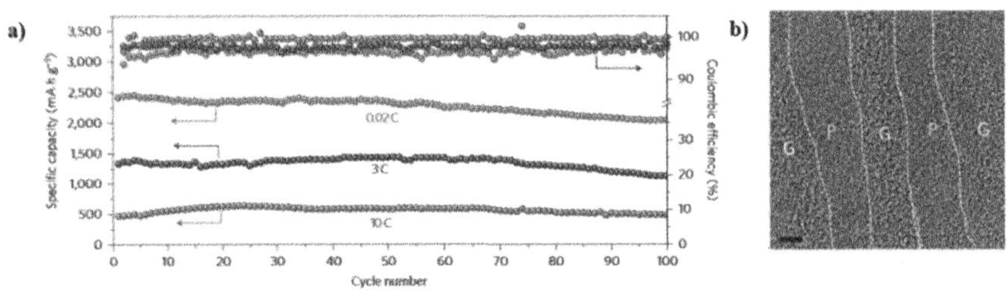

Figure 4.13. *Example of a phosphorene-graphene hybrid material as a high capacity anode material for Na-ion batteries (adapted from Sun* et al. *2015). For a color version of this figure, see www.iste.co.uk/monconduit/batteries.zip*

As seen above (see siloxene, section 4.3.1), *p*-bloc elements have recently emerged as very promising 2D material able to buffer volume expansion during alkali insertion (Loaiza *et al.* 2020a). Layered black P, with a considerable in-plane electrical conductivity, showed to be highly beneficial to enhance power densities. Phosphorene synthesized by an exfoliation method showed rapid charge transfer between less conductive phosphorene interlayers. In the same spirit, layered graphene/phosphorene electrodes achieved a reversible capacity of 2,440 mAh g^{-1} (/g of P) over 100 cycles (Figure 4.13) (Sun *et al.* 2015).

In summary, coupling nanostructuration with the presence of appropriate conductive additives is definitively a promising direction, even though the carbon portion has to be reduced to not penalize the energy density. Furthermore, safety issues is still a limitation for P electrodes, as toxic white P formed during the synthesis may be present in the materials.

Finally, the mechanism of sodiation/desodiation has been unclear until not so long ago, limiting further optimization. The partial amorphization of pristine P limits the use of XRD to follow the electrochemical sodiation processes. Grey and co-workers used a combined experimental/theoretical approach to provide molecular level insight into the sodiation/desodiation pathways in black P anodes for NIB (Marbella *et al.* 2018). During sodiation, ^{31}P chemical shift anisotropies in NMR data reveals P helices and P at the end of chains as the primary structural components of amorphous Na_xP (Figure 4.14). XRD and ^{23}Na magic-angle spinning NMR support the formation during the discharge of a new Na_3P polymorph, also predicted by DFT. Upon charge, P helices are reformed in the amorphous disordered intermediates. The pristine material is not reformed back at the end of desodiation, and this might be connected to the irreversibility observed in the Na-P system.

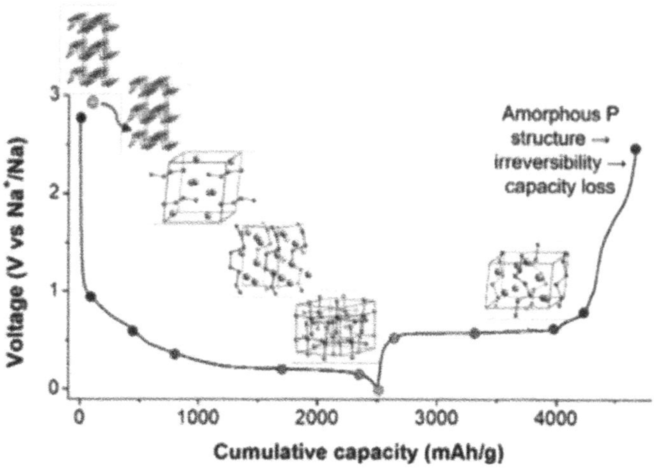

Figure 4.14. *Proposed mechanism for the sodiation/desodiation of black P (adapted from Marbella* et al. *2018). For a color version of this figure, see www.iste.co.uk/monconduit/batteries.zip*

4.3.4. *Antimony*

Sb was first studied in NIB as a carbon/Sb nanocomposite, which demonstrated an initial capacity of 610 mAh g^{-1} (/g of Sb) (Qian *et al.* 2012). Darwiche *et al.* showed an impressive stable cycling with a raw Sb rich electrode (70% wt), maintaining a stable capacity of 580 mAh g^{-1} at a moderate rate (C/2) over 160 cycles with a coulombic efficiency over 99% (Figure 4.15). The mechanism of sodiation was deeply investigated by *in situ* XRD (Darwiche *et al.* 2012), showing that Na reacts with crystalline Sb to form intermediate amorphous Na_xSb phases. While X-ray absorption spectroscopy (XAS) was unable to provide valuable information (Darwiche *et al.* 2018), *operando* PDF analysis and *ex situ* ^{23}Na magic-angle spinning (MAS) solid-state NMR spectroscopy provided significant insight into the alloying mechanism of Sb with Na (Figure 4.16) (Allan *et al.* 2016).

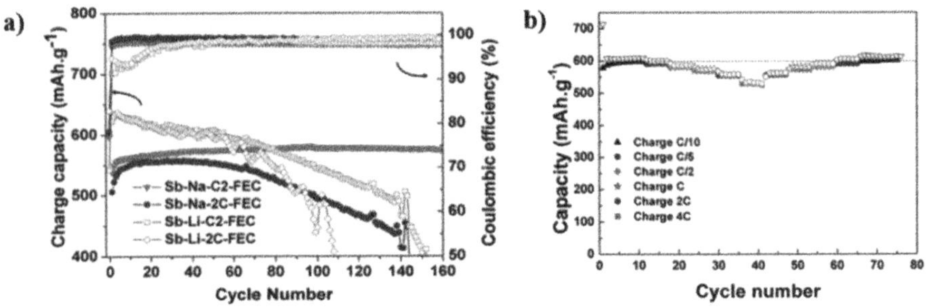

Figure 4.15. *(a) Capacity retention of Sb/Li and Sb/Na half-cells cycled at C/2 and 2 C. (b) Rate capability of Sb/Na from C/10 to 4 C (adapted from Darwiche* et al. *2012). For a color version of this figure, see www.iste.co.uk/monconduit/batteries.zip*

While only two sodiated antimonides, NaSb and Na_3Sb, are known in the Na–Sb phase diagram, two new intermediate species formed electrochemically were identified: a-$Na_{3-x}Sb$ ($x \approx 0.4$–0.5), with a local structure similar to crystalline Na_3Sb (c-Na_3Sb) but with a significant number of sodium vacancies, and a-$Na_{1.7}Sb$, a highly amorphous structure featuring some Sb-Sb dumbbell species. The first sodiation was shown to break down crystalline antimony to form first a-$Na_{3-x}Sb$, and finally, crystalline Na_3Sb. Desodiation results in a composite of crystalline and amorphous antimony networks, likely able to react at different voltages, through different pathways, and thus to minimize strain in the electrode. These intermediates, mostly amorphous, seem to minimize the adverse effects of volumetric changes in electrochemical mechanisms.

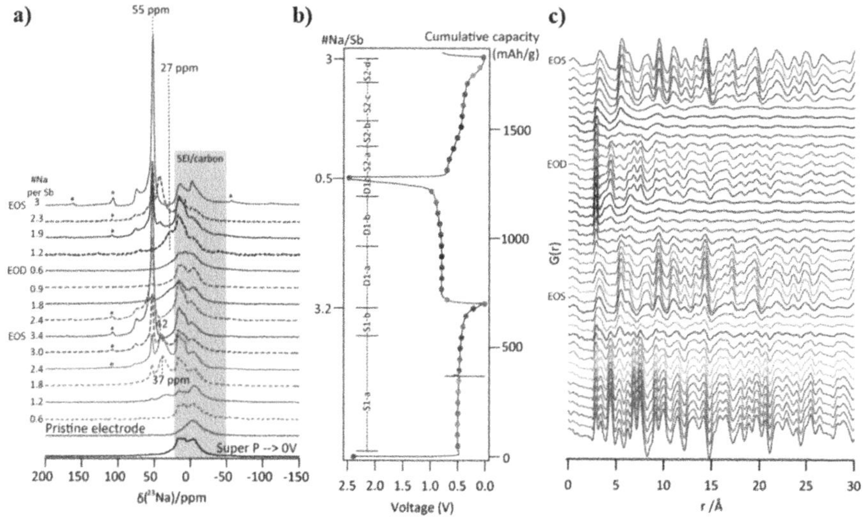

Figure 4.16. *(a) Ex situ ^{23}Na NMR spectra of cycled Sb electrodes at different states of charge. The shaded region marks the dominant contribution of Na in CMC binder, SEI and conductive carbon. *Spinning sidebands. (b) Discharge–charge curves obtained for Sb during the in situ PDF measurements. (c) Selected PDFs obtained by Fourier transform the total scattering X-ray data during the first discharge, first charge and second discharge (from Allan* et al. *2016). For a color version of this figure, see www.iste.co.uk/monconduit/batteries.zip*

A great deal of work has been devoted to further improving the performance of Sb-based electrodes, with the preparation of nanocomposites with Sb nanoparticles embedded in conductive carbon matrices, where carbon can accommodate the large volume expansion. Authors claimed improved capacity retention, attributed to the formation of stable SEI and improved conductivity. Sb/C nanofibers were prepared by electrospinning and subsequent calcination treatments (Wu *et al.* 2014). These Sb nanoparticles (20 nm diameters) uniformly dispersed in carbon nanofibers exhibited a capacity of 631 mAh g^{-1} at 40 mA g^{-1}. Ultrasmall Sb nanoparticles (≈ 10 nm) embedded into porous carbon spheres (with a correct 69% in weight Sb content in the composite) were prepared via an aerosol spray route (Palomares *et al.* 2012). Such spherical Sb/C nanocomposite exhibits excellent rate capability and capacity retention of 88.5% after 500 cycles at 100 mA g^{-1}, however the Sb loading is very low. The addition of highly stretchy and robust graphene layers, in good contact with the Sb particles, was suggested to buffer the strain associated with the alloying process and to improve the conductivity (Li *et al.* 2015a). A flexible and binder free full cell was achieved with Sb/rGO and $Na_3V_2(PO_4)_3$/rGO as the negative and positive electrodes, respectively, delivering a reversible capacity of ≈ 400 mAh g^{-1} (/g for the negative electrode) at 100 mA g^{-1} after 100 cycles (Cheng *et al.* 2018).

The design of M–Sb intermetallics, with M being either an electrochemically inactive or active component, was also proposed to alleviate volume changes during cycling. The additional element can also increase the electric conductivity of the system M–Sb (M = Mo, Bi, Al, Cu, Zn, Fe) (Baggetto *et al.* 2013b; Darwiche *et al.* 2015). In this case, the reaction mechanism is a combination of conversion (section 4.4) and alloying reactions. When M is electrochemically active versus Na, it can work as mutual buffer with Sb to alleviate the volume expansion (Darwiche *et al.* 2013; Baggetto *et al.* 2014; Zhao and Manthiram 2015b). SnSb, for instance, appropriately formulated with a carboxymethyl cellulose binder and cycled in fluoroethylene carbonate containing electrolyte, sustains a reversible capacity largely exceeding 525 mAh g^{-1} over more than 125 cycles at a rate of C/2 (55 mA g^{-1}), with a satisfactory coulombic efficiency of >97% (Darwiche *et al.* 2013). The reversible sodiation–desodiation reaction of SnSb was investigated by simultaneous *operando* Sn and Sb K-edge XAS along with *operando* ^{119}Sn Mössbauer spectroscopy. The sodiation reaction is a two-step process clearly distinct from the reaction of SnSb versus Li. First, the sodiation of SnSb starts by a conversion step producing Na_3Sb and an intermediate semiamorphous metallic tin identified as nanosized α-Sn, commonly unstable at ambient conditions. During the second step, nanosized α-Sn is fully sodiated to form $Na_{15}Sn_4$, as rarely observed for pure Sn-based electrodes. Finally, the analysis of the XAS spectra proved that the amorphous SnSb phase formed after one complete cycle is clearly distinct from the pristine material. Nonetheless, in this case, both conversion and alloying reactions are reversible. The key to the enhanced cycle life and capacity retention lies in the gradual formation of amorphous, nanoconfined intermediates and correlated elastic softening of highly sodiated tin and antimony phases, producing a system with enhanced ability to absorb and mitigate the strong volume changes occurring upon sodiation and desodiation (Fehse *et al.* 2018).

When M is electrochemically inactive versus Na, it can however bring a benefit to the system, as in the case of $FeSb_2$. By *operando* Sb K-edge XAS combined with ^{57}Fe synchrotron Mössbauer spectroscopy, the role of antimony and iron in the electrochemical reaction mechanism of $FeSb_2$ versus Na was clarified. The reaction of Na with $FeSb_2$ leads to the formation of Na_3Sb along with superparamagnetic Fe nanoparticles, which contain small amounts of Sb dissolved in their lattice. During the following desodiation, the pristine material $FeSb_2$ is not recovered while iron nanoparticles grow in size, and continue growing also along the following discharge. In this case, the conversion reaction is not reversible and limited to the first sodiation. Even though such iron nanoparticles remain electrochemically inactive, they play a key role in the reduction and stabilization of the limited polarization as well as in the reversibility of the electrochemical sodiation of antimony (Fehse *et al.* 2019). The use of $TiSb_2$ was also reported as anode material for NIB. $TiSb_2$ electrode shows stable cycling with a capacity limited to 225 mAh g^{-1} for 200 cycles, and an excellent rate capability with an energy density of 132 Wh kg^{-1} at 114 W kg^{-1} (Arnaiz *et al.* 2018).

4.3.5. *Other post-transition metal elements*

Lead and bismuth achieve initial sodiation close to the theoretical values. However, after the first cycle, Bi electrodes show a very poor cyclability probably related to its volume expansion of 365%. Surprisingly Pb, despite an elevated volume expansion (490%), shows a correct cyclability for 50 cycles (Figure 4.17).

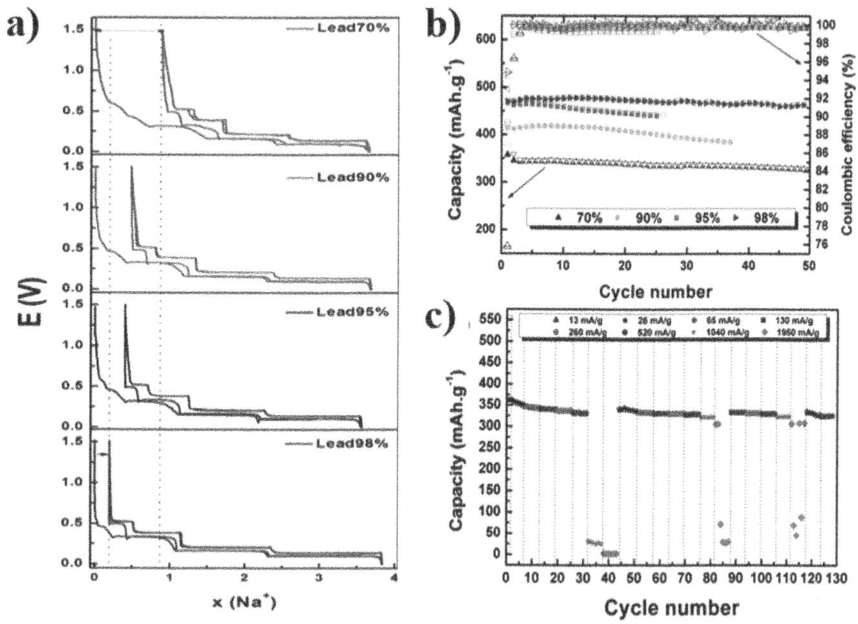

Figure 4.17. *(a) Composition–voltage profile for Pb/Na cells cycled at C/10 (top to bottom: 70%, 90%, 95% and 98% wt. lead electrodes). (b) Specific capacity and coulombic efficiency versus the cycle number for the four electrode formulations tested. (c) Sodiation–desodiation capacity of 70% wt. lead electrode plotted versus cycle number at different current densities (C/10 to 15 C rate) (adapted from Darwiche* et al. *2016). For a color version of this figure, see www.iste.co.uk/monconduit/batteries.zip*

Electrodes fabricated with a high Pb loading (90% wt. of Pb in the electrode) showed a capacity retention of 464 mAh g^{-1} after 50 cycles, representing a high volumetric capacity of 5,289 mAh cm^{-3} due to the high density of Pb (Darwiche *et al.* 2016). These results demonstrated the potential of a Pb anode for NIB in terms of electrochemical performance. Although it is highly recyclable, Pb is toxic and environmentally unfriendly, and probably does not represent a promising negative electrode material for the practical commercialization of NIB in the long run.

4.4. Negative electrode materials based on conversion reactions

Beside intercalation or alloy materials, CM have been considered as potential candidates for all metal-ion batteries. In the case of NIB, CM seem to have more chance since till now carbon based materials are not as efficient as graphite for LIB. CM were introduced first at high temperature by the end of 1970s (iron oxides vs. Li at 420°C) (Thackeray and Coetzer 1981) and later at room temperature based on copper or tin oxides (Ikeda and Narukawa 1983; Novák and Podhájecký 1991; Courtney and Dahn 1997; Idota *et al.* 1997; Huggins 1997). This class of materials has been made famous and named "conversion" materials by Tarascon's group who reported excellent cycle stability for TMO (Poizot *et al.* 2000; Cabana *et al.* 2010; Reddy *et al.* 2013; Zhao *et al.* 2017). After that, CM then promptly extended to other families such as sulfides (Kim *et al.* 2005), phosphates (Son *et al.* 2004), carbonates (Shao *et al.* 2013; Cabana *et al.* 2010) and more recently carbodiimides (Sougrati *et al.* 2016a). In the case of NIB, as early as 1985, Novak showed that sodium can react with cupric oxide and suggested that it forms copper metal and Na_2O (Novák 1985). In 2002, Tirado and co-workers successfully cycled $NiCo_2O_4$ in half-cells and full cells versus Na_xCoO_2 (Alcántara *et al.* 2002). After that, a large number of CM has been reported as illustrated in Figure 4.18. The timeline of the development of CM (Figure 4.18(a)) shows that metal oxides and sulfides are the most studied negative CM electrodes. The fast development of novel CM for NIB has undoubtedly benefited from the solid experience accumulated with LIB. Among the strategies transposed from LIB to NIB, already described in the previous sections on intercalation and alloy materials, the most popular is the use of CM composites with carbon materials such as graphene (Zhou *et al.* 2014a), rGO (Kollu *et al.* 2015) and CNT (Zhang *et al.* 2017a).

As a general rule, and if one excludes sodium metal and phosphorus, CM hold by far the highest theoretical and practical gravimetric/volumetric capacities (see Figures 4.18(b) and (c)) (Wu *et al.* 2018). However, in some cases such as titanates and vanadates, the practical capacities are significantly lower than theoretical ones (Brown *et al.* 2015). In addition, the high theoretical and practical capacities are associated with unsolved obstacles such as the low initial coulombic efficiency (Figure 4.18(d)), poor kinetics and some parasitic reactions generally attributed to SEI formation and electrolyte degradation (Nayak *et al.* 2018; Wang *et al.* 2019b).

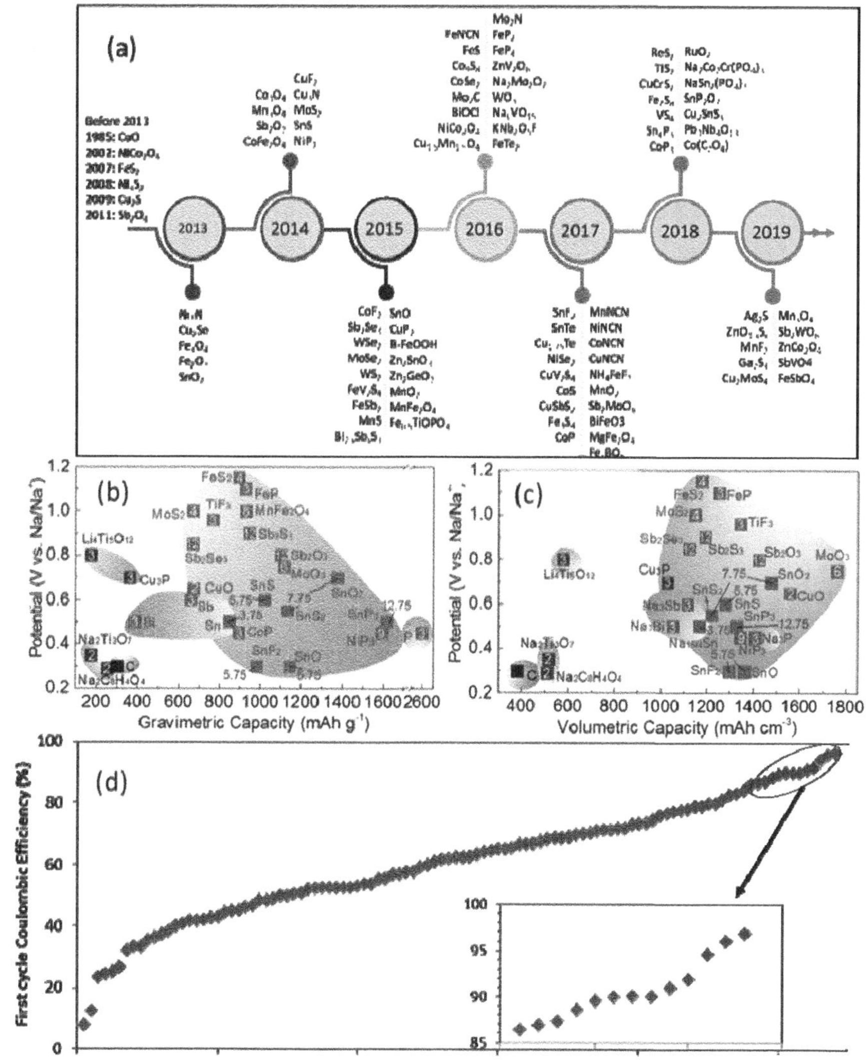

Figure 4.18. *(a) Timeline of conversion materials successfully tested versus Na (excluding composites). (b and c) Average operating potentials and specific gravimetric and volumetric capacity (adapted from Wu* et al. *2018). (d) Initial coulombic efficiency for materials compiled in Table 4.1. For a color version of this figure, see www.iste.co.uk/monconduit/batteries.zip*

Compound	Cap ($mAhg^{-1}$) /Cycle number	Ref.	Compound	Cap ($mAhg^{-1}$) /Cycle number	Ref.	Compound	Cap ($mAhg^{-1}$) /Cycle number	Ref.
Oxides			Sulphides			Tellurides		
CuO	100/40	1	FeS_2	85/50	2	SnTe	300/100	3
Sb_2O_4	720/20	4	Ni_3S_2	350/15	5	$Cu_{1.75}Te$	178/500	6
Fe_3O_4	250/10	7	Cu_2S	261/20	8	$FeTe_2$	300/80	9
Fe_2O_3	800/50	10	CuS		11	Phosphides		
SnO_2	700/100	12	MoS_2	400/100	13	NiP_3	900/15	14
α-MoO_3	140/500	15	MoS_2	600/1000	16	CoP	400/2000	17
Co_3O_4	450/50	18	WS_2	300/1400	19	FeP_2 FeP_4	20/30 1000/30	20
Sb_2O_3	180/30	21	MnS	308/130		CuP_2	450/30	22
Sb_2O_3	900/100	23	$Bi_{2-x}SbxS_3$	380/200	24	Sn_4P_3	540/150	25
FeOOH	300/60	26	FeS	354/500	27	CoP_3	212/80	28
FeOOH	-	26	Co_9S_8	320/30	29	Fluorides		
MnO_2	200/500	30	FeV_2S_4	500/25	31	SnF_2	337/50	32
SnO	530/50	33	SnS	1050/50	34	CoF_2	200/100	35
WO_3	-	36	SnS_2	500/50	34	CuF_2	49/50	37
MnO	191/100	38	SnS	364/100	39	NH_4FeF_3	175/100	40
RuO_2	236/100	41	CuV_2S_4	580/300	42	MnF_2	320/60	43
Mn_3O_4	770/200	44	$CuSbS_2$	200/20	45	Carbodiimides		
Mn_3O_4	100/100	46	Fe_3S_4	275/3500	47	FeNCN	400/200	48
$Cu_{1.5}Mn_{1.5}O_4$	608/100	49	CoS	542/2000	50	CuNCN	268/100	51
$CoFe_2O_4$	850/50	52	CoS	231/100	53	MnNCN	175/100	54
$ZnSnO_3$	320/20	55	Fe_7S_8	380/3500	56	NiNCN	-	54
$NiCo_2O_4$	1346/60	57	VS_4	453/42	58	CoNCN	-	54
$MnFe_2O_4$	280/50	59	$CuCrS_2$	424/200	60	Phosphates		
Na_6VO_{15} NaV_2O_5	1000/1000	61	ReS_2	375/100	62	$Fe_{0.5}TiOPO_4$	100/25	63
ZnV_2O_6	250/120	64	Cu_2SnS_3	178/50	65	$Na_2Co_2Cr(PO_4)_3$	-	63
$Na_2Mo_2O_7$	200/50		TiS_2	740/9000	66	$NaSn_2(PO_4)_3$	400/80	67
Zn_2GeO_4	342/100	68	Ag_2S	189/100	69	Others		
Zn_2SnO_4	306/100	68	Ga_2S_3	476/100	70	Mo_2N	180/200	71
Sb_2MoO_6	372/220	72	Cu_2MoS_4	206/2000	73	Ni_3N	126/30	74
$BiFeO_3$	180/100	75	Selenides			Cu_3N	80/50	37
$MgFe_2O_4$	135/150	76	WSe_2	180/30	77	SnP_2O_7	200/200	78
Fe_3BO_6	444/130	79	$MoSe_2$	350/50	80	$Co(C_2O_4)$	247/200	81
$SbVO_4$	446/1500	82	Cu_2Se	114/100	5	$ZnO_{1-x}S_x$	173/500	83
$FeSbO_4$	180/200	84	Sb_2Se_3	600/100	85	$FeSb_2$	500/140	86
Sb_2WO_6	350/100	87	$CoSe_2$	420/1800	88	$MnSn_2$	400/50	89
$ZnCo_2O_4$	564/200	90	$CoSe_2$	400/800	91	BiOCl	-	92
$Pb_3Nb_4O_{13}$	61/400	93	$NiSe_2$	346/1000	94	KNb_2O_5F	100/140	95

Table 4.2. *List of selected CM investigated in NIB based on their anteriority. C($mAhg^{-1}$)/N is the capacity maintained after N cycles*

Table 4.2 gives a non-exhaustive summary of the investigated CM for NIB classified by the nature of active anion. The largest CM family is that of oxides (Deng *et al.* 2018) followed by sulfides (Liu *et al.* 2016c; Zhang and Huang 2017;

Zhang *et al.* 2017b; Xia *et al.* 2018). These two families have attracted the attention of researchers owing to their intrinsic properties such as chemical stability, the abundance of their constituents and easiness of their synthesis in a large variety of sizes and shapes. Other CM families such as phosphates (Bleith *et al.* 2015; Hadouchi *et al.* 2018), phosphides (Fullenwarth *et al.* 2014; Qian *et al.* 2014; Zhao *et al.* 2015a, 2018; Zhang *et al.* 2016d; Zhu *et al.* 2017), fluorides (Li *et al.* 2014b; Tan *et al.* 2015; Ali *et al.* 2017; Kong *et al.* 2017; Grenier *et al.* 2019), carbodiimides (Sougrati *et al.* 2016b; Eguía-Barrio *et al.* 2017a, 2017b) and nitrides (Li *et al.* 2013, 2014b; Liu *et al.* 2016a) were investigated as well.

While using affordable CM based on metals such as iron, manganese and copper can be justified by their economical or environmental interest, many other less affordable or less abundant metals were also explored (e.g., Ru (Lu *et al.* 2018), Te (Park and Park 2017), Se (Share *et al.* 2015), Mo (Wang *et al.* 2015a), Sb (Sun *et al.* 2011), Sn (Li *et al.* 2015c), Ge (Lim *et al.* 2016) and Bi (Zhang *et al.* 2017b)) for the fundamental interest in understanding their reaction mechanism. Less well-known materials than the CM anode can also show voltages higher enough to make them usable as cathode materials for NIB. Typically, metal chlorides and fluorides operate between 2.5 and 4.4 V (Kim *et al.* 2018). Good performance has been reported for CuCl, $CuCl_2$ (Klein *et al.* 2013), CuF_2, $CuBr_2$ (Kim *et al.* 2016a) and $Cu_3(PO_4)_2$ (Zhao *et al.* 2016).

4.4.1. *Reaction mechanisms of CM*

Detailed reviews on the reaction mechanism were reported recently (Cheng *et al.* 2018; Kim and Kim 2018; Wu *et al.* 2018; Yu *et al.* 2018; Zhang *et al.* 2018a; Wang *et al.* 2019b). In general and by analogy with LIB (Ikeda and Narukawa 1983; Kim *et al.* 2008), it is commonly consented that CM follow a general reaction mechanism in NIB that can be simplified as (Klein *et al.* 2013; Wang *et al.* 2018):

$$M_aX_b + (bc)\cdot Na \rightleftarrows aM + bNa_cX$$

where M = Ti, V, Cr, Mn, Fe, Co, Ni, Cu, Mo, W and Zn; X = Sn, Sb, P, O, N, F, S, P, CO_3, C_2O_4, PO_4, and NCN, and c is the formal number of the maximum achievable negative charge of X (Cabana *et al.* 2010). M can also be a mixture of two metals or even more (Zhao *et al.* 2017). Additionally, M can undergo an alloying reaction once the conversion reaction is completed (Huggins 1997). This implies that the operational potential of the conversion reaction is higher than that of the alloying one. This higher operating voltage can be seen in a dual manner: on the one hand, it limits the risk of sodium plating (Zhang *et al.* 2018a), and, on the other hand, it reduces the overall achievable energy due to the higher operating voltage. During sodiation, the active metal is reduced by gaining *n* electrons while going from M^{n+} to the metallic state. Some exceptions to this general mechanism have been reported such as NH_4FeF_3, which follows a slightly different reaction path forming NH_4F with metallic iron and lithium fluoride (Kong *et al.* 2017) or β–FeOOH, which reacts as a cathode material with sodium to form FeO_2 + NaOH at a voltage higher than 1 V (Zhang *et al.* 2018b). At lower potentials, β–FeOOH undergoes a classical conversion reaction leading to metallic iron and Na_2O (Yu *et al.* 2015). It is worth noting that some electrode materials can combine up to three different reaction mechanisms depending on the potential range. SnTe is a typical example starting by the intercalation of sodium to form $Na_{\sim 1.5}SnTe$ that undergoes a conversion reaction to form Na_2Te and Sn, which in its turns undergoes the alloying processes (Figure 4.19(a)).

The reversibility of the conversion mechanism must be considered only after the formation step (first sodiation), which is often irreversible. In some cases, the conversion reaction can be limited to the formation step with all the subsequent cycling exclusively based on alloying reaction as illustrated in Figure 4.19(b) for $ZnSnO_3$. Even in favorable cases where the conversion reaction is totally reversible, the voltage profile of the formation step is drastically different from those of the succeeding sodiations as illustrated in the case of the model material RuO_2 (Lu *et al.* 2018). This characteristic difference arises from the fundamentally differences in terms of particle size, morphology and SEI layer between the fresh CM anode and that formed after a sodiation/desodiation process leading to nanostructured, SEI-"coated" material. This transformation is associated with sodium trapping in the SEI, electrolyte degradation, partial disconnection of active material particles due to large volume changes, implying generally high first cycle irreversible capacity (a low initial coulombic efficiency) (Figure 4.19(c)). The first cycle irreversible capacity is even exacerbated for nanostructured electrodes, in which higher surface areas lead to boosted electrolyte decomposition (Zhang *et al.* 2018a). In case of metal sulfides, the irreversible capacity was found to be caused by cracks and pulverization of the active material upon the first cycle and inherent slow sodium ion diffusion kinetics and mobility at the electrode interface (Figure 4.19(c); Wang *et al.* 2018).

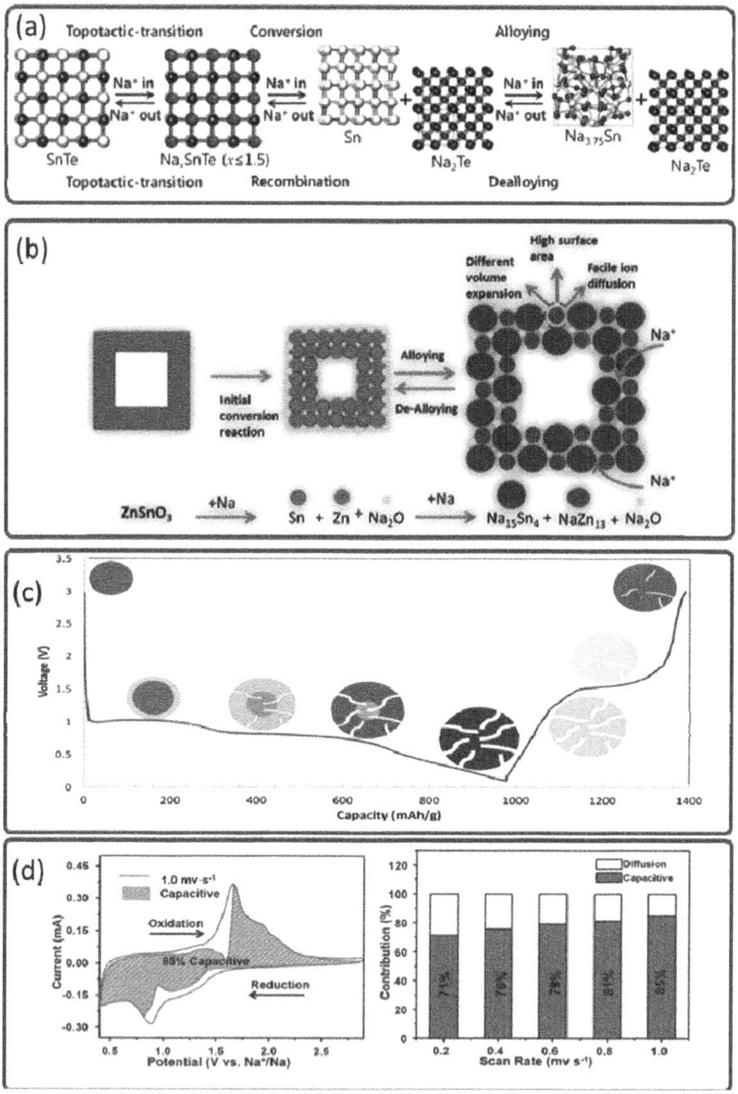

Figure 4.19. *(a) SnTe, a typical multistep reaction mechanism including conversion (Park and Park 2017). (b) Illustration of the irreversibility of the formation step in the case of $ZnSnO_3$ with a totally irreversible initial sodiation and reversible alloying processes (Wang* et al. *2015b). (c) Schematic representation of the active particle's cracking in the case of nickel sulfide anode for NIB (Wang* et al. *2018). (d) Cyclic voltammogram of NiS_2 nanospheres with the capacitive contribution to the total current shown by the shaded region, and its evolution at different scan rates (Sun* et al. *2017). For a color version of this figure, see www.iste.co.uk/monconduit/batteries.zip*

The pure conversion mechanism is sometimes accompanied with significant pseudo-capacitive contribution. In the case of mesoporous NiS_2 nanospheres (Sun *et al.* 2017), such capacitive contribution was found to represent 70% of the total capacity or even higher at low scan rates (Figure 4.18(d)).

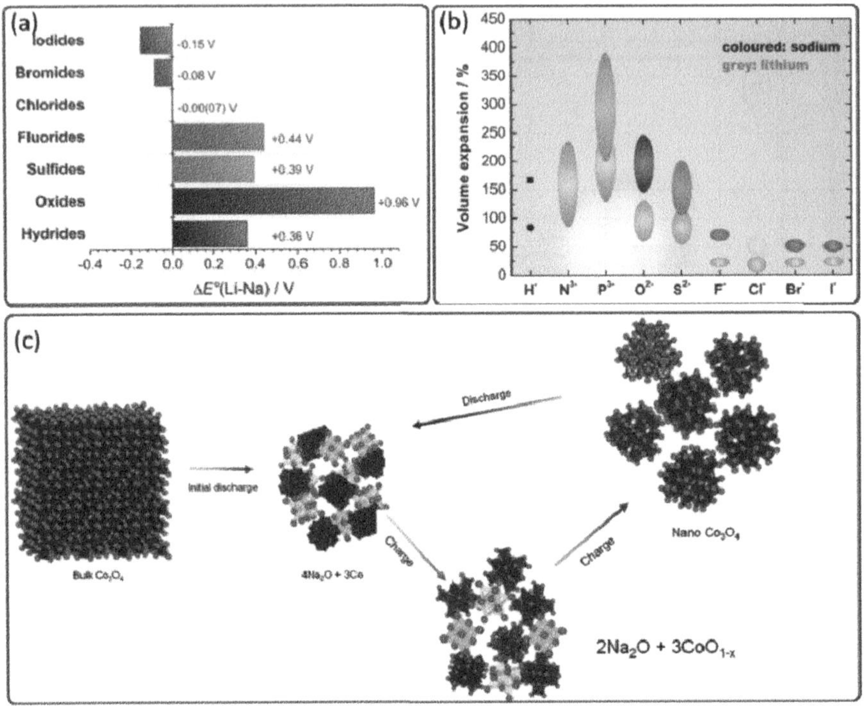

Figure 4.20. *Calculated differences in (a) cell potential and (b) volume change for conversion reactions with Na and Li. Schematic of conversion and reconversion reaction mechanisms for Co_3O_4 in NIB (adapted from Klein* et al. *2013). For a color version of this figure, see www.iste.co.uk/monconduit/batteries.zip*

Compared to LIB, CM for NIB show some differences from the thermodynamic point of view. Theoretical calculation of cell voltage for CM in the case of oxides, sulfides, hydrides and fluorides showed that the cell voltage expected versus Na is generally lower than that obtained versus Li (Figure 4.20(a)) (Klein *et al.* 2013). In the same study, the authors revealed that volume expansion is expected to be higher in the case of sodium CM as illustrated in Figure 4.20(b). These DFT predictions need however to be conciliated with some experimental observations such as the *in situ* study showing that the fracture of particles are less probable in the case of CM for NIB than for LIB (Boebinger *et al.* 2018). The later study revealed, using *in situ* TEM, that different nanoscale reaction pathways are followed during lithiation and sodiation of FeS_2. Moreover, chemomechanical modeling of stress evolution and mechanical testing shows that, despite significantly larger volume changes during sodiation, particles were observed to fracture only during lithiation but not during sodiation. These counterintuitive results suggest that larger volume changes in battery materials do not necessarily lead to larger stress and more significant mechanical degradation. Hence, other structural consideration such as reaction fronts must be taken into consideration in mechanical failures. Therefore, even though high-capacity CM undergo larger volume variation in NIB compared with LIB, these volume changes may be manageable and do not unavoidably lead to particle failure (Boebinger *et al.* 2018). Another aspect of CM anodes is the high-voltage hysteresis (polarization). The origin of this drawback was explained in the case of cobalt oxide mainly by the difference in reaction pathways between sodiation and desodiation with some additional intermediate steps observed in the desodiation process (Figure 4.20(c)) (Kim *et al.* 2016b). Finally, since CM operate at low voltage, the sodiation is generally accompanied by the decomposition of electrolyte components. These side reactions may provide additional partially reversible capacity (Cabana *et al.* 2010).

4.4.2. *Approaches toward efficient anode CM for NIB*

After intense exploration of the negative electrode CM, it comes out that they have a real chance to enter the application step only if their main drawbacks (high volume expansion, low initial coulombic efficiency, low conductivities and high polarization) are overcome. To this end, researchers developed a large arsenal of strategies, and most efforts are in particular dedicated to mitigate the impact of volume expansion.

Among the selected strategies, the most explored one is CM carbon composites. The idea of synergetic effects between the high capacity of CM and the good mechanical and electrical properties of carbon has oriented many studies and led to a considerable number of composites with all the CM families shown in Table 4.2. Significant examples were recently reported for metal oxide/carbon composites (Jian *et al.* 2014a, 2014b; Zhou *et al.* 2014b, 2015b; Zhang *et al.* 2015a; Dou *et al.* 2017). For example, γ-Fe_2O_3@C composites with innerconnected nanochannels embedded in a porous carbon matrix (Figures 4.21(a–d)) were found to deliver high rate capability and long-term cyclability (Figures 4.21(e) and (f)). The comparison between 20 nm and 5 nm γ-Fe_2O_3@C composites clearly evidences that the particle size affects only slightly the rate capability. The excellent performance was attributed to the synergy between carbon and γ-Fe_2O_3 that reduces mechanical stress, helps accommodating volume expansion, prevents nanoparticle aggregation and facilitates charge transfer (Zhang *et al.* 2015a).

Nanoarchitecture design including nanosheet assembly and nanoarray construction is also an effective strategy to improve the electrochemical performance of CM anodes for NIB (Zhang *et al.* 2017b). In this case, the key is to provide sufficient active sites, open space, short ion-diffusion length and strong connections between the electrode components. One common way for doing so is to use 1D (Jin *et al.* 2018) and 2D materials (Peng *et al.* 2018). Here again, metal oxides (Mei *et al.* 2017) and chalcogenides (Xie *et al.* 2017) are the most commonly studies 1D/2D materials. Inspired by graphene and hybrid 2D chemistries, both top-down and bottom-up approaches are used for the synthesis of 2D metal oxides with the promise of high yield and cost-effective production. A typical example of improved performance for 2D oxides is that of Co_3O_4 nanosheets obtained via a surfactant-assisted self-assembly approach. The nanosheets grown on stainless steel meshes deliver high capacity with excellent stability and outperform conventionally prepared electrodes (Figures 4.21(g–j)) (Dou *et al.* 2017). In such 2D structures, the contribution of capacitive storage (Figure 4.21(i)) is dominant, thus also allowing appealing rate capability performance.

The impact of morphology and porosity of the CM particles on the performance as an NIB anode has also attracted great attention (Ryu *et al.* 2016; Yang *et al.* 2016; Dong *et al.* 2017; Liu *et al.* 2018b). Yang *et al.* (2016) used mesoporous silica as a template (Figures 4.22(a) and (b)) to prepare cobalt oxide with spherical mesopores connected by nanoporous channels. The cycle life is greatly improved compared to bulk cobalt oxide electrodes (Figure 4.22(d)). However, the low initial coulombic efficiency problem is not solved since irreversible capacity is still high because of electrolyte degradation at the particles surface. It is worth noting the positive impact of electrolyte additives (FEC) on cycle life. Similar performance was reported for stannous oxide electrodes (Shimizu *et al.* 2014).

Finally, electrode formulation was found to play a significant role in improving the electrochemical performance of CM-based electrodes for NIB. In a systematic work, Ma *et al.* examined six different binders in the formulation of SnS_2 electrodes keeping all the other cycling conditions identical (Ma *et al.* 2018). The studied binders are sodium carboxymethyl cellulose (CMC-Na), sodium polyacrylate (PAA-Na), CMC-Na/PAA-Na (1:1, wt%, denoted as PAA-CMC), sodium alginate (ALG-Na), PVDF and PTFE. Figure 4.23(a) compares the obtained capacities and their cycle stability. The differences in performance can vary by more than 50%. Among these binders, PAA-CMC does not deliver the highest capacity during the early cycles, but shows the most stable cycle life and rate capability (Figure 4.23(b)).

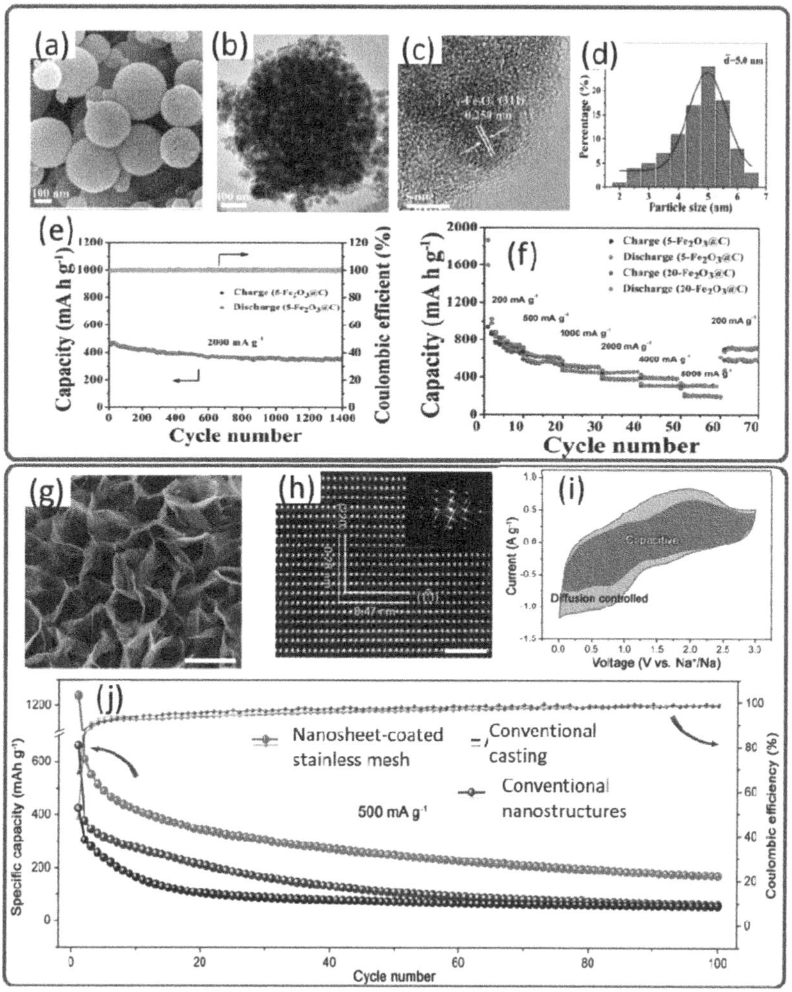

Figure 4.21. *(a) SEM and (b and c) TEM images of 5 nm Fe_2O_3. (d) Particle size distribution. (e) Long-term cycling stability of γ-Fe_2O_3@C. (f) Rate performance of γ-Fe_2O_3@C and α-Fe_2O_3@C at different current densities (adapted from Zhang* et al. *2015a). (g) SEM image showing the coated ultrathin Co_3O_4 nanosheets. (h) HRTEM image of Co_3O_4 nanosheets and its corresponding FFT pattern (inset). (i) Separation of the capacitive current from the diffusion-controlled current in a Co_3O_4 nanosheets-coated stainless steel mesh at a scan rate of 2 mVs^{-1}. (j) Cycle performance of Co_3O_4 nanosheet-coated stainless mesh prepared with conventional casting and as a conventional nanostructure (adapted from Dou* et al. *2017). For a color version of this figure, see www.iste.co.uk/monconduit/batteries.zip*

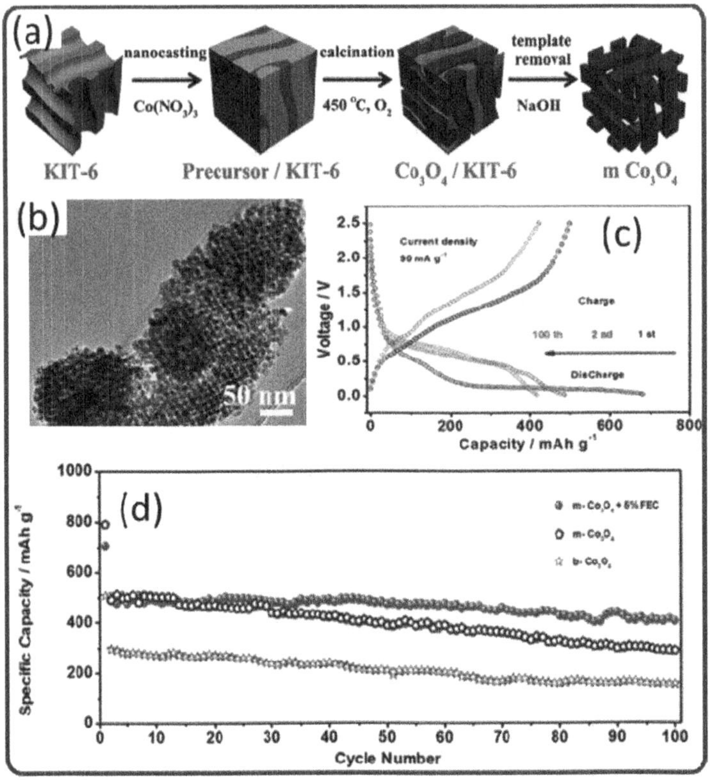

Figure 4.22. *(a) Preparation and (b) TEM image of mesoporous Co_3O_4 preparation via KIT-6 templating method. (c) Galvanostatic discharge/charge profiles of the mesoporous Co_3O_4 electrode at a current density of 90 mA g^{-1}. (d) Capacity retention of electrodes containing bulk or mesoporous Co_3O_4 with or without FEC in the electrolyte (adapted from Yang* et al. *2016). For a color version of this figure, see www.iste.co.uk/monconduit/batteries.zip*

This stability seems to correlate with the fraction of carboxylate and hydroxyl groups, which leads to stronger hydrogen and/or covalent bonds between the carbon black additive and the active materials. The electrode morphology made with different binders after 100 cycles is shown in Figures 4.23(c–h). Electrodes based on PAA-CMC, CMC-Na and PAA-Na behave significantly better than those based on ALG-Na, PVDF and PTFE. In addition, the ALG-Na electrode is severely cracked and this might be the reason for its poor performance, whereas electrodes based on PVDF and PTFE seem to suffer from particle agglomeration. Unfortunately, for all these binders, the initial coulombic efficiency is still considerably low.

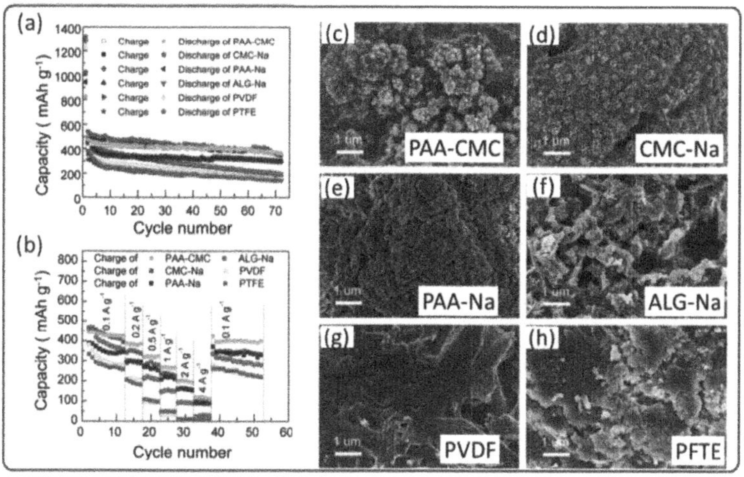

Figure 4.23. *(a) Cycling stability of SnS_2 for six different electrode formulations. (b) Their rate capabilities and (c-d-e-f-g-h) the surface morphologies after 100 cycles (adapted from Ma* et al. *2018). For a color version of this figure, see www.iste.co.uk/monconduit/batteries.zip*

4.5. Conclusion

As for LIB, insertion material have advantages for NIB in terms of long cycle performance, however for limited capacity and lower voltage in full cells than in LIB.

The screening of the reaction toward Na of the main posttransition metal elements showed that most of them can form Na-rich phases, whereas Si does not belong to this category due to a diffusion barrier that is too high for Na.

It is interesting to see that the sodiation/desodiation mechanisms associated with the good performance of posttransition metal elements are far from being directly deduced from those observed in LIB and sometimes hold good surprise. Indeed, the higher expected volume expansion during the alloying of main group elements (X = P, Sb, Ge, etc.) with Na does not hamper the cycling performance because of the higher flexibility of Na-X than Li-X bonds. Moreover, it was shown that, unlikely LIB, intermediate amorphous phases are formed during the sodiation/desodiation of X, which are able to better buffer the volume changes.

However, as described above, the main challenge for alloy and CM is still the huge volume expansion, which results in continuous pulverization of these electrode materials and then a steep deterioration of electrochemical performance. The consequences of the huge volume expansion are worsen by a continuous electrolyte reduction, which affects both coulombic efficiency and cycle life. The strategies employed to improve the electrochemical performance are (1) nanosizing of the active materials particles to shorten the Na diffusion and buffer the volume changes and (2) association with robust templates, usually carbon based, to suppress active material aggregation and improve conductivity.

These strategies have led to great progress in the electrochemical performance in terms of cycle life in the recent years, even though unavoidably lower volumetric energy densities are to deplore. This is, nevertheless, not a true issue since the main application foreseen for NIB is stationary energy storage. It has pushed the researchers to go one step forward toward commercial application through the use of CM and alloy negative electrodes in full cells versus well-established cathode materials. Full cell combinations have been reported (for CM, see Chen *et al.* 2018a and references therein). However, the reported results show that, up to now, CM as well as alloy anodes do not surpass the carbon-based anodes. This is partially due to the large amount of cathode excess required to compensate for the large initial irreversible capacity, and the low coulombic efficiency in cycling. Therefore, one can conclude that the primary challenge for conversion or alloy materials is still to be solved (Chen *et al.* 2018a). Hard carbon (Chapter 3) remains the most relevant candidate, however its cost is much higher than graphite. Among alloy materials, antimony is likely the most promising one, possessing a large gravimetric charge storage capacity (660 mAh g^{-1}) combined with good rate capability and cycle life. Such excellent performance, higher than in LIB, should give it a good opportunity when used in composites with HC. Indeed, a good compromise between energy and life span-cost can in principle be attained.

4.6. References

Abel, P.R., Fields, M.G., Heller, A., and Mullins, C.B. (2014). Tin-germanium alloys as anode materials for sodium-ion batteries. *ACS Appl. Mater. Interfaces*, 6(18), 15860–15867.

Alcántara, R., Jaraba, M., Lavela, P., and Tirado, J.L. (2002). $NiCo_2O_4$ spinel: First report on a transition metal oxide for the negative electrode of sodium-ion batteries. *Chem. Mater.*, 14(7), 2847–2848.

Ali, G., Lee, J.-H., Oh, S.H., Jung, H.-G., and Chung, K.Y. (2017). Elucidating the reaction mechanism of SnF_2@C nanocomposite as a high-capacity anode material for Na-ion batteries. *Nano Energy*, 42, 106–114.

Allan, P.K., Griffin, J.M., Darwiche, A., Borkiewicz, O.J., Wiaderek, K.M., Chapman, K.W., Morris, A.J., Chupas, P.J., Monconduit, L., and Grey, C.P. (2016). Tracking sodium-antimonide phase transformations in sodium-ion anodes: Insights from operando pair distribution function analysis and solid-state NMR spectroscopy. *J. Am. Chem. Soc.*, 138(7), 2352–2365.

Aravindan, V., Lee, Y.-S., Yazami, R., and Madhavi, S. (2015). TiO_2 polymorphs in 'rocking-chair' Li-ion batteries. *Mater. Today*, 18(6), 345–351.

Arnaiz, M., Gómez-Cámer, J.L., Ajuria, J., Bonilla, F., Acebedo, B., Jáuregui, M., Goikolea, E., Galceran, M., and Rojo, T. (2018). High performance titanium antimonide $TiSb_2$ alloy for Na-ion batteries and capacitors. *Chem. Mater.*, 30(22), 8155–8163.

Baggetto, L., Keum, J.K., Browning, J.F., and Veith, G.M. (2013a). Germanium as negative electrode material for sodium-ion batteries. *Electrochem. Commun.*, 34, 41–44.

Baggetto, L., Marszewski, M., Górka, J., Jaroniec, M., and Veith, G.M. (2013b). AlSb thin films as negative electrodes for Li-ion and Na-ion batteries. *J. Power Sources*, 243, 699–705.

Baggetto, L., Hah, H.Y., Jumas, J.-C., Johnson, C.E., Johnson, J.A., Keum, J.K., Bridges, C.A., and Veith, G.M. (2014). The reaction mechanism of SnSb and Sb thin film anodes for Na-ion batteries studied by X-ray diffraction, 119Sn and 121Sb Mössbauer spectroscopies. *J. Power Sources*, 267, 329–336.

Bleith, P., Kaiser, H., Novák, P., and Villevieille, C. (2015). In situ X-ray diffraction characterisation of $Fe_{0.5}TiOPO_4$ and $Cu_{0.5}TiOPO_4$ as electrode material for sodium-ion batteries. *Electrochim. Acta*, 176, 18–21.

Boebinger, M.G., Yeh, D., Xu, M., Miles, B.C., Wang, B., Papakyriakou, M., Lewis, J.A., Kondekar, N.P., Cortes, F.J.Q., Hwang, S., Sang, X., Su, D., Unocic, R.R., Xia, S., Zhu, T., and McDowell, M.T. (2018). Avoiding fracture in a conversion battery material through reaction with larger ions. *Joule*, 2(9), 1783–1799.

Brown, Z.L., Smith, S., and Obrovac, M.N. (2015). Mixed transition metal titanate and vanadate negative electrode materials for Na-ion batteries. *Journal of The Electrochemical Society*, 162(1), A15–A20.

Bu, F., Zagho, M.M., Ibrahim, Y., Ma, B., Elzatahry, A., and Zhao, D. (2020). Porous MXenes: Synthesis, structures, and applications. *Nano Today*, 30, 100803.

Cabana, J., Monconduit, L., Larcher, D., and Palacín, M.R. (2010). Beyond intercalation-based Li-ion batteries: The state of the art and challenges of electrode materials reacting through conversion reactions. *Adv. Mater.*, 22, E170–E192.

Chen, C., Wen, Y., Hu, X., Ji, X., Yan, M., Mai, L., Hu, P., Shan, B., and Huang, Y. (2015). Na+ intercalation pseudocapacitance in graphene-coupled titanium oxide enabling ultra-fast sodium storage and long-term cycling. *Nature Commun.*, 6(1), 6929.

Chen, J., Mohrhusen, L., Ali, G., Li, S., Chung, K.Y., Al-Shamery, K., and Lee, P.S. (2019). Electrochemical mechanism investigation of Cu_2MoS_4 hollow nanospheres for fast and stable sodium ion storage. *Advanced Functional Materials*, 29(7), 1807753.

Chen, J., Zhou, X., Mei, C., Xu, J., and Wong, C.-P. (2017). Improving the sodiation performance of $Na_2Ti_3O_7$ through Nb-doping. *Electrochim. Acta*, 224, 446–451.

Chen, L., Fiore, M., Wang, J.E., Ruffo, R., Kim, D.-K., and Longoni, G. (2018a). Readiness level of sodium-ion battery technology: A materials review. *Adv. Sus. Syst.*, 2(3), 1700153.

Chen, Z., Cheng, X., Yu, H., Zhu, H., Zheng, R., Liu, T., Zhang, J., Shui, M., and Shu, J. (2018b). Lithium, sodium and potassium storage behaviors of $Pb_3Nb_4O_{13}$ nanowires for rechargeable batteries. *Ceramics International*, 44(14), 17094–17101.

Cheng, D.-L., Yang, L.-C., and Zhu, M. (2018). High-performance anode materials for Na-ion batteries. *Rare Metals*, 37(3), 167–180.

Chevrier, V.L. and Ceder, G. (2011). Challenges for Na-ion negative electrodes. *Journal of the Electrochemical Society*, 158(9), A1011.

Cho, J.S., Lee, S.Y., Lee, J.-K., and Kang, Y.C. (2016). Iron telluride-decorated reduced graphene oxide hybrid microspheres as anode materials with improved Na-ion storage properties. *ACS Applied Materials & Interfaces*, 8(33), 21343–21349.

Chou, C.-Y., Lee, M., and Hwang, G.S. (2015). A comparative first-principles study on sodiation of silicon, germanium, and tin for sodium-ion batteries. *J. Phys. Chem. C*, 119(27), 14843–14850.

Courtney, I.A. and Dahn, J.R. (1997). Electrochemical and in situ X-ray diffraction studies of the reaction of lithium with tin oxide composites. *J. Electrochem. Soc.*, 144(6), 2045–2052.

Cui, J., Yao, S., and Kim, J.-K. (2017). Recent progress in rational design of anode materials for high-performance Na-ion batteries. *Energy Storage Mater.*, 7, 64–114.

Dahbi, M., Yabuuchi, N., Kubota, K., Tokiwa, K., and Komaba, S. (2014). Negative electrodes for Na-ion batteries. *Phys. Chem. Chem. Phys.*, 16(29), 15007–15028.

Darwiche, A., Marino, C., Sougrati, M.T., Fraisse, B., Stievano, L., and Monconduit, L. (2012). Better cycling performances of bulk Sb in Na-ion batteries compared to Li-ion systems: An unexpected electrochemical mechanism. *J. Am. Chem. Soc.*, 134(51), 20805–20811.

Darwiche, A., Sougrati, M.T., Fraisse, B., Stievano, L., and Monconduit, L. (2013). Facile synthesis and long cycle life of SnSb as negative electrode material for Na-ion batteries. *Electrochem. Commun.*, 32, 18–21.

Darwiche, A., Toiron, M., Sougrati, M.T., Fraisse, B., Stievano, L., and Monconduit, L. (2015). Performance and mechanism of $FeSb_2$ as negative electrode for Na-ion batteries. *J. Power Sources*, 280, 588–592.

Darwiche, A., Dugas, R., Fraisse, B., and Monconduit, L. (2016). Reinstating lead for high-loaded efficient negative electrode for rechargeable sodium-ion battery. *J. Power Sources*, 304, 1–8.

Darwiche, A., Fehse, M., Mahmoud, A., Fontaine, C. La, Fraisse, B., Hermann, R., Doublet, M.-L., Monconduit, L., Sougrati, M.T., Ben Yahia, M., and Stievano, L. (2018). The electrochemical sodiation of Sb investigated by operando X-ray absorption and 121Sb Mössbauer spectroscopy: What does one really learn? *Batteries*, 4(2), 25.

Dawson, J.A. and Robertson, J. (2016). Improved calculation of Li and Na intercalation properties in anatase, rutile, and TiO_2(B). *J. Phys. Chem.*, 120(40), 22910–22917.

Deng, X., Chen, Z., and Cao, Y. (2018). Transition metal oxides based on conversion reaction for sodium-ion battery anodes. *Materials Today Chemistry*, 9, 114–132.

Dong, S., Shen, L., Li, H., Nie, P., Zhu, Y., Sheng, Q., and Zhang, X. (2015). Pseudocapacitive behaviours of $Na_2Ti_3O_7$@CNT coaxial nanocables for high-performance sodium-ion capacitors. *J. Mater. Chem. A*, 3(42), 21277–21283.

Dong, S., Li, C., Ge, X., Li, Z., Miao, X., and Yin, L. (2017). ZnS-Sb_2S_3@C core-double shell polyhedron structure derived from metal–organic framework as anodes for high performance sodium ion batteries. *ACS Nano*, 11(6), 6474–6482.

Dou, Y., Wang, Y., Tian, D., Xu, J., Zhang, Z., Liu, Q., Ruan, B., Ma, J., Sun, Z., and Dou, S.X. (2017). Atomically thin Co_3O_4 nanosheet-coated stainless steel mesh with enhanced capacitive Na+ storage for high-performance sodium-ion batteries. *2D Materials*, 4(1), 015022.

Eames, C. and Islam, M.S. (2014). Ion intercalation into two-dimensional transition-metal carbides: Global screening for new high-capacity battery materials. *J. Am. Chem. Soc.*, 136(46), 16270–16276.

Edison, E., Gogoi, P.K., Zheng, Y., Sreejith, S., Pennycook, S.J., Lim, C.T., and Srinivasan, M. (2019). Electrochemically induced amorphization and unique lithium and sodium storage pathways in $FeSbO_4$ nanocrystals. *ACS Appl. Mater. Interfaces*, 11(22), 20082–20090.

Eguía-Barrio, A., Castillo-Martinez, E., Liu, X., Dronskowski, R., Lezama, L., Armand, M., and Rojo, T. (2017a). Advanced anode materials for sodium ion batteries: Carbodiimides. *MRS Adv.*, 2(21–22), 1165–1176.

Eguía-Barrio, A., Castillo-Martínez, E., Klein, F., Pinedo, R., Lezama, L., Janek, J., Adelhelm, P., and Rojo, T. (2017b). Electrochemical performance of CuNCN for sodium ion batteries and comparison with ZnNCN and lithium ion batteries. *J. Power Sources*, 367, 130–137.

Ellis, L.D., Hatchard, T.D., and Obrovac, M.N. (2012). Reversible insertion of sodium in tin. *J. Electrochem. Soc.*, 159(11), A1801–A1805.

Fang, S., Bresser, D., and Passerini, S. (2020). Transition metal oxide anodes for electrochemical energy storage in lithium- and sodium-ion batteries. *Adv. Energy Mater.*, 10(1), 1902485.

Fehse, M., Sougrati, M.T., Darwiche, A., Gabaudan, V., La Fontaine, C., Monconduit, L., and Stievano, L. (2018). Elucidating the origin of superior electrochemical cycling performance: New insights on sodiation–desodiation mechanism of SnSb from operando spectroscopy. *J. Mater. Chem. A*, 6(18), 8724–8734.

Fehse, M., Monconduit, L., Fischer, F., Tessier, C., and Stievano, L. (2014). Study of the insertion mechanism of lithium into anatase by operando X-ray diffraction and absorption spectroscopy. *Solid State Ionics*, 268, 252–255.

Fehse, M., Bessas, D., Darwiche, A., Mahmoud, A., Rahamim, G., La Fontaine, C., Hermann, R.P., Zitoun, D., Monconduit, L., Stievano, L., and Sougrati, M.T. (2019). The electrochemical sodiation of $FeSb_2$: New insights from operando 57 Fe synchrotron Mössbauer and X-ray absorption spectroscopy. *Batteries & Supercaps*, 2(1), 66–73.

Fei, J., Cui, Y., Li, J., Xu, Z., Yang, J., Wang, R., Cheng, Y., and Hang, J. (2017). A flexible Sb_2O_3/carbon cloth composite as a free-standing high performance anode for sodium ion batteries. *Chemical Communications*, 53(98), 13165–13167.

Fukunishi, M., Yabuuchi, N., Dahbi, M., Son, J.-Y., Cui, Y., Oji, H., and Komaba, S. (2016). Impact of the cut-off voltage on cyclability and passive interphase of Sn-polyacrylate composite electrodes for sodium-ion batteries. *J. Phys. Chem. C*, 120(28), 15017–15026.

Fullenwarth, J., Darwiche, A., Soares, A., Donnadieu, B., and Monconduit, L. (2014). NiP 3: A promising negative electrode for Li- and Na-ion batteries. *J. Mater. Chem. A*, 2(7), 2050–2059.

Gao, P., Wang, L., Zhang, Y., Huang, Y., and Liu, K. (2015). Atomic-scale probing of the dynamics of sodium transport and intercalation-induced phase transformations in MoS_2. *ACS Nano*, 9(11), 11296–11301.

González, J.R., Alcántara, R., Tirado, J.L., Fielding, A.J., and Dryfe, R.A.W. (2017). Electrochemical interaction of few-layer molybdenum disulfide composites vs sodium: New insights on the reaction mechanism. *Chem. Mater.*, 29(14), 5886–5895.

Grenier, A., Porras-Gutierrez, A.-G., Desrues, A., Leclerc, S., Borkiewicz, O.J., Groult, H., and Dambournet, D. (2019). Synthesis and optimized formulation for high-capacity manganese fluoride (MnF_2) electrodes for lithium-ion batteries. *Journal of Fluorine Chemistry*, 224, 45–51.

Guo, Y., Zhu, Y., Yuan, C., and Wang, C. (2017). $MgFe_2O_4$ hollow microboxes derived from metal-organic-frameworks as anode material for sodium-ion batteries. *Materials Letters*, 199, 101–104.

Hadouchi, M., Assani, A., Saadi, M., Saadoune, I., Lahmar, A., Bouyanfif, H., El Marssi, M., and El Ammari, L. (2018). Synthesis, crystal structure and properties of a new phosphate, $Na_2Co_2Cr(PO_4)_3$. *Journal of Inorganic and Organometallic Polymers and Materials*, 28(6), 2854–2864.

Han, Y., Yang, M., Zhang, Y., Xie, J., Yin, D., and Li, C. (2016). Tetragonal tungsten bronze framework as potential anode for Na-ion batteries. *Chem. Mater.*, 28(9), 3139–3147.

Hao, W., Si, H., Li, W., Zhang, C., Zhu, W., and Qiu, X. (2019). Sodium storage performance and mechanism of Ag_2S nanospheres as electrode material for sodium-ion batteries. *Solid State Ionics*, 343, 115071.

Hariharan, S., Kuppan, S., Ramar, V., and Balaya, P. (2013). A rationally designed dual role anode material for lithium-ion and sodium-ion batteries: Case study of eco-friendly Fe_3O_4. *Phys. Chem. Chem. Phys.*, 15, 2945–2953

Hariharan, S., Saravanan, K., and Balaya, P. (2013). α-MoO_3: A high performance anode material for sodium-ion batteries. *Electrochem. Commun.*, 31, 5–9.

Huang, J.P., Yuan, D.D., Zhang, H.Z., Cao, Y.L., Li, G.R., Yang, H.X., and Gao, X.P. (2013). Electrochemical sodium storage of TiO_2(B) nanotubes for sodium ion batteries. *RSC Adv.*, 3(31), 12593.

Huggins, R.A. (1997). Alloy negative electrodes for lithium batteries formed in-situ from oxides. *Ionics*, 3(3–4), 245–255.

Idota, Y., Kubota, T., Matsufuji, A., Maekawa, Y., and Miyasaka, T. (1997). Tin-based amorphous oxide: A high-capacity lithium-ion-storage material. *Science*, 276(5317), 1395–1397.

Ikeda, H. and Narukawa, S. (1983). Behaviour of various cathode materials for non-aqueous lithium cells. *Journal of Power Sources*, 9(3), 329–334.

Irisarri, E., Amini, N., Tennison, S., Ghimbeu, C.M., Gorka, J., Vix-Guterl, C., Ponrouch, A., and Palacin, M.R. (2018). Optimization of large scale produced hard carbon performance in Na-ion batteries: Effect of precursor, temperature and processing conditions. *J. Electrochem. Soc.*, 165(16), A4058–A4066.

Irisarri, E., Ponrouch, A., and Palacin, M.R. (2015). Review—Hard carbon negative electrode materials for sodium-ion batteries. *J. Electrochem. Soc.*, 162(14), A2476–A2482.

Jian, Z., Liu, P., Li, F., Chen, M., and Zhou, H. (2014a). Monodispersed hierarchical Co_3O_4 spheres intertwined with carbon nanotubes for use as anode materials in sodium-ion batteries. *Journal of Materials Chemistry A*, 2(34), 13805–13809.

Jian, Z., Zhao, B., Liu, P., Li, F., Zheng, M., Chen, M., Shi, Y., and Zhou, H. (2014b). Fe_2O_3 nanocrystals anchored onto graphene nanosheets as the anode material for low-cost sodium-ion batteries. *Chem. Commun.*, 50(10), 1215–1217.

Jiang, Y., Hu, M., Zhang, D., Yuan, T., Sun, W., Xu, B., and Yan, M. (2014). Transition metal oxides for high performance sodium ion battery anodes. *Nano Energy*, 5, 60–66.

Jin, T., Han, Q., Wang, Y., and Jiao, L. (2018). 1D nanomaterials: Design, synthesis, and applications in sodium-ion batteries. *Small*, 14(2), 1703086.

Jo, C.H., Yashiro, H., Yuan, S., Shi, L., and Myung, S.T. (2018). Conversion chemistry of cobalt oxalate for sodium storage. *ACS Appl. Mater. Interfaces*, 10(47), 40523–40530.

Jung, S.C., Jung, D.S., Choi, J.W., and Han, Y.-K. (2014). Atom-level understanding of the sodiation process in silicon anode material. *J. Phys. Chem. Lett.*, 5(7), 1283–1288.

Kajiyama, S., Szabova, L., Sodeyama, K., Iinuma, H., Morita, R., Gotoh, K., Tateyama, Y., Okubo, M., and Yamada, A. (2016). Sodium-ion intercalation mechanism in MXene nanosheets. *ACS Nano*, 10(3), 3334–3341.

Kim, B.-C., Takada, K., Ohta, N., Seino, Y., Zhang, L., Wada, H., and Sasaki, T. (2005). All solid state Li-ion secondary battery with FeS anode. *Solid State Ionics*, 176(31–34), 2383–2387.

Kim, B.-R., Jeong, G., Kim, A., Kim, Y., Kim, M.G., Kim, H., and Kim, Y.-J. (2016a). High performance Na-$CuCl_2$ rechargeable battery toward room temperature ZEBRA-type battery. *Advanced Energy Materials*, 6(20), 1600862.

Kim, H., Kim, Hyunchul, Kim, Hyungsub, Kim, J., Yoon, G., Lim, K., Yoon, W.S., and Kang, K. (2016b). Understanding origin of voltage hysteresis in conversion reaction for Na rechargeable batteries: The case of cobalt oxides. *Advanced Functional Materials*, 26(28), 5042–5050.

Kim, J.-H. and Kim, D.K. (2018). Conversion-alloying anode materials for Na-ion batteries: Recent progress, challenges, and perspective for the future. *Journal of the Korean Ceramic Society*, 55(4), 307–324.

Kim, J., Kim, H., and Kang, K. (2018). Conversion-based cathode materials for rechargeable sodium batteries. *Adv. Energy Mater.*, 8(17), 1702646.

Kim, J.S., Ahn, H.J., Ryu, H.S., Kim, D.J., Cho, G.B., Kim, K.W., Nam, T.H., and Ahn, J.H. (2008). The discharge properties of Na/Ni_3S_2 cell at ambient temperature. *Journal of Power Sources*, 178(2), 852–856.

Kim, J.S., Kim, D.Y., Cho, G.B., Nam, T.H., Kim, K.W., Ryu, H.S., Ahn, J.H., and Ahn, H.J. (2009). The electrochemical properties of copper sulfide as cathode material for rechargeable sodium cell at room temperature. *Journal of Power Sources*, 189(1), 864–868.

Kim, S.-W., Seo, D.-H., Ma, X., Ceder, G., and Kang, K. (2012). Electrode materials for rechargeable sodium-ion batteries: Potential alternatives to current lithium-ion batteries. *Adv. Energy Mater.*, 2(7), 710–721.

Kim, T.B., Choi, J.W., Ryu, H.S., Cho, G.B., Kim, K.W., Ahn, J.H., Cho, K.K., and Ahn, H.J. (2007). Electrochemical properties of sodium/pyrite battery at room temperature. *Journal of Power Sources*, 174(2), 1275–1278.

Kim, Y., Park, Y., Choi, A., Choi, N.-S., Kim, J., Lee, J., Ryu, J.H., Oh, S.M., and Lee, K.T. (2013). An amorphous red phosphorus/carbon composite as a promising anode material for sodium ion batteries. *Adv. Mater.*, 25(22), 3045–3049.

Klein, F., Jache, B., Bhide, A., and Adelhelm, P. (2013). Conversion reactions for sodium-ion batteries. *Physical Chemistry Chemical Physics*, 15(38), 15876–15887.

Kollu, P., Kumar, P.R., Santosh, C., Kim, D.K., and Grace, A.N. (2015). A high capacity MnFe2O4/rGO nanocomposite for Li and Na-ion battery applications. *RSC Advances*, 5(78), 63304–63310.

Komaba, S., Matsuura, Y., Ishikawa, T., Yabuuchi, N., Murata, W., and Kuze, S. (2012). Redox reaction of Sn-polyacrylate electrodes in aprotic Na cell. *Electrochem. Commun.*, 21, 65–68.

Kong, M., Liu, K., Ning, J., Zhou, J., and Song, H. (2017). Perovskite framework NH_4FeF_3 /carbon composite nanosheets as a potential anode material for Li and Na ion storage. *Journal of Materials Chemistry A*, 5(36), 19280–19288.

Krengel, M., Hansen, A.-L., Kaus, M., Indris, S., Wolff, N., Kienle, L., Westfal, D., and Bensch, W. (2017). CuV2S4: A high rate capacity and stable anode material for sodium ion batteries. *ACS Appl. Mater. Interfaces*, 9(25), 21283–21291.

Krengel, M., Hansen, A.-L., Hartmann, F., van Dinter, J., and Bensch, W. (2018). Elucidation of the sodium–Copper extrusion mechanism in $CuCrS_2$: A high capacity, long-life anode material for sodium-ion batteries. *Batteries & Supercaps*, 1(5), 176–183.

Krengel, M., Adelhelm, P., Klein, F., and Bensch, W. (2015). FeV2S4 as a high capacity electrode material for sodium-ion batteries. *Chem. Commun.*, 51(70), 13500–13503.

Legrain, F., Malyi, O.I., and Manzhos, S. (2013). Comparative computational study of the diffusion of Li, Na, and Mg in silicon including the effect of vibrations. *Solid State Ionics*, 253, 157–163.

Li, F., Ma, J., Ren, H., Wang, H., and Wang, G. (2017a). Fabrication of MnO nanowires implanted in graphene as an advanced anode material for sodium-ion batteries. *Materials Letters*, 206, 132–135.

Li, K., Su, D., Liu, H., and Wang, G. (2015a). Antimony-carbon-graphene fibrous composite as freestanding anode materials for sodium-ion batteries. *Electrochim. Acta*, 177, 304–309.

Li, K., Liu, H., and Wang, G. (2014a). Sb_2O_3 nanowires as anode material for sodium-ion battery. *Arabian Journal for Science and Engineering*, 39(9), 6589–6593.

Li, L., Zheng, Y., Zhang, S., Yang, J., Shao, Z., and Guo, Z. (2018a). Recent progress on sodium ion batteries: Potential high-performance anodes. *Energy Environ. Sci.*, 11(9), 2310–2340.

Li, M., Xiao, X., Fan, X., Huang, X., Liu, Y., and Chen, L. (2017b). Carbon coated sodium-titanate nanotube as an advanced intercalation anode material for sodium-ion batteries. *J. Alloys Compd.*, 712, 365–372.

Li, Q., Wei, Q., Zuo, W., Huang, L., Luo, W., An, Q., Pelenovich, V.O., Mai, L., and Zhang, Q. (2016). Greigite Fe_3S_4 as a new anode material for high-performance sodium-ion batteries. *Chemical Science*, 8(1), 160–164.

Li, W., Fukunishi, M., Morgan, B.J., Borkiewicz, O.J., Chapman, K.W., Pralong, V., Maignan, A., Lebedev, O.I., Ma, J., Groult, H., Komaba, S., and Dambournet, D. (2017c). A reversible phase transition for sodium insertion in anatase TiO_2. *Chem. Mater.*, 29(4), 1836–1844.

Li, W., Zhou, M., Li, H., Wang, K., Cheng, S., and Jiang, K. (2015b). Carbon-coated Sb2Se3 composite as anode material for sodium ion batteries. *Electrochem. Commun.*, 60, 74–77.

Li, W., Hu, S., Luo, X., Li, Z., Sun, X., Li, M., Liu, F., and Yu, Y. (2017d). Confined amorphous red phosphorus in MOF-derived N-doped microporous carbon as a superior anode for sodium-ion battery. *Advanced Materials*, 29(16), 1605820.

Li, W., Huang, J., Feng, L., Cao, L., Liu, Y., and Pan, L. (2018b). VS4 microspheres winded by (1 1 0)-oriented nanotubes with high rate capacities as sodium-ion battery anode. *Materials Letters*, 230, 105–108.

Li, X., Hasan, M.M., Hector, A.L., and Owen, J.R. (2013). Performance of nanocrystalline Ni3N as a negative electrode for sodium-ion batteries. *Journal of Materials Chemistry A*, 1(21), 6441–6445.

Li, X., Hector, A.L., and Owen, J.R. (2014b). Evaluation of Cu_3N and CuO as negative electrode materials for sodium batteries. *J. Phys. Chem. C*, 118(51), 29568–29573.

Li, Y., Wang, R., Guo, Z., Xiao, Z., Wang, H., Luo, X., and Zhang, H. (2019). Emerging two-dimensional noncarbon nanomaterials for flexible lithium-ion batteries: Opportunities and challenges. *J. Mater. Chem. A*, 7(44), 25227–25246.

Li, Z., Ding, J., and Mitlin, D. (2015c). Tin and tin compounds for sodium ion battery anodes: Phase transformations and performance. *Acc. Chem. Res.*, 48(6), 1657–1665.

Liang, Y., Lai, W.-H., Miao, Z., and Chou, S.-L. (2018). Nanocomposite materials for the sodium-ion battery: A review. *Small*, 14(5), 1702514.

Lignesh Durai, B., Moorthy, B., Issac Thomas, C., Kyung Kim, D., and Kamala Bharathi, K. (2017). Electrochemical properties of $BiFeO_3$ nanoparticles: Anode material for sodium-ion battery application. *Materials Science in Semiconductor Processing*, 68, 165–171.

Lim, Y.R., Jung, C.S., Im, H.S., Park, K., Park, J., Cho, W. Il, and Cha, E.H. (2016). Zn_2GeO_4 and Zn_2SnO_4 nanowires for high-capacity lithium- and sodium-ion batteries. *Journal of Materials Chemistry A*, 4(27), 10691–10699.

Ling, L., Bai, Y., Li, Y., Ni, Q., Wang, Z., Wu, F., and Wu, C. (2017). Quick activation of nanoporous anatase TiO_2 as high-rate and durable anode materials for sodium-ion batteries. *ACS Appl. Mater. Interfaces*, 9(45), 39432–39440.

Ling, L., Bai, Y., Wang, Z., Ni, Q., Chen, G., Zhou, Z., and Wu, C. (2018). Remarkable effect of sodium alginate aqueous binder on anatase TiO_2 as high-performance anode in sodium ion batteries. *ACS Appl. Mater. Interfaces*, 10(6), 5560–5568.

Liu, Q., Yan, Y., Chu, X., Zhang, Y., Xue, L., and Zhang, W. (2017). Graphene-induced growth of single crystalline Sb_2MoO_6 sheets and their sodium storage performance. *Journal of Materials Chemistry A*, 5(40), 21328–21333.

Liu, S.-L., Huang, J., Liu, J., Lei, M., Min, J., Li, S., and Liu, G. (2016a). Porous Mo_2N nanobelts as a new anode material for sodium-ion batteries. *Materials Letters*, 172, 56–59.

Liu, S., Lei, W., Liu, Y., Qiao, Q., and Zhang, W.-H. (2018a). Hierarchical nanosheet-based MS_2 (M = Re, Mo, W) nanotubes prepared by templating sacrificial Te nanowires with superior lithium and sodium storage capacity. *ACS Applied Materials & Interfaces*, 10(43), 37445–37452.

Liu, X., Liu, H., Zhao, Y., Dong, Y., Fan, Q., and Kuang, Q. (2016b). Synthesis of the carbon-coated nanoparticle Co_9S_8 and its electrochemical performance as an anode material for sodium-ion batteries. *Langmuir*, 32(48), 12593–12602.

Liu, Y., Li, Y., Kang, H., Jin, T., and Jiao, L. (2016c). Design, synthesis, and energy-related applications of metal sulfides. *Materials Horizons*, 3(5), 402–421.

Liu, Y., Qiao, Y., Zhang, W., Hu, P., Chen, C., Li, Z., Yuan, L., Hu, X., and Huang, Y. (2014). Facile fabrication of CuO nanosheets on Cu substrate as anode materials for electrochemical energy storage. *Journal of Alloys and Compounds*, 586, 208–215.

Liu, Y., Yu, J., Guo, D., Li, Z., and Su, Y. (2020). Ti_3C_2Tx MXene/graphene nanocomposites: Synthesis and application in electrochemical energy storage. *J. Alloys Compd.*, 815, 152403.

Liu, Y., Zhang, N., Jiao, L., Tao, Z., and Chen, J. (2015a). Ultrasmall Sn nanoparticles embedded in carbon as high-performance anode for sodium-ion batteries. *Advanced Functional Materials*, 25(2), 214–220.

Liu, Y., Zhang, N., Kang, H., Shang, M., Jiao, L., and Chen, J. (2015b). WS2 nanowires as a high-performance anode for sodium-ion batteries. *Chem. Eur. J.*, 21(33), 11878–11884.

Liu, Z., Yue, C., Chen, C., Xiang, J., Hu, F., Lee, D., Shin, D., Sun, S., Hu, L., and Song, T. (2018b). A self-buffering structure for application in high-performance sodium-ion batteries. *Energy Storage Materials*, 15, 242–248.

Loaiza, L.C., Monconduit, L., and Seznec, V. (2020a). Si and Ge-based anode materials for Li-, Na-, and K-ion batteries: A perspective from structure to electrochemical mechanism. *Small*, 16(5), 1905260.

Loaiza, L.C., Monconduit, L., and Seznec, V. (2019). Siloxene: A potential layered silicon intercalation anode for Na, Li and K ion batteries. *J. Power Sources*, 417(February), 99–107.

Loaiza, L.C., Monconduit, L., and Seznec, V. (2020b). Towards germanium layered materials as superior negative electrodes for Li-, Na-, and K-ion batteries. *Batteries & Supercaps*, 30 January 2020. https://onlinelibrary.wiley.com/doi/abs/10.1002/batt.201900198.

Longoni, G., Pena Cabrera, R.L., Polizzi, S., D'Arienzo, M., Mari, C.M., Cui, Y., and Ruffo, R. (2017). Shape-controlled TiO_2 nanocrystals for Na-Ion battery electrodes: The role of different exposed crystal facets on the electrochemical properties. *Nano Lett.*, 17(2), 992–1000.

Lu, F., Liu, S., Zeng, G., Li, J., and Yang, Y. (2018). Probing the electrochemical reaction mechanism and crystallinity effect of RuO_2 for sodium storage. *J. Electrochem. Soc.*, 165(11), A2897–A2903.

Lu, X., Adkins, E.R., He, Y., Zhong, L., Luo, L., Mao, S.X., Wang, C.-M., and Korgel, B.A. (2016). Germanium as a sodium ion battery material: In situ TEM reveals fast sodiation kinetics with high capacity. *Chem. Mater.*, 28(4), 1236–1242.

Lu, Y.C., Ma, C., Alvarado, J., Kidera, T., Dimov, N., Meng, Y.S., and Okada, S. (2015). Electrochemical properties of tin oxide anodes for sodium-ion batteries. *J. Power Sources*, 284, 287–295.

Ma, Z., Lyu, Y., Yang, H., Li, Q., Guo, B., and Nie, A. (2018). Systematic investigation of the Binder's role in the electrochemical performance of tin sulfide electrodes in NIBs. *Journal of Power Sources*, 401, 195–203.

Malyi, O., Kulish, V. V., Tan, T.L., and Manzhos, S. (2013). A computational study of the insertion of Li, Na, and Mg atoms into Si(111) nanosheets. *Nano Energy*, 2(6), 1149–1157.

Marbella, L.E., Evans, M.L., Groh, M.F., Nelson, J., Griffith, K.J., Morris, A.J., and Grey, C.P. (2018). Sodiation and desodiation via helical phosphorus intermediates in high-capacity anodes for sodium-ion batteries. *J. Am. Chem. Soc.*, 140(25), 7994–8004.

Marino, C., Debenedetti, A., Fraisse, B., Favier, F., and Monconduit, L. (2011). Activated-phosphorus as new electrode material for Li-ion batteries. *Electrochem. Commun.*, 13(4), 346–349.

Marino, C., Block, T., Pöttgen, R., and Villevieille, C. (2017). $CuSbS_2$ as a negative electrode material for sodium ion batteries. *J. Power Sources*, 342, 616–622.

Mei, J., Liao, T., Kou, L., and Sun, Z. (2017). Two-dimensional metal oxide nanomaterials for next-generation rechargeable batteries. *Adv. Mater.*, 29(48), 1700176.

Mitra, S., Veluri, P.S., Chakraborthy, A., and Petla, R.K. (2014). Electrochemical properties of spinel cobalt ferrite nanoparticles with sodium alginate as interactive binder. *ChemElectroChem*, 1(6), 1068–1074.

Naguib, M., Mochalin, V.N., Barsoum, M.W., and Gogotsi, Y. (2014). 25th Anniversary Article: MXenes: A new family of two-dimensional materials. *Adv. Mater.*, 26(7), 992–1005.

Natarajan, S., Subramanyan, K., and Aravindan, V. (2019). Focus on spinel Li4Ti5O12 as insertion type anode for high-performance Na-ion batteries. *Small*, 15(49), 1904484.

Nayak, P.K., Yang, L., Brehm, W., and Adelhelm, P. (2018). From lithium-ion to sodium-ion batteries: Advantages, challenges, and surprises. *Angew. Chem. Int. Ed.*, 57(1), 102–120.

Ni, J. and Li, L. (2018). Self-supported 3D array electrodes for sodium microbatteries. *Advanced Functional Materials*, 28(3), 1704880.

Ni, J., Li, L., and Lu, J. (2018). Phosphorus: An anode of choice for sodium-ion batteries. *ACS Energy Lett.*, 3(5), 1137–1144..

Nithya, C. and Gopukumar, S. (2015). Sodium ion batteries: A newer electrochemical storage. *WIREs Energy Environ.*, 4(3), 253–278.

Novák, P. (1985). CuO cathode in lithium cells-II. Reduction mechanism of CuO. *Electrochimica Acta*, 30(12), 1687–1692.

Novák, P. and Podhájecký, P. (1991). A high-temperature lithium-copper oxide cell with a solid polymer electrolyte. *Journal of Power Sources*, 35(3), 235–247.

Opra, D.P., Gnedenkov, S. V., and Sinebryukhov, S.L. (2019). Recent efforts in design of TiO2(B) anodes for high-rate lithium-ion batteries: A review. *J. Power Sources*, 442(September), 227225.

Ou, X., Li, J., Zheng, F., Wu, P., Pan, Q., Xiong, X., Yang, C., and Liu, M. (2017a). In situ X-ray diffraction characterization of $NiSe_2$ as a promising anode material for sodium ion batteries. *Journal of Power Sources*, 343, 483–491.

Ou, X., Liang, X., Zheng, F., Wu, P., Pan, Q., Xiong, X., Yang, C., and Liu, M. (2017b). In situ X-ray diffraction investigation of $CoSe_2$ anode for Na-ion storage: Effect of cut-off voltage on cycling stability. *Electrochimica Acta*, 258, 1387–1396.

Palomares, V., Serras, P., Villaluenga, I., Hueso, K.B., Carretero-González, J., and Rojo, T. (2012). Na-ion batteries, recent advances and present challenges to become low cost energy storage systems. *Energy Environ. Sci.*, 5(3), 5884–5901.

Palomares, V., Casas-Cabanas, M., Castillo-Martínez, E., Han, M.H., and Rojo, T. (2013). Update on Na-based battery materials. A growing research path. *Energy Environ. Sci.*, 6(8), 2312–2337.

Pan, H., Hu, Y.-S., and Chen, L. (2013). Room-temperature stationary sodium-ion batteries for large-scale electric energy storage. *Energy Environ. Sci.*, 6(8), 2338–2360.

Pan, J., Zhang, Y., Li, L., Cheng, Z., Li, Y., Yang, X., Yang, J., and Qian, Y. (2019). Polyanions enhance conversion reactions for lithium/sodium-ion batteries: The case of $SbVO_4$ nanoparticles on reduced graphene oxide. *Small Methods*, 3(10), 1900231.

Pan, J., Chen, S., Zhang, D., Xu, X., Sun, Y., Tian, F., Gao, P., and Yang, J. (2018). SnP_2O_7 covered carbon nanosheets as a long-life and high-rate anode material for sodium-ion batteries. *Advanced Functional Materials*, 28(43), 1804672.

Park, A.R. and Park, C.M. (2017). Cubic crystal-structured SnTe for superior Li- and Na-ion battery anodes. *ACS Nano*, 11(6), 6074–6084.

Park, Y., Shin, D.-S., Woo, S.H., Choi, N.S., Shin, K.H., Oh, S.M., Lee, K.T. and Hong, S.Y. (2012). Sodium terephthalate as an organic anode material for sodium ion batteries. *Adv. Mater.*, 24(26), 3562–3567.

Pei, L., Zhao, Q., Chen, C., Liang, J. and Chen, J. (2015). Phosphorus nanoparticles encapsulated in graphene scrolls as a high-performance anode for sodium-ion batteries. *ChemElectroChem*, 2(11), 1652–1655.

Peng, L., Fang, Z., Zhu, Y., Yan, C., and Yu, G. (2018). Holey 2D nanomaterials for electrochemical energy storage. *Advanced Energy Materials*, 8(9), 1702179.

Persson, P.O.Å. and Rosen, J. (2019). Current state of the art on tailoring the MXene composition, structure, and surface chemistry. *Curr. Opin. Solid State Mater. Sci.*, 23(6), 100774.

Poizot, P., Laruelle, S., Grugeon, S., Dupont, L., and Tarascon, J.-M. (2000). Nano-sized transition-metal oxides as negative-electrode materials for lithium-ion batteries. *Nature*, 407(September), 496–499.

Qian, J., Chen, Y., Wu, L., Cao, Y., Ai, X., and Yang, H. (2012). High capacity Na-storage and superior cyclability of nanocomposite Sb/C anode for Na-ion batteries. *Chem. Commun.*, 48(56), 7070–7072.

Qian, J., Xiong, Y., Cao, Y., Ai, X., and Yang, H. (2014). Synergistic Na-storage reactions in Sn_4P_3 as a high-capacity, cycle-stable anode of Na-ion batteries. *Nano Lett.*, 14(4), 1865–1869.

Rahman, M.M., Glushenkov, A.M., Ramireddy, T., and Chen, Y. (2014). Electrochemical investigation of sodium reactivity with nanostructured Co_3O_4 for sodium-ion batteries. *Chem. Commun.*, 50(39), 5057–5060.

Reddy, M. V., Subba Rao, G. V., and Chowdari, B.V.R. (2013). Metal oxides and oxysalts as anode materials for Li ion batteries. *Chem. Rev.*, 113(7), 5364–5457.

Rudola, A., Saravanan, K., Mason, C.W., and Balaya, P. (2013). $Na_2Ti_3O_7$: An intercalation based anode for sodium-ion battery applications. *J. Mater. Chem. A*, 1(7), 2653–2662.

Ryu, W.H., Wilson, H., Sohn, S., Li, J., Tong, X., Shaulsky, E., Schroers, J., Elimelech, M., and Taylor, A.D. (2016). Heterogeneous WSx/WO_3 Thorn-Bush nanofiber electrodes for sodium-ion batteries. *ACS Nano*, 10(3), 3257–3266.

Santhosha, A.L., Das, S.K., and Bhattacharyya, A.J. (2016). Tungsten trioxide (WO3) nanoparticles as a new anode material for sodium-ion batteries. *Journal of Nanoscience and Nanotechnology*, 16(4), 4131–4135.

Schon, T.B., McAllister, B.T., Li, P.-F., and Seferos, D.S. (2016). The rise of organic electrode materials for energy storage. *Chem. Soc. Rev.*, 45(22), 6345–6404.

Senguttuvan, P., Rousse, G., Seznec, V., Tarascon, J.-M., and Palacín, M.R. (2011). $Na_2Ti_3O_7$: Lowest voltage ever reported oxide insertion electrode for sodium ion batteries. *Chem. Mater.*, 23(18), 4109–4111.

Shao, L., Ma, R., Wu, K., Shui, M., Lao, M., Wang, D., Long, N., Ren, Y., and Shu, J. (2013). Metal carbonates as anode materials for lithium ion batteries. *J. Alloys Compd.*, 581, 602–609.

Share, K., Lewis, J., Oakes, L., Carter, R.E., Cohn, A.P., and Pint, C.L. (2015). Tungsten diselenide (WSe_2) as a high capacity, low overpotential conversion electrode for sodium ion batteries. *RSC Advances*, 5(123), 101262–101267.

Shi, L., Wang, W., Wu, C., Ding, J., and Li, Q. (2017). Synthesis of Cu_2SnS_3 nanosheets as an anode material for sodium ion batteries. *Journal of Alloys and Compounds*, 699, 517–520.

Shimizu, M., Usui, H., and Sakaguchi, H. (2014). Electrochemical Na-insertion/extraction properties of SnO thick-film electrodes prepared by gas-deposition. *J. Power Sources*, 248, 378–382.

Sinha, S., Didwal, P.N., Nandi, D.K., Verma, R., Cho, J.Y., Kim, S.H., Park, C.J., and Heo, J. (2019). Revealing the simultaneous effects of conductivity and amorphous nature of atomic-layer-deposited double-anion-based zinc oxysulfide as superior anodes in Na-ion batteries. *Small*, 15(37), 1900595.

Slater, M.D., Kim, D., Lee, E., and Johnson, C.S. (2012). Sodium-ion batteries. *Adv. Funct. Mater.*, 23(8), 947–958.

Son, D., Kim, E., Kim, T.-G., Kim, M.G., Cho, J., and Park, B. (2004). Nanoparticle iron-phosphate anode material for Li-ion battery. *Appl. Phys. Lett.*, 85(24), 5875–5877.

Song, J., Yu, Z., Gordin, M.L., Li, X., Peng, H., and Wang, D. (2015). Advanced sodium ion battery anode constructed via chemical bonding between phosphorus, carbon nanotube, and cross-linked polymer binder. *ACS Nano*, 9(12), 11933–11941.

Sougrati, M.T., Darwiche, A., Liu, X., Mahmoud, A., Hermann, R.P., Jouen, S., Monconduit, L., Dronskowski, R., and Stievano, L. (2016a). Batteries transition-metal carbodiimides as molecular negative electrode materials for lithium- and sodium-ion batteries with excellent cycling properties. *Angew. Chem. Int. Ed.*, 55(16), 5090–5095.

Sougrati, M.T., Darwiche, A., Monconduit, L., Stievano, L., Hermann, R.P., Mahmoud, A., Herlitschke, M., Dronskowski, R., and Liu, X. (2016b). Metal carbodiimides and metal cyanamides as electrode materials. Patent, WO/2016/198550.

Su, D., Ahn, H.-J., and Wang, G. (2013). SnO_2@graphene nanocomposites as anode materials for Na-ion batteries with superior electrochemical performance. *Chem. Commun.*, 49(30), 3131–3133.

Su, D., Dou, S., and Wang, G. (2015). Anatase TiO_2: Better anode material than amorphous and rutile phases of TiO_2 for Na-ion batteries. *Chem. Mater.*, 27(17), 6022–6029.

Su, H., Jaffer, S., and Yu, H. (2016). Transition metal oxides for sodium-ion batteries. *Energy Storage Mater.*, 5, 116–131.

Sun, C., Chen, Z., Lu, L., and Li, N. (2019). $Na_2Ti_3O_7$ nanotubes as an anode for sodium ion batteries and self-powered system. *ChemElectroChem*, 6(12), 3085–3090.

Sun, J., Lee, H.-W., Pasta, M., Yuan, H., Zheng, G., Sun, Y., Li, Y., and Cui, Y. (2015). A phosphorene–graphene hybrid material as a high-capacity anode for sodium-ion batteries. *Nature Nanotechnology*, 10(11), 980–985.

Sun, Q., Ren, Q.-Q., Li, H., and Fu, Z.-W. (2011). High capacity Sb2O4 thin film electrodes for rechargeable sodium battery. *Electrochem. Commun.*, 13(12), 1462–1464.

Sun, R., Liu, S., Wei, Q., Sheng, J., Zhu, S., An, Q., and Mai, L. (2017). Mesoporous NiS_2 nanospheres anode with pseudocapacitance for high-rate and long-life sodium-ion Battery. *Small*, 13(39), 1701744.

Sun, W., Liu, J., Wu, H., Yue, X., Wang, Z., Rooney, D., Feng, J., and Sun, K. (2016a). Facile synthesis of copper-manganese spinel anodes with high capacity and cycling performance for lithium-ion batteries. *Materials Letters*, 182, 147–150.

Sun, Y., Li, C.-S., Yang, Q.-R., Chou, S.-L., and Liu, H.-K. (2016b). Electrochemically active, novel layered m-ZnV2O6 nanobelts for highly rechargeable Na-ion energy storage. *Electrochimica Acta*, 205, 62–69.

Tan, J., Liu, L., Guo, S., Hu, H., Yan, Z., Zhou, Q., Huang, Z., Shu, H., Yang, X., and Wang, X. (2015). The electrochemical performance and mechanism of cobalt (II) fluoride as anode material for lithium and sodium ion batteries. *Electrochimica Acta*, 168, 225–233.

Tao, H., Zhou, M., Wang, R., Wang, K., Cheng, S., and Jiang, K. (2018). TiS_2 as an advanced conversion electrode for sodium-ion batteries with ultra-high capacity and long-cycle life. *Advanced Science*, 5(11), 1801021.

Thackeray, M.M. and Coetzer, J. (1981). A preliminary investigation of the electrochemical performance of α-Fe_2O_3 and Fe_3O_4 cathodes in high-temperature cells. *Materials Research Bulletin*, 16(5), 591–597.

Tian, J., Wang, B., Zhao, F., Ma, X., Liu, Y., Liu, H.K., and Huang, Z. (2017). Highly active Fe_3BO_6 as an anode material for sodium-ion batteries. *Chem. Commun.*, 53(34), 4698–4701.

Veluri, P.S. and Mitra, S. (2013). Enhanced high rate performance of α-Fe_2O_3 nanotubes with alginate binder as a conversion anode. *RSC Advances*, 3(35), 15132.

Verma, R., Raman, R.K., and Varadaraju, U.V. (2016). Disodium dimolybdate: A potential high-performance anode material for rechargeable sodium ion battery applications. *J. Solid State Electrochem.*, 20(5), 1501–1505.

Vogt, L.O. and Villevieille, C. (2016). $MnSn_2$ negative electrodes for Na-ion batteries: A conversion-based reaction dissected. *Journal of Materials Chemistry A*, 4(48), 19116–19122.

Wang, B., Li, F., Wang, X., Wang, G., Wang, H., and Bai, J. (2019a). Mn_3O_4 nanotubes encapsulated by porous graphene sheets with enhanced electrochemical properties for lithium/sodium-ion batteries. *Chemical Engineering Journal*, 364, 57–69.

Wang, H., Lan, X., Jiang, D., Zhang, Y., Zhong, H., Zhang, Z., and Jiang, Y. (2015a). Sodium storage and transport properties in pyrolysis synthesized $MoSe_2$ nanoplates for high performance sodium-ion batteries. *Journal of Power Sources*, 283, 187–194.

Wang, J.W., Liu, X.H., Mao, S.X., and Huang, J.Y. (2012). Microstructural evolution of tin nanoparticles during in situ sodium insertion and extraction. *Nano Lett.*, 12(11), 5897–5902.

Wang, L., Światowska, J., Dai, S., Cao, M., Zhong, Z., Shen, Y., and Wang, M. (2019b). Promises and challenges of alloy-type and conversion-type anode materials for sodium–ion batteries. *Materials Today Energy*, 11, 46–60.

Wang, L., Wang, Jiajun, Guo, F., Ma, L., Ren, Y., Wu, T., Zuo, P., Yin, G., and Wang, J. (2018). Understanding the initial irreversibility of metal sulfides for sodium-ion batteries via operando techniques. *Nano Energy*, 43, 184–191.

Wang, L.P., Zhao, Y., Wei, C., Wong, C., Srinivasan, M., and Xu, Z.J. (2015b). Polycrystalline zinc stannate as an anode material for sodium-ion batteries. *Journal of Materials Chemistry A*, 3(26), 14033–14038.

Wang, P., Liu, M., Mo, F., Long, Z., Fang, F., Sun, D., Zhou, Y.N., and Song, Y. (2019c). Exploring the sodium ion storage mechanism of gallium sulfide (Ga_2S_3): A combined experimental and theoretical approach. *Nanoscale*, 11(7), 3261–3267.

Wang, P., Xie, S., She, Y.Y., Fan, W., Leung, M.K.H., and Wang, H. (2019d). Microwave-hydrothermal synthesis of hierarchical Sb_2WO_6 nanostructures as a new anode material for sodium storage. *ChemistrySelect*, 4(3), 1078–1083.

Wang, S., Wang, L., Zhu, Z., Hu, Z., Zhao, Q., and Chen, J. (2014a). All organic sodium-ion batteries with $Na_4C_8H_2O_6$. *Angewandte Chemie–International Edition*, 53(23), 5892–5896.

Wang, W., Yu, C., Lin, Z., Hou, J., Zhu, H., and Jiao, S. (2013a). Microspheric $Na_2Ti_3O_7$ consisting of tiny nanotubes: An anode material for sodium-ion batteries with ultrafast charge–discharge rates. *Nanoscale*, 5(2), 594–599.

Wang, X., Shen, X., Wang, Z., Yu, R., and Chen, L. (2014b). Atomic-scale clarification of structural transition of MoS_2 upon sodium intercalation. *ACS Nano*, 8(11), 11394–11400.

Wang, X., Kajiyama, S., Iinuma, H., Hosono, E., Oro, S., Moriguchi, I., Okubo, M., and Yamada, A. (2015c). Pseudocapacitance of MXene nanosheets for high-power sodium-ion hybrid capacitors. *Nat. Commun.*, 6(1), 6544.

Wang, X., Zheng, S., Wang, S., Zhang, Y., and Du, H. (2016). Self-anchoring dendritic ternary vanadate compound on graphene nanoflake as high-performance conversion-type anode for lithium ion batteries. *Nano Energy*, 22, 179–188.

Wang, Y., Yu, X., Xu, S., Bai, J., Xiao, R., Hu, Y.-S., Li, H., Yang, X.-Q., Chen, L., and Huang, X. (2013b). A zero-strain layered metal oxide as the negative electrode for long-life sodium-ion batteries. *Nat. Commun.*, 4(1), 2365.

Wang, Y.X., Seng, K.H., Chou, S.L., Wang, J.Z., Guo, Z., Wexler, D., Liu, H.K., and Dou, S.X. (2014c). Reversible sodium storage via conversion reaction of a MoS_2-C composite. *Chemical Communications*, 50(73), 10730–10733.

Weng, Y.T., Huang, T.Y., Lim, C.H., Shao, P.S., Hy, S., Kuo, C.Y., Cheng, J.H., Hwang, B.J., Lee, J.F., and Wu, N.L. (2015). An unexpected large capacity of ultrafine manganese oxide as a sodium-ion battery anode. *Nanoscale*, 7(47), 20075–20081.

Wu, D., Li, X., Xu, B., Twu, N., Liu, L., and Ceder, G. (2015). NaTiO2: a layered anode material for sodium-ion batteries. *Energy Environ. Sci.*, 8(1), 195–202.

Wu, F., Zhao, C., Chen, S., Lu, Y., Hou, Y., Hu, Y.S., Maier, J., and Yu, Y. (2018). Multi-electron reaction materials for sodium-based batteries. *Materials Today*, 21(9), 960–973.

Wu, L., Hu, X., Qian, J., Pei, F., Wu, F., Mao, R., Ai, X., Yang, H., and Cao, Y. (2014). Sb–C nanofibers with long cycle life as an anode material for high-performance sodium-ion batteries. *Energy Environ. Sci.*, 7(1), 323–328.

Wu, Z.G., Li, J.T., Zhong, Y.J., Liu, J., Wang, K., Guo, X.D., Huang, L., Zhong, B.H., and Sun, S.G. (2016). Synthesis of FeS@C-N hierarchical porous microspheres for the applications in lithium/sodium ion batteries. *Journal of Alloys and Compounds*, 688, 790–797.

Xia, D., Gong, F., Pei, X., Wang, W., Li, H., Zeng, W., Wu, M., and Papavassiliou, D. V (2018). Molybdenum and tungsten disulfides-based nanocomposite films for energy storage and conversion: A review. *Chemical Engineering Journal*, 348, 908–928.

Xiao, Y., Lee, S.H., and Sun, Y.-K. (2017). The application of metal sulfides in sodium ion batteries. *Adv. Energy Mater.*, 7(3), 1601329.

Xie, X., Wang, S., Kretschmer, K., and Wang, G. (2017). Two-dimensional layered compound based anode materials for lithium-ion batteries and sodium-ion batteries. *Journal of Colloid and Interface Science*, 499, 17–32.

Xu, J., Ma, C., Balasubramanian, M., and Meng, Y.S. (2014). Understanding $Na_2Ti_3O_7$ as an ultra-low voltage anode material for a Na-ion battery. *Chem. Commun.*, 50(83), 12564–12567.

Xu, X., Ji, S., Gu, M., and Liu, J. (2015). In situ synthesis of MnS hollow microspheres on reduced graphene oxide sheets as high-capacity and long-life anodes for Li- and Na-ion batteries. *ACS Applied Materials & Interfaces*, 7(37), 20957–20964.

Xu, Y., Memarzadeh Lotfabad, E., Wang, H., Farbod, B., Xu, Z., Kohandehghan, A., and Mitlin, D. (2013). Nanocrystalline anatase TiO_2: A new anode material for rechargeable sodium ion batteries. *Chem. Commun.*, 49(79), 8973–8975.

Xu, Y., Swaans, E., Basak, S., Zandbergen, H.W., Borsa, D.M., and Mulder, F.M. (2016). Reversible Na-ion uptake in Si nanoparticles. *Advanced Energy Materials*, 6(2), 1501436.

Yang, J., Zhou, T., Zhu, R., Chen, X., Guo, Z., Fan, J., Liu, H.K., and Zhang, W.X. (2016). Highly ordered dual porosity mesoporous cobalt oxide for sodium-ion batteries. *Advanced Materials Interfaces*, 3(3), 1500464.

Yang, Z., Choi, D., Kerisit, S., Rosso, K.M., Wang, D., Zhang, J., Graff, G.L., and Liu, J. (2009). Nanostructures and lithium electrochemical reactivity of lithium titanites and titanium oxides: A review. *J. Power Sources*, 192(2), 588–598.

Yi, T.-F., Yang, S.-Y., and Xie, Y. (2015). Recent advances of $Li_4Ti_5O_{12}$ as a promising next generation anode material for high power lithium-ion batteries. *J. Mater. Chem. A*, 3(11), 5750–5777.

Yu, H., Yang, J., Geng, H., and Li, C.C. (2017). Facile preparation of carbon wrapped copper telluride nanowires as high performance anodes for sodium and lithium ion batteries. *Nanotechnology*, 28(14), 145403.

Yu, L., Wang, L.P., Liao, H., Wang, J., Feng, Z., Lev, O., Loo, J.S.C., Sougrati, M.T., and Xu, Z.J. (2018). Understanding fundamentals and reaction mechanisms of electrode materials for Na-ion batteries. *Small*, 14(16), 1703338.

Yu, L., Wang, L.P., Xi, S., Yang, P., Du, Y., Srinivasan, M., and Xu, Z.J. (2015). β-FeOOH: An earth-abundant high-capacity negative electrode material for sodium-ion batteries. *Chem. Mater.*, 27(15), 5340–5348.

Yu, Y.-X. (2016). Prediction of mobility, enhanced storage capacity, and volume change during sodiation on interlayer-expanded functionalized Ti_3C_2 MXene anode materials for sodium-ion batteries. *J. Phys. Chem. C*, 120(10), 5288–5296.

Zaghib, K., Simoneau, M., Armand, M., and Gauthier, M. (1999). Electrochemical study of $Li_4Ti_5O_{12}$ as negative electrode for Li-ion polymer rechargeable batteries. *J. Power Sources*, 81–82, 300–305.

Zhang, C., Wang, X., Liang, Q., Liu, X., Weng, Q., Liu, J., Yang, Y., Dai, Z., Ding, K., Bando, Y., Tang, J., and Golberg, D. (2016a). Amorphous phosphorus/nitrogen-doped graphene paper for ultrastable sodium-ion batteries. *Nano Letters*, 16(3), 2054–2060.

Zhang, D., Sun, W., Chen, Z., Zhang, Y., Luo, W., Jiang, Y., and Dou, S.X. (2016b). Two-dimensional cobalt-/nickel-based oxide nanosheets for high-performance sodium and lithium storage. *Chemistry–A European Journal*, 22(50), 18060–18065.

Zhang, H., Hasa, I., and Passerini, S. (2018a). Beyond insertion for Na-ion batteries: Nanostructured alloying and conversion anode materials. *Adv. Energy Mater.*, 8(17), 1702582.

Zhang, K., Park, M., Zhou, L., Lee, G.-H., Li, W., Kang, Y.-M., and Chen, J. (2016c). Urchin-like $CoSe_2$ as a high-performance anode material for sodium-ion batteries. *Advanced Functional Materials*, 26(37), 6728–6735.

Zhang, L. and Zhu, S. (2019). Hollow polyhedral $ZnCo_2O_4$ anode materials for lithium-ion batteries. *Materials Letters*, 236, 337–341.

Zhang, N., Han, X., Liu, Y., Hu, X., Zhao, Q., and Chen, J. (2015a). 3D porous γ-Fe_2O_3@C nanocomposite as high-performance anode material of Na-ion batteries. *Adv. Energy Mater.*, 5(5), 1401123.

Zhang, S., Yue, L., Zhao, H., Wang, Z., and Mi, J. (2017a). Mwcnts wrapped flower-like SnS composite as anode material for sodium-ion battery. *Materials Letters*, 209, 212–215.

Zhang, W.-J. and Huang, K.-J. (2017). A review of recent progress in molybdenum disulfide-based supercapacitors and batteries. *Inorganic Chemistry Frontiers*, 4(10), 1602–1620.

Zhang, W., Dahbi, M., Amagasa, S., Yamada, Y., and Komaba, S. (2016d). Iron phosphide as negative electrode material for Na-ion batteries. *Electrochem. Commun.*, 69, 11–14.

Zhang, Y., Pu, X., Yang, Y., Zhu, Y., Hou, H., Jing, M., Yang, X., Chen, J., and Ji, X. (2015b). An electrochemical investigation of rutile TiO_2 microspheres anchored by nanoneedle clusters for sodium storage. *Phys. Chem. Chem. Phys.*, 17(24), 15764–15770.

Zhang, Y., Lu, S., Wang, M.-Q., Niu, Y., Liu, S., Li, Y., Wu, X., Bao, S.-J., and Xu, M. (2016e). Bismuth oxychloride ultrathin nanoplates as an anode material for sodium-ion batteries. *Materials Letters*, 178, 44–47.

Zhang, Y., Zhou, Q., Zhu, J., Yan, Q., Dou, S.X., and Sun, W. (2017b). Nanostructured metal chalcogenides for energy storage and electrocatalysis. *Adv. Funct. Mater.*, 27(35), 1702317.

Zhang, Y., Guo, L., and Yang, S. (2014). Three-dimensional spider-web architecture assembled from $Na_2Ti_3O_7$ nanotubes as a high performance anode for a sodium-ion battery. *Chem. Commun.*, 50(90), 14029–14032.

Zhang, Z., Qian, J., Lu, W., Chan, C.H., Lau, S.P., and Dai, J.Y. (2018b). In situ TEM study of the sodiation/desodiation mechanism of MnO_2 nanowire with gel-electrolytes. *Energy Storage Materials*, 15, 91–97.

Zhao, F., Han, N., Huang, W., Li, J., Ye, H., Chen, F., and Li, Y. (2015a). Nanostructured CuP2/C composites as high-performance anode materials for sodium ion batteries. *Journal of Materials Chemistry A*, 3(43), 21754–21759.

Zhao, W., Ma, X., Wang, G., Long, X., Li, Y., Zhang, W., and Zhang, P. (2018). Carbon-coated CoP3 nanocomposites as anode materials for high-performance sodium-ion batteries. *Applied Surface Science*, 445, 167–174.

Zhao, Wengao, Zhong, G., McDonald, M.J., Gong, Z., Liu, R., Bai, J., Yang, C., Li, S., Zhao, Weimin, Wang, H., Fu, R., Jiang, Z., and Yang, Y. (2016). $Cu_3(PO_4)_2/C$ composite as a high-capacity cathode material for rechargeable Na-ion batteries. *Nano Energy*, 27, 420–429.

Zhao, Y., Wang, L.P., Sougrati, M.T., Feng, Z., Leconte, Y., Fisher, A., Srinivasan, M., and Xu, Z. (2017). A review on design strategies for carbon based metal oxides and sulfides nanocomposites for high performance Li and Na ion battery anodes. *Adv. Energy Mater.*, 7(9), 1601424.

Zhao, Y., Feng, Z., and Xu, Z.J. (2015b). Yolk-shell $Fe_2O_3 \odot C$ composites anchored on MWNTs with enhanced lithium and sodium storage. *Nanoscale*, 7(21), 9520–9525.

Zhao, Y. and Manthiram, A. (2015a). $Bi_{0.94}Sb_{1.06}S_3$ nanorod cluster anodes for sodium-ion batteries: Enhanced reversibility by the synergistic effect of the Bi_2S_3–Sb_2S_3 solid solution. *Chem. Mater.*, 27(17), 6139–6145.

Zhao, Y. and Manthiram, A. (2015b). High-capacity, high-rate Bi–Sb alloy anodes for lithium-ion and sodium-ion batteries. *Chem. Mater.*, 27(8), 3096–3101.

Zhou, L., Zhang, K., Sheng, J., An, Q., Tao, Z., Kang, Y.M., Chen, J., and Mai, L. (2017). Structural and chemical synergistic effect of CoS nanoparticles and porous carbon nanorods for high-performance sodium storage. *Nano Energy*, 35, 281–289.

Zhou, T., Pang, W.K., Zhang, C., Yang, J., Chen, Z., Liu, H.K., and Guo, Z. (2014a). Enhanced sodium-ion battery performance by structural phase transition from two-dimensional hexagonal-SnS_2 to orthorhombic-SnS. *ACS Nano*, 8(8), 8323–8333.

Zhou, X., Liu, X., Xu, Y., Liu, Y., Dai, Z., and Bao, J. (2014b). An SbOx/reduced graphene oxide composite as a high-rate anode material for sodium-ion batteries. *J. Phys. Chem. C*, 118(41), 23527–23534.

Zhu, C., Mu, X., van Aken, P.A., Yu, Y., and Maier, J. (2014). Single-layered ultrasmall nanoplates of MoS_2 embedded in carbon nanofibers with excellent electrochemical performance for lithium and sodium storage. *Angewandte Chemie*, 126(8), 2184–2188.

Zhu, K., Liu, J., Li, S., Liu, L., Yang, L., Liu, S., Wang, H., and Xie, T. (2017). Ultrafine cobalt phosphide nanoparticles embedded in nitrogen-doped carbon matrix as a superior anode material for lithium ion batteries. *Advanced Materials Interfaces*, 4(19), 1700377.

Zhu, Y., Wen, Y., Fan, X., Gao, T., Han, F., Luo, C., Liou, S.-C., and Wang, C. (2015). Red phosphorus-single-walled carbon nanotube composite as a superior anode for sodium ion batteries. *ACS Nano*, 9(3), 3254–3264.

5

Electrolytes for Sodium Batteries

Faezeh MAKHLOOGHIAZAD[1], **Cristina POZO-GONZALO**[1], **Patrik JOHANSSON**[2,3] **and Maria FORSYTH**[1]

[1] *Institute for Frontier Materials, Deakin University, Burwood, Australia*

[2] *Dept. of Physics, Chalmers University of Technology, Gothenburg, Sweden*

[3] *ALISTORE-European Research Institute, CNRS FR 3104, Amiens, France*

5.1. Introduction

Alternatives to fossil fuel usage have been the goal of research for decades to overcome the associated depletion of natural resources and environmental pollution. Cleaner and renewable energy from wind and solar power have emerged as alternative sources, which, however, do require large-scale energy storage systems to ensure a balance between energy supply and demand. Additionally, the ever-increasing demand on portable, entertainment and computing equipment has for over two decades required light-weight and reliable energy storage technologies (Tarascon and Armand 2001).

Since the 1990s, lithium ion batteries (LIBs) have been widely used in portable devices, and the demand continues to increase following the trends in advanced electronic devices. LIBs are also the technology of choice to power electric vehicles. This is mostly due to their superior performance (e.g. operating voltage (>3.7 V), energy density (>250 Wh/kg at cell level) and cyclability (>5,000 cycles)) (Nishi 2001; Dunn *et al.* 2011).

Na-ion Batteries,
coordinated by Laure MONCONDUIT and Laurence CROGUENNEC.

However, the scarcity of the raw materials required for LIBs, such as lithium and cobalt, together with the high cost and potential toxicity of cobalt, is jeopardizing large-scale device fabrication and availability. Therefore, new chemistries are required to produce reliable, cost-effective and sustainable batteries that can meet the increasing global energy demands. Sodium is the fourth most abundant element in the Earth's crust and has a projected low production cost, providing a compelling rationale to investigate sodium ion batteries (NIBs) as an alternative to LIBs. The similar intercalation chemistry required for both technologies enables much of the knowledge already gained on LIBs, in terms of mechanisms and chemistry, to be translated to NIBs, while considering differences in the radii of the ions, i.e. Na^+ 1.02 Å and Li^+ 0.76 Å.

Additionally, design features such as the use of aluminum as the material for the current collectors at both sides of electrodes for NIBs, as opposed to the copper required at the anode side of the LIB, will ensure both a decrease in the cost of one of the battery components and an improvement in cell energy density.

Start-up energy companies including Tiamat and Faradion[1,2] have already demonstrated the promising performance and commercialization potential of NIBs, which has encouraged further research in this technology. Still, however, the specific energy of NIBs is lower than that of LIBs. The main hurdles for the realization of sodium-based technologies is the lack of an anode with an appropriate potential, and to achieve large reversible capacity, and high structural stability in the NIBs.

The natural alternative used in LIB technology, graphite, is not a suitable anode option for NIBs due to the poor intercalation of sodium ions. Notably, both Li and K can be readily intercalated in graphite to reach LiC_6 and KC_8, respectively, but for Na this is not possible without solvent co-intercalation. The most prominent example is perhaps $Na(diglyme)_2C_{20}$, which exhibits very good reversibility and a reversible capacity close to 100 mAh/g for >2,500 cycles (Jache and Adelhelm 2014; Kim *et al.* 2015b, 2015a). However, this is still low compared to LIB anodes and the volumetric capacity is also negatively affected. Instead, other different carbonaceous materials have been employed as hosts for the Na-ions – most notably hard carbons (HCs).

HCs have strong cross-links created during high-temperature (>800°C) synthesis, immobilizing the carbon layers. This prevents any true graphitic structure to evolve, instead resulting in a complex structure of lateral extent and stacking of

1 See http://www.tiamat-energy.com/.
2 See https://www.faradion.co.uk/.

graphene sheets. As a result, HCs generally have two different modes of Na insertion and associated potential regions; one sloping extending to ca. 0.2 V and one plateau very close to 0 V versus Na. The practical capacity can reach up to 300 mAh/g if the organic liquid electrolyte is optimized (Ponrouch *et al.* 2013). The main practical drawback of HCs being used for NIB anode material is the large irreversible capacity loss – hence loss of cyclable Na, and the margin needed when building a cell to ensure that no Na plating will occur carries a severe penalty on the practical NIB cell energy density.

Among the many possible alternative anode materials, it is obvious that Na metal possesses the most fundamental promises with the lowest redox potential, highest theoretical capacity (1,166 mAh/g), an abundance without any need for (high-temperature) synthesis and a demonstrated feasibility in other electrochemical systems including batteries – most notably the Na-S technology. In fact, the use of Na metal in sodium-based batteries would advance the specific capacity beyond competitive values versus LIBs.

However, the reactive Na metal also brings a different set of issues such as dendrite growth, low melting point, traditional organic liquid electrolyte solvents decomposition and unstable passivation layers (Hwang *et al.* 2017). Many of these issues are, however, not only related to the Na metal anode as such but also to the (type of) electrolyte employed. Stable performance has already been demonstrated using ionic liquid (IL) based electrolytes as well as plastic crystal electrolytes, both well suited for efficient plating and stripping of sodium metal (Basile *et al.* 2017).

In this chapter, we start with an overview of different liquid and solid electrolytes employed for sodium(-ion) batteries, before we cover the basics in more depth and discuss the current status of IL-based electrolytes. Finally, we outline the challenges that remain to be solved to enable the realization of sodium batteries based on such electrolytes.

5.2. Liquid and solid electrolytes for sodium batteries

In all battery technologies, the electrolyte plays a critical role for both performance and safety. The role of the electrolyte is basically to separate the electrodes electrically and provide proper ion transfer between electrodes during the charge and discharge processes.

Hence, the electrolyte needs to be electronically insulating but provide a high ionic conductivity. It must also have high electrochemical stability to avoid oxidation or reduction during battery operation, and this is especially a key property of lithium and sodium batteries given their high voltage. An ideal electrolyte should be thermally stable in a wide range of temperatures, and present low volatility and flammability to ensure overall safety. It should be chemically stable and ideally cost effective, and based on abundant and non-toxic elements (Ponrouch *et al.* 2015).

In a real battery (full cell), the electrodes and the electrolyte are in direct contact, which means that there are chemical interactions occurring between them. As most (liquid) electrolytes are in fact meta-stable, the cell stability and performance such as rate capability and cyclability all depend on the properties of the solid electrolyte interphase (SEI) formed on the anode by partial electrolyte decomposition. The equivalent on the cathode side is the cathode electrolyte interphase (CEI) or sometimes the "solid polymer interphase" (SPI). The SEI is electrically insulating but allows for ion conduction to support the required electrochemical processes required. Since the SEI is formed as a result of partial electrolyte decomposition, the electrolyte composition has a crucial effect on its properties and optimizing electrolytes must be made taking into account the specific electrode in mind to improve the battery stability, safety and electrochemical performance (Komaba *et al.* 2011).

Electrolytes for batteries can basically be liquids or solids; both have their own merits and drawbacks. Low viscosity, high ionic conductivity and low cost are advantages of liquid electrolytes. On the other hand, some are flammable and volatile. In contrast, solid electrolytes show non-flammability and non-volatility, thus solve many safety-related problems. However, they often display low ionic conductivity and some show poor mechanical flexibility, i.e. they are brittle (Kundu *et al.* 2015). Below, the research and development of different kinds of electrolytes are outlined, including organic liquid electrolytes, IL-based electrolytes and hybrids thereof, followed by different solid electrolytes, organic ionic plastic crystals (OIPCs), solid polymer electrolytes (SPEs) and gel polymer electrolytes (GPEs). While we do cover both the materials constituting the electrolytes and the intrinsic resulting properties of the electrolytes herein, the focus is on their application in sodium(-ion) batteries, with a particular emphasis on the properties of the IL-based electrolytes.

5.2.1. *Organic liquid electrolytes*

Organic liquid electrolytes for sodium batteries typically consist of one or more sodium salts dissolved in one or more organic solvents. The properties of the solvent(s) and the dissolved sodium salt(s) play a pivotal role in determining cell performance. Typical solvent properties that must be taken into consideration are as follows: a large permittivity/high dielectric constant to dissociate and solvate the sodium salt(s), a low viscosity to increase ion mobility and improve the wetting of electrodes, and wide chemical and electrochemical stabilities, as well as low-melting and high-boiling temperatures – all to enable wide and safe operation conditions. The organic solvent of interest for sodium batteries are typically carbonate ester-type and ether-type compounds such as ethylene carbonate (EC), propylene carbonate (PC), di-methyl carbonate (DMC), di-ethyl carbonate (DEC), ethyl-methyl carbonate (EMC), di-methoxyethane (DME), di-ethyleneglycol di-methylether (diglyme) and tri-ethylene glycol di-methyl ether (triglyme).

The Na salts should present a high solubility in the solvent(s) chosen to enable electrolytes with a sufficient concentration of charge carriers, and to be safe, non-toxic and chemically stable during cell cycling. The sodium salts of main interest for sodium batteries are based on the same weakly coordinating anions (WCAs), i.e. with highly delocalized negative charges, as employed in LIB electrolytes: ClO_4^-, PF_6^-, BF_4^-, TFSI, FSI and OTf.

$NaClO_4$ and $NaPF_6$ are by far the most studied salts for sodium batteries; however, they both have severe safety concerns. The perchlorate anion (ClO_4^-) is toxic, can create explosive compounds in combination with organic solvents and is highly oxidative. The salt is also difficult and time consuming to dry. For the PF_6^- anion, the safety issue is the auto-decomposition leading to hydrofluoric acid in the presence of traces of water (Krause *et al.* 1997; Ponrouch *et al.* 2015). Nevertheless, $NaClO_4$ is still the most commonly used electrolyte salt for basic materials research studies, largely due to its low cost.

Other and safer salt alternatives are sodium bis(fluorosulfonyl)imide (NaFSI) and sodium bis(trifluoromethylsulfonyl)imide (NaTFSI), which both present high thermal stabilities and yield non-toxic electrolytes with high ionic conductivity. However, they are both more expensive than $NaClO_4$.

Sodium dicyanamide (NaDCA) is a non-fluorinated salt based on the electron withdrawing nitrile group –CN (a strong donor ligand) (Scheers *et al.* 2014). However, the low anodic stability of the DCA anion limits its application to cells with medium voltage cathode materials.

Moving on from the separate components to exemplify how they are used to create NIB electrolytes, we have chosen to highlight just a few of many systematic studies of several electrolytes that have been made. The first study includes the following electrolytes: $NaClO_4$ in PC, $NaBF_4$ in PC, $NaPF_6$ in EC:DEC, NaFSI in EC:DEC and NaFSI in PC (Jian *et al.* 2013). As these carbonate-based electrolytes were tested against NASICON electrodes ($Na_3V_2(PO_4)_3/C$), the NaFSI-based electrolytes resulted in much better coulombic efficiencies, close to 100%, with a stable and high capacity retention. The authors suggested that either electrolyte decomposition and/or the presence of side-reactions could be responsible for the lower coulombic efficiencies observed for the other electrolytes (Jian *et al.* 2013).

The above results were also confirmed by Komaba *et al.* (2011) who investigated the effect of the various solvents in a NIB using a HC electrode and employing 1 M $NaClO_4$ in a series of electrolyte solvents including EC, PC and butylene carbonate (BC), with the two former yielding the best performance in terms of capacity retention. Furthermore, the electrochemical performance of the 1 M $NaClO_4$ in EC:DMC, EC:EMC and EC:DEC showed the latter electrolyte to outperform all the others (Dahbi *et al.* 2014). Thus, PC- and EC:DEC-based electrolytes are the most suitable for both $NaClO_4$ and NaFSI salts.

Oh *et al.* (2014) introduced another type of organic solvent electrolyte, ethyl methanesulfonate (EMS), which outperformed the conventional PC solvent in terms of higher ionic conductivity (6 mS cm^{-1} vs. 5.7 mS cm^{-1} in PC) and a wider anionic stability (5.6 V compared to 4.5 V vs. Na/Na^+) when 1 M $NaClO_4$ electrolytes were formed by each (Figure 5.1).

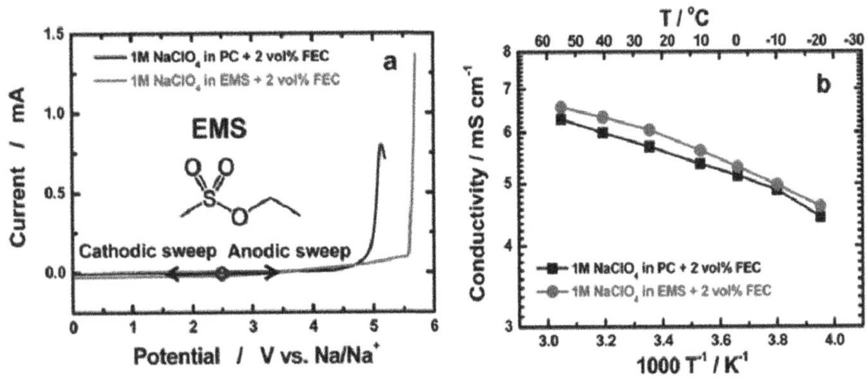

Figure 5.1. *(a) Linear sweep voltammetry of a sodium cell with two electrolytes: 1 M $NaClO_4$ in PC + 2 vol% FEC and 1 M $NaClO_4$ in EMS + 2 vol% FEC and (b) the ionic conductivity of the same electrolytes. Reprinted with permission from (Oh* et al. *2014). Copyright (2014) American Chemical Society. For a color version of this figure, see www.iste.co.uk/monconduit/batteries.zip*

5.2.2. *IL-based electrolytes*

A great deal of effort has been devoted to addressing the safety issues arising from the use of organic liquid electrolytes, due to their low thermal stabilities, high flammability and large volatilities (Yang *et al.* 2018). IL-based electrolytes can overcome some of the abovementioned issues in terms of safety, since they can be designed to be both thermally very stable, non-flammable and non-volatile.

ILs are composed exclusively of ions and most consist of large organic cations and small inorganic anions (MacFarlane *et al.* 2014; Welton 2018) (Table 5.1). When these ILs are mixed with a salt containing the charge carrier, e.g. Na salts (mostly with the same anion in both the salt and IL), multifunctional electrolytes are formed that are suitable for sodium battery applications. Herein, we highlight both the basics and the recent developments in the use of IL-based electrolytes for sodium metal batteries and NIBs.

Cation structures		
	Abbreviation	Cation structure
Quaternary cations Alkylmethylimidazolium	$[C_{nm}im]^+$	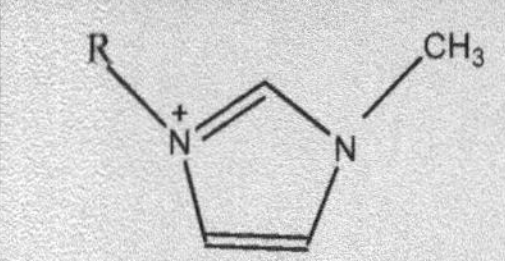
Alkylmethylpyrrolidinium	$[C_{nm}pyr]^+$	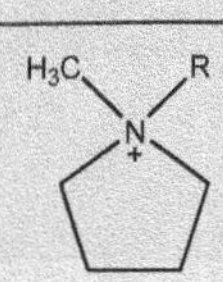
Ammonium (ether functionalized)	$[N_{n,n,n,n}\,O1]^+$	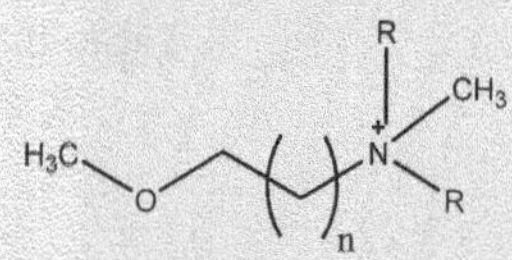
Ammonium	$[N_{n,n,n,n}]^+$	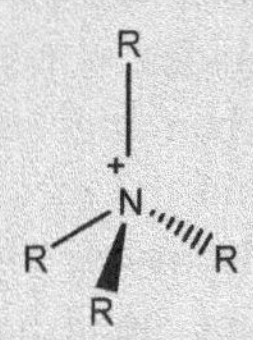
Phosphonium	$[P_{n,n,n,n}]^+$	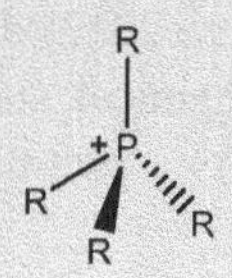

Anion structures		
Dicyanamide	DCA	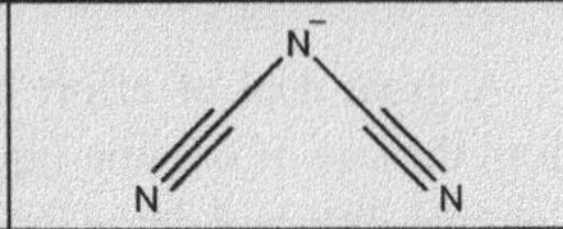
Trifluoromethanesulfonate	TFSI	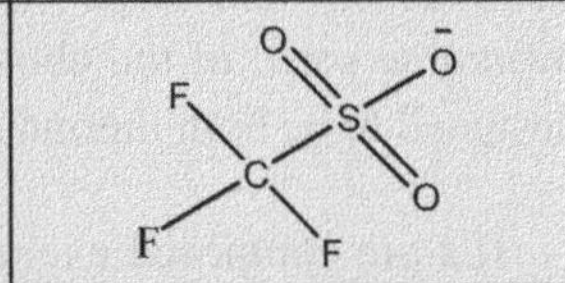
Bis(trifluoromethanesulfonyl)imide	TFSI	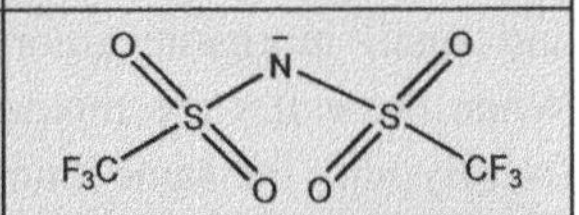
Bis(fluorosulfonyl)imide	FSI	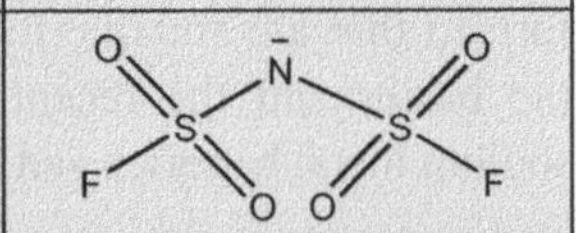
Tetrafluoroborate	BF_4^-	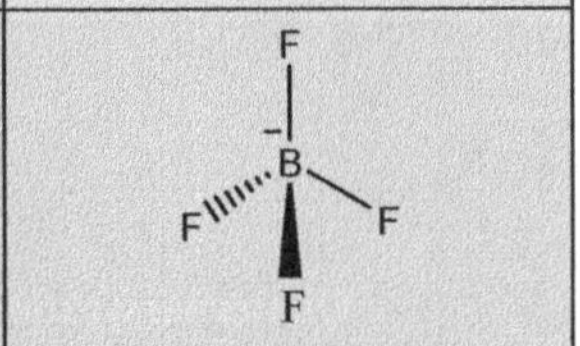
Hexafluorophosphate	PF_6^-	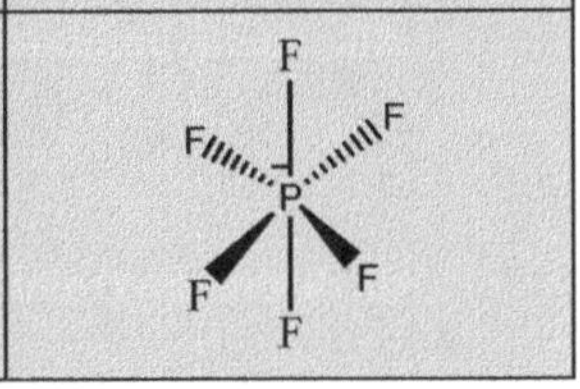

Table 5.1. *Structure and abbreviations of common IL and plastic crystal cations and anions*

Organic liquid carbonate based electrolytes are not compatible with sodium metal anodes, which are important for the formation of high energy density anodes due to their high reactivity, dendrite formation and an unstable passivation layer formed on the metal surface (Hwang *et al.* 2017). In contrast, a wide range of IL-based electrolyte chemistries have been reported to successfully enable the cycling of sodium metal batteries (e.g. pyrrolidinium and phosphonium-based ILs) (Mohd Noor *et al.* 2013;Monti *et al.* 2014; Yoon *et al.* 2014b). They also yield high anodic electrochemical stabilities (e.g. 5.2 V for NaFSI in [C_3mpyr][FSI]) (Ding *et al.* 2013). As a prominent example, reversible sodium stripping and plating has been attained for a 0.25 mol kg^{-1} (~0.25 M) NaTFSI in [C_3mpyr][TFSI] electrolyte at room temperature (Yoon *et al.* 2014a).

To exemplify the important role that IL-based electrolytes play in the performance of sodium metal based batteries, full cell cycling with a sodium metal anode and a $Na_{0.45}Ni_{0.22}Co_{0.66}O_2$ cathode was performed with a "traditional" organic liquid electrolyte, 0.5 M $NaPF_6$ in PC versus a 10 mol% (~0.45 M) NaTFSI in [C_4mpyr][FSI] electrolyte (Chagas *et al.* 2014). The latter cell presented superior cycling performance, ascribed to the higher electrochemical stability of the IL-based electrolyte. Furthermore, the formation of a uniform and thin SEI layer at the surface of the cathode material was confirmed via SEM imaging, which was responsible for the enhanced cycling performance of the cells with the IL-based electrolyte (Chagas *et al.* 2014).

Overall, ILs are more viscous than conventional organic solvents. This should be detrimental to the mass transport properties and therefore long-term battery performance, as well as the rate performance. However, this can be mitigated by the use of highly concentrated electrolytes (HCEs); up to 1:1 Na salt:IL have been reported to present high Na-ion transference numbers leading to higher reversibility in Na-based batteries (Forsyth *et al.* 2016; Girard *et al.* 2017). For example, Hilder *et al.* reported on sodium metal cells using an IL-based HCE, 2.3 M NaFSI in [P_{111i4}][FSI] in combination with a cathode composed of a polyanionic compound (P_2- and O_3-$Na_{2/3}[Fe_{2/3}Mn_{1/3}]O_2$). These cells outperformed cells that used a 0.5 M $NaPF_6$ in EC/DMC (1:1) electrolyte – the capacity retention after 15 cycles was 86% using the IL-based electrolyte versus 77% for the carbonate-based electrolyte (Hilder *et al.* 2017). Acceptable rate capability and cycling performance have also been reported using an HCE composed of 40 mol% NaFSI in [C_3mpyr][FSI] in a Na/$NaCrO_2$ cell (Ding *et al.* 2014).

While the cycling stability of Na metal batteries is the main argument for the introduction of IL-based electrolytes, cells using HC anodes have also been studied. For example, 10 mol% NaFSI in [C_3mpyr][FSI] was demonstrated to be stable even at 90°C in a Na/HC cell (Fukunaga *et al.* 2014). This same electrolyte also resulted in a high performant full cell, which exhibited excellent electrochemical properties at the laboratory scale (1.5 mAh coin-type) – the HC/$NaCrO_2$ cell cycled at 60 and 90°C showed 96.9% and 97.5% coulombic efficiency, respectively, and under the latter conditions, a specific capacity of 102 mAh/$g_{\text{-NaCrO2}}$, maintaining 90% of the initial discharge capacity after 1,000 cycles at 90°C (Fukunaga *et al.* 2016). The specific capacity of this NIB is comparable to the sodium metal cell with the same cathode and electrolyte (106 mAh $g_{\text{-NaCrO2}}$) (Ding *et al.* 2013).

Fukunaga *et al.* used a similar IL composition (NaFSI in [C_3mpyr][FSI]) and sodium chemistry (HC/$NaCrO_2$ anode and cathode electrode, respectively) for the successful fabrication of large-scale NIBs (27 Ah capacity) for the first time in 2016. Their results indicated that the performance of a large-sized NIB is comparable with that of the laboratory-scale coin-cell (96.9% coulombic efficiency at 60°C and 90% capacity retention after 1,000 cycles) (Fukunaga *et al.* 2016).

For applications requiring room temperature battery operation, MacFarlane's research group investigated a full NIB cell, composed of a carbon coated sodium vanadium phosphate (NVP/C) cathode and a HC anode using 1 M NaFSI in ([C_3mpyr][FSI]), at room temperature. They demonstrated a safe, high energy density (368 W h kg^{-1}) NIB, which is higher in energy density than that reported with $NaCrO_2$ cathode material (102 mAh $g^{-1}{}_{-NaCrO2}$). This cell showed 75% capacity retention after 100 cycles with high rate capability, and it outperforms that with a commercial organic solvent electrolyte (1 M $NaClO_4$ in EC:PC (1:1) v/v), which showed only 57% capacity retention. The good performance of this cell was attributed to the ability of the IL to form conductive, stable and uniform SEI/CEI layers on the surfaces of the electrodes (Figure 5.2; Manohar *et al.* 2018a).

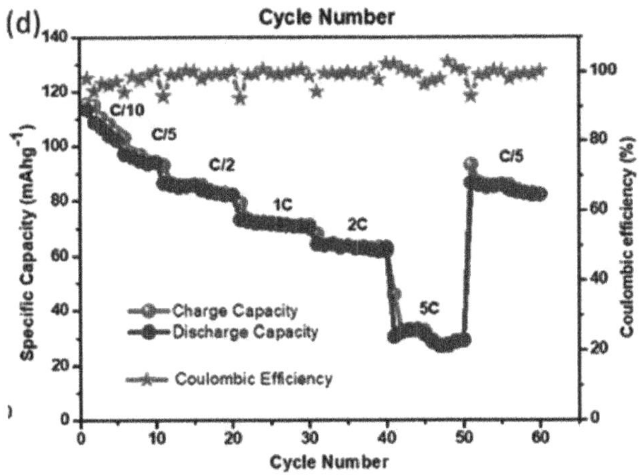

Figure 5.2. *Cycling performance of the NVP/C/IL/HC at different current densities. Reprinted with permission from Manohar* et al. *(2018a). Copyright (2018) Royal Society of Chemistry. For a color version of this figure, see www.iste.co.uk/monconduit/batteries.zip*

5.2.3. *Hybrid electrolytes*

ILs have paved the way to improve the cyclability and safety of batteries, particularly those operating above room temperature (Ding *et al.* 2013; Yoon *et al.* 2013, 2014b). However, depending on the cation and anion combination, IL-based electrolytes have some limitations in terms of high viscosity and lower ionic conductivity compared to conventional organic solvent electrolytes (Tsunashima and Sugiya 2007). Therefore, employing ILs as co-solvents in organic solvents to form hybrid electrolytes has been a rational way to enhance mass transport and electrochemical performance in NIBs (Sato *et al.* 2004; Guerfi *et al.* 2010). Other benefits of the hybrid electrolytes are that ILs can act as fire retardants, to otherwise flammable organic solvents, increasing safety (Plylahan *et al.* 2016) and lowering the cost of IL-based electrolytes.

The physicochemical behavior and electrochemical performance of hybrid electrolytes depend on the concentration and chemical composition of each component. For instance, the addition of 10–25 wt% of IL ([C_4mim][TFSI] or [C_3mpyr][TFSI]) to EC:PC barely affects the ionic conductivity of the pure organic solvent (Monti *et al.* 2016). However, increasing the IL content by too much, relative to the organic solvent, results in electrolytes with inferior physicochemical properties (Monti *et al.* 2016; Manohar *et al.* 2018b).

As previously mentioned, electrolytes that have a high viscosity are not always directly related to poor battery performance, and great cycling performance has also been reported by using hybrid electrolytes in NIBs. Therefore, it is important to understand the structure–property relationships of each system.

As an example, Monti *et al.* (2016) showed a significant specific capacity (182 mAh/g at C/10 after 40 cycles) using HC anode cells and a 0.80 M NaTFSI in [C_3mpyr][TFSI]/EC:PC (10:90 wt%) electrolyte. This result is comparable with organic solvent electrolytes such as 1 M $NaPF_6$ or $NaClO_4$ in EC:PC (Ponrouch *et al.* 2012), with specific capacities of 200 mAh g^{-1} at C/10 after 180 cycles.

Higher IL content based hybrid electrolyte compositions (1–75 vol%) have been studied by Manohar *et al.* (e.g. 1 M NaFSI in different volume ratios of 1:1 by volume [C_3mpyr][TFSI]/EC:PC). Superior performance was attained for the hybrid electrolyte containing 50 and 75 vol% IL (Figure 5.3). The main reason for this optimum performance was attributed to the formation of a stable and conductive passivation layer at the surface of the cathode (Manohar *et al.* 2018b).

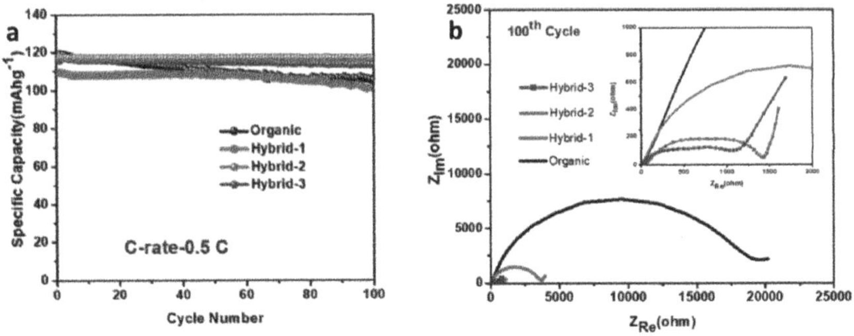

Figure 5.3. *(a) Cycling performance of sodium metal cells with an NVP/C cathode containing organic solvent and the IL hybrid electrolytes at 0.5 C and (b) EIS measurements for Na/Na symmetrical cells after 100 cycles. Hybrids 1, 2 and 3 are 1 M NaFSI in 25, 50 or 75 vol% [C3mpyr][TFSI] mixtures with organic solvent (EC:PC 1:1). Reprinted with permission from Manohar* et al. *(2018b). Copyright (2018) Royal Society of Chemistry. For a color version of this figure, see www.iste.co.uk/monconduit/batteries.zip*

5.2.4. *Effects of additives and impurities*

Water is a common impurity in electrolytes that has a major impact on the performance of high-voltage batteries. Water reacts with metals, forming oxides that block the surface and also degrade the electrolyte, as reported by Xiong's group while exploring $NaPF_6$ in various organic solvents (EC, PC, DMC, EMC and DEC). They reported the formation of HF in battery-grade electrolyte with less than 20 ppm water content (Barnes *et al.* 2020). However, it is important to note that this may not be entirely relevant to the carbonate-based solvents, as $NaPF_6$ is prone to hydrolyze in the presence of water.

Among the different carbonate-based electrolytes, the degradation rate depends on the chemical nature of the carbonate. For instance, EC shows a faster rate of decarboxylation compared to PC, whereas among acyclic carbonate solvents the rate of degradation of DMC was slower than DEC, probably related to the length of the alkyl chain and hence the nucleophilic character of anions generated through degradation (Barnes *et al.* 2020).

Interestingly, the presence of water in IL-based electrolytes has been shown to improve the cycling performance of lithium and sodium batteries through the formation of a more conductive and uniform SEI layer on sodium metal, which leads to stable stripping and plating of sodium. Basile *et al.* (2016) studied the effect of water in a [C_3mpyr][DCA]-based electrolyte and concentrations less than 100 ppm had a significant beneficial effect concentrations >200 ppm presented a detrimental effect on cell cycling.

Sodium cells with FSI-based IL electrolytes (e.g. 50 mol% NaFSI in [C_3mpyr][FSI]) present a higher water tolerance; an improved ionic conductivity and a decrease in the polarization potential at high current density (1 mA cm^{-2}) were attained in symmetrical cells with concentrations up to 10,000 ppm water (Basile *et al.* 2018a).

Finally, the coulombic efficiency of a full Na | IL-based | $NaFePO_4$ cell in the presence of 1,000 ppm water was higher (99%) than for the dry electrolyte system (80%). This confirms that water is a convenient additive for IL-based electrolytes for sodium battery applications (Ferdousi *et al.* 2019).

The purity of the chemicals used for battery preparation is also paramount to sodium battery performance and can affect the coulombic efficiency, capacity and cycling stability. Komaba's research group focused on the influence of the purity of 1 M $NaPF_6$ electrolytes, battery-grade versus commercial, on the electrochemical performance of Na/HC cells, and found the former to be superior (Kubota and Komaba 2015).

Poor electrochemical performance of HC/$Na_3V_2(PO_4)_2F_3$ cells were attained with 1 M $NaPF_6$ dissolved in $EC_{0.45}$:$PC_{0.45}$:$DMC_{0.1}$ as compared to 1 M $NaClO_4$ in the same solvent mixture. The conclusion was that the lower purity of $NaPF_6$ influenced the electrochemical performance (Kubota and Komaba 2015).

5.2.5. *Solid-state electrolytes*

5.2.5.1. *Organic ionic plastic crystals*

OIPCs, the solid-state analogues of ILs, are emerging solid-state electrolytes (SSEs) that have advantageous properties, similar to ILs. These materials form several crystal structures at temperatures, below their melting point, and show plastic behavior with high ion conduction derived from rotational motion of the ions within their 3D lattice crystal structure (Pringle *et al.* 2010). This plasticity is also desired to promote a good contact between the electrodes and the OIPC (Shekibi *et al.* 2007, 2009; Jin *et al.* 2011).

Numerous studies of OIPC-based SSEs in Li batteries have been reported (Zhou and Matsumoto 2007; Lin *et al.* 2015; Zhou *et al.* 2017a, 2017b, 2018), while, however, the use of OIPCs for sodium battery application is in the very early stages of development.

Determining OIPC phase diagrams identifies the microstructure, composition and temperature where a material presents plastic behavior. The phase diagram investigation of a mixture of Na salts in OIPCs was done for the first time for NaTFSI in the OIPC [C_2mpyr][TFSI] and showed a eutectic composition for 15 mol% NaTFSI in [C_2mpyr][TFSI], which melts at 63°C. Also, a significant increase of three orders of magnitude in ionic conductivity compared to the pure OIPC (~10^{-7} S cm^{-1}) was achieved for the 40 mol% NaTFSI in [C_2mpyr][TFSI], while still maintaining plastic behavior (Forsyth *et al.* 2014).

Subsequently, Forsyth's group has studied other OIPCs, which has highlighted the importance of the anion and cation chemistry. For example, by changing the anion to DCA (NaDCA in [C_1mpyr][DCA]), the phase behavior, crystal structure, dynamics and ionic conductivity of the binary system changed; the electrolyte yielded a wider temperature range where it maintained its plastic behavior (e.g. 20 mol% NaDCA is a eutectic composition with a eutectic transition at 89°C) (Chimdi *et al.* 2015).

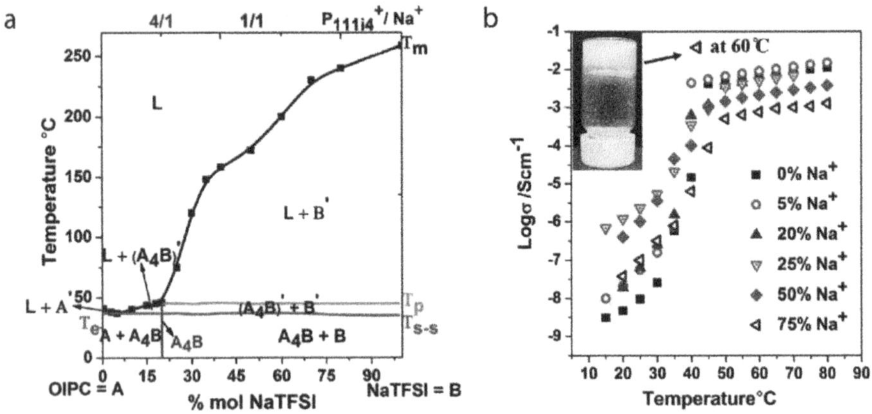

Figure 5.4. *(a) Phase diagram of NaTFSI in [P_{111i4}][TFSI] binary systems and (b) ionic conductivity of pure OIPC and mixtures with different concentrations of NaTFSI. Reproduced from Makhlooghiazad* et al. *(2017a). For a color version of this figure, see www.iste.co.uk/monconduit/batteries.zip*

A new solid phase was attained while investigating the phase diagram of NaTFSI in short alkyl chain phosphonium, [P_{111i4}][TFSI]; an OIPC with 20 mol% NaTFSI (4:1, i.e. [$Na_{0.2}(P_{111i4})_{0.8}$][TFSI]). All compositions studied revealed high room temperature ionic conductivity (>10^{-5} S cm^{-1}) (Figure 5.4; Makhlooghiazad *et al.* 2017a, 2017b).

A combination of phosphonium cations with smaller alkyl chain length and a small and delocalized anion leads to lower melting temperature with higher ionic conductivity and superior electrochemical behavior (Makhlooghiazad *et al.* 2018). The effect of mixed sodium salts with different anions, NaFSI, NaTFSI and $NaPF_6$ in $[P_{1i444}][FSI]$ OIPC, was explored. The $[P_{1i444}][FSI]/NaPF_6$ mixture results in a new phase compared to neat $[P_{1i444}][FSI]$. The physicochemical and electrochemical behaviors of NaFSI and NaTFSI systems are very similar. For the $NaPF_6$ mixture, however, preconditioning procedures such as the need to apply higher current density were required to form a stable and smooth SEI layer and to allow for stable stripping and plating of Na (Makhlooghiazad *et al.* 2017c).

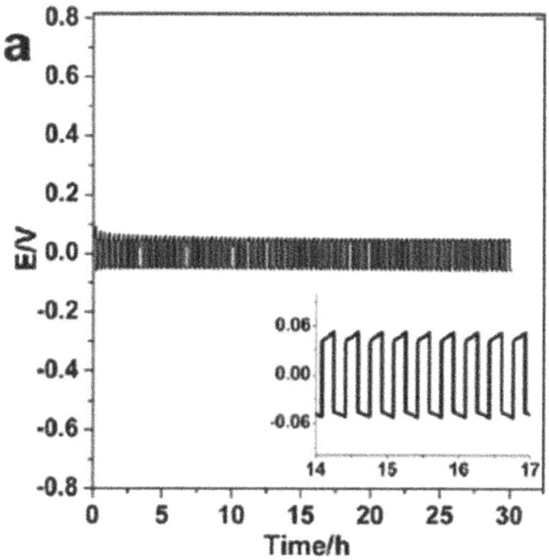

Figure 5.5. *Galvanostatic cycling of sodium symmetrical cell with 90 mol% NaFSI in $[P_{1i444}][FSI]$ at 50°C at 0.05 mA cm^{-2} (polarization period of 10 min). Reproduced from Makhlooghiazad* et al. *(2017b) with permission from The Royal Society of Chemistry*

Sodium symmetrical cells containing high concentration of NaFSI (90 mol%) in $[P_{1i444}][FSI]$ showed very low, reversible and stable polarization potential at 50 and 90°C. The intermediate composition (60 mol% NaFSI) was also cycled at room temperature at 0.25 mA cm^{-2} with a very stable and low cell polarization (200 mV) (Figure 5.5) (Makhlooghiazad *et al.* 2017b). These results highlight the possibility of using OIPC SSEs in sodium batteries.

5.2.5.2. *Solid polymer electrolytes*

SPEs for sodium battery applications contain a sodium salt dissolved in a polymer matrix and are highly desirable for energy storage device applications because they are light weight, cheap, resistant to volume change of the electrodes upon charge/discharge processing and enable desired flexible design of batteries without any leakage issue, hence enhancing safety of the batteries. Due to their mechanical flexibility and good interfacial contact with the electrodes, dendrite formation is suppressed in batteries employing SPEs (Long *et al.* 2016).

Since the major disadvantage of SPEs is their low ionic conductivity at room temperature, they are typically used at elevated temperatures to achieve enough ionic conductivity to be applied in cells. However, working at high temperatures can be detrimental for Na metal cells, as the metal melts at 95°C. Therefore, SPEs, mostly polyethylene glycol (PEO) based, have been studied in NIBs at temperature below the sodium melt temperature. PEO readily forms SPEs due to its intrinsic properties that makes it a great solvent for a variety of alkali salts (Armand 1983) (Wright 1998; Mohapatra *et al.* 2007; Colò *et al.* 2017). Furthermore, PEO shows high electrochemical stability compared to other polyethers and copolymers (Ni'mah *et al.* 2015). A few examples from the literature are described below.

$NaTFSI:PEO_{20}$ SPEs were studied by Moreno *et al.* and showed high ionic conductivity around 10^{-3} S cm^{-1} above 70°C. Additionally, improved mechanical and mass transport properties were attained in the presence of 5 wt% SiO_2, with the sodium transference number increased from 0.39 to 0.51 (Serra Moreno *et al.* 2014). In fact, inorganic nanoparticles have been widely used as fillers in polymeric electrolyte materials to improve the conductivity and mechanical properties (Weston and Steele 1982; Croce 1998; Klongkan and Pumchusak 2015; Liu *et al.* 2015). Those improvements are due to promotion of the cation–anion dissociation at the surface region of the nanoparticles, together with cross-linking of the polymer backbone by the nanoparticles, or by physical entanglements (Croce *et al.* 2001; Xie *et al.* 2007).

A study on the ionic conductivity and charge carriers in PEO-based SPEs using different sodium salts, $NaFSI:PEO_n$ and $NaTFSI:PEO_n$ (n = 6, 9, 20), showed that the NaFSI-based SPEs have lower ionic conductivities for more salt concentrated systems (n = 6), which was attributed to higher aggregation due to stronger cation–anion interactions, resulting in less free anions. Furthermore, NaTFSI-based SPEs prevent crystallization because of its large size and charge delocalization, which lead to higher ionic conductivity (Boschin and Johansson 2015).

Later, Qi *et al.* reported the high ionic conductivity of 4.1×10^{-4} S cm^{-1} at 80°C in NaFSI:PEO_{20} SPEs. Na symmetrical cells using this SPE showed excellent Na stripping and plating with low polarization potential, mainly due to the interfacial stability of the SPE and Na electrodes. These SPEs also showed great cycling performance in a sodium metal cell with $Na_{0.67}Ni_{0.33}Mn_{0.67}O_2$ cathode material at 80°C (Qi *et al.* 2016).

Alternative polymer host materials to PEO have also been investigated for sodium batteries. For example, Mindemark's research group showed that a high molecular weight poly(tri-methylene carbonate) (PTMC) polymer matrix can dissolve a large amount of NaFSI (up to 1:1) with a room temperature ionic conductivity of 5×10^{-5} S cm^{-1}. Comparable discharge capacities to those for PEO-based batteries were attained (90 mAh g^{-1} discharge capacity at 60°C) in a Na|5:1 PTMC:NaFSI|($Na_xFe[Fe(CN)_6]$) cell after 80 cycles (Sångeland *et al.* 2019a).

Brandell's research group used a polycaprolactone-poly(tri-methylene carbonate (PCL–PTMC) copolymer with NaTFSI and fabricated a full cell with HC and Prussian blue analogue (PBA) electrodes and cycled it at room temperature (22°C, that is impossible using PEO-based electrolytes) for 120 cycles (Sångeland *et al.* 2019b). Furthermore, a high transference number of 0.5 at 80°C was obtained for the SSPE.

5.2.5.3. *Gel polymer electrolytes*

SPEs can suffer from low room temperature ionic conductivity, which can be addressed by incorporating a solvent and forming GPEs. The solvents can be either an organic solvent (section 5.2.1) or an IL (section 5.2.2) – resulting in ionogels. In general, GPEs combine the benefits of both SPEs and liquid electrolytes, offering higher ionic conductivity and better interfacial contact with electrodes than an SPE (Fdz De Anastro *et al.* 2019a, 2019b; Mendes *et al.* 2019).

GPEs composed of poly(vinylidinefluoride-co-hexafluoropropylene) (PVdF-HFP) have been intensively studied due to their low crystallinity and high free volume that can accommodate and immobilize liquid electrolyte. This co-polymer contains both an amorphous phase and a crystalline phase due to the nature of the co-polymers (HFP and PVDF, respectively). In that way, a large amount of liquid electrolyte can be entrapped due to the nature of HFP, while PVDF supports the mechanical stability of the co-polymer (Deepak Kumar 2010). Isa *et al.* showed that a GPE composed of 20 wt% NaTfO in the organic solvent EC:PC incorporated in PVdF-HFP have thermal stability up to 145°C with a room temperature ionic conductivity of 2.5×10^{-3} S cm^{-1}; that is higher than SPEs, and one order of magnitude higher than the organic electrolyte solvent 1 M $NaClO_4$ in EC:DEC:DMC (e.g. 6×10^{-4} S cm^{-1} at room temperature).

The addition of polycrystalline beta alumina nanowires (Al_2O_3) to GPEs is a rational way to achieve higher mechanical stability of GPEs. By incorporating Al_2O_3 into the abovementioned GPE (1 M $NaClO_4$ in EC:DEC with 5 wt% FEC in PVdF–HFP), an ionic conductivity of 7.13×10^{-4} at 25°C was achieved. This composite can effectively entrap a large amount of liquid electrolyte and form dense and uniform solid–liquid hybrid Na-ion transportation channels. As a result, long-term and stable symmetrical Na cells at a high current density of 2 mA cm^{-2} were demonstrated, likely supported by the formation of a stable and uniform SEI on the sodium metal surface. The Na| composite| $Na_3V_2(PO_4)_3$ (NVP) full cell cycled for 1,000 cycles at 1C rate of charge/discharge and 60°C and delivered a capacity retention of 78.8% (Lei *et al.* 2019).

GPEs based on carbonate electrolytes encounter a number of drawbacks, such as the reactivity of the electrolyte with Na metal resulting in poor cell performance. Furthermore, the ionic conductivity of these GPEs may decrease due to solvent evaporation. The significant problem of flammability and volatility exists in GPEs using organic solvent plasticizers. As mentioned earlier, to overcome these issues, non-flammable and non-volatile ILs have been used, which produces more stable and safer GPEs for all solid-state Li and NIBs. The substitution of an organic solvent by an IL, depending on the loading, would lead to an increase in ionic conductivity due to the inherent nature of the IL.

Boschin *et al.* investigated the plasticizing effect of [C_3mpyr][TFSI] or [C_3mpyr][FSI] to $NaTFSI(PEO)_n$ or $NaFSI(PEO)_n$ SPEs (n = 6, 9, 20), respectively. The addition of IL can hinder PEO crystallization, especially in the presence of FSI anions. The ionic conductivity in GPEs with either anion is higher than SPEs based on PEO, due to more conduction pathways and larger amount of charge carriers present (Boschin and Johansson 2016). Further studies of GPEs composed of [C_3mpyr][FSI] in $NaClO_4$:PEO SPEs by infrared spectroscopy and density functional theory (DFT) calculations revealed the plasticizing role of IL for the PEO backbone through interaction of FSI anions with the PEO backbone, resulting in decreased crystallinity of PEO and increased dynamics of the PEO chains. Furthermore, this GPE shows a Na^+ transference number of 0.44 with great electrochemical stability up to 4.4 V versus Na/Na^+. The cell Na|GPE|NVP shows a reversible discharge capacity of 100 mAh g^{-1} with a capacity retention of 86.7% after 70 cycles at 0.1 C rate. Even at the higher rate of 2 C, this cell delivers a discharge capacity of 94 mAh g^{-1} at 60 °C (Chen *et al.* 2019).

MacFarlane's research group synthesized an ionogel by incorporating an IL-based electrolyte, NaFSI in [C_3mpyr][FSI], into a PVdF–HFP host, which was tested in a full battery with a sodium metal anode and an NVP cathode. The battery exhibited a capacity retention of 92% with a coulombic efficiency of 99.9% after 150 cycles at a current density of 1 C, which is higher than for the same polymer backbone with an EC:DEC electrolyte incorporated (Mendes *et al.* 2019).

Fdz De Anastro *et al.* (2019b) prepared a GPE based on incorporation of the IL-based electrolyte, NaFSI in [C_3mpyr][FSI] into another commercially available polymer matrix, poly(ethylene glycol) diacrylate (PEGDA), by fast (<1 min) UV photo-polymerization. The GPE containing 90 wt% IL with 20 and 50 mol% NaFSI showed high ionic conductivity (6.5×10^{-3} and 1.5×10^{-3} $S \cdot cm^{-1}$, respectively) at 50°C with high t_{Na}^{+} of 0.3 and 0.49. They reported better electrochemical performance of GPE containing higher concentrations of NaFSI in both a sodium symmetrical cell and a full cell with a Na metal anode and $NaFePO_4$ cathode.

5.3. Properties of IL-based electrolytes for Na batteries

As mentioned above, the strong case for IL-based electrolytes lies in the many unique intrinsic properties of the ILs themselves that fulfill battery electrolyte requirements, such as low vapor pressure, excellent thermal and electrochemical stability and low flammability. Furthermore, these properties as well as those that are more related to electrolyte performance, such as ionic conductivity, polarity and fluidity, can be tailored by the IL composition due to the large variety of cation–anion configurations that can be employed and also further widened by engineering their chemical structures (Yang *et al.* 2018).

5.3.1. *Physical properties*

High viscosity and hence low ionic conductivity is normally one of the main drawbacks of IL-based electrolytes, since it is detrimental toward the rapid ion transport between the two electrodes. This is the result of the strong intermolecular interactions arising from the coulombic interactions existing in ILs due to their intrinsic ionic nature. In principle, smaller ions should enhance the fluidity, but this comes at the expense of stronger interactions and hence delocalization of the ionic charge over a large ion, or other strategies that weaken the intermolecular interactions, which are key strategies to reduce the viscosity of the electrolytes.

Different strategies have been explored to increase the fluidity and thus optimize the transport properties of IL-based electrolytes. For instance, the chemical structure of the cation and anion can be tuned to decrease the viscosity of the electrolyte. The inclusion of oligo-ether side chains into a quaternary ammonium cation has been reported to lead to more fluid electrolytes by weakening the electrostatic interactions

(Kar *et al.* 2016). The planarity of the cationic core is also key to improve the fluidity where tetraalkylammonium < *N,N*-dialkylpyrrolidinium < *N,N*-dialkylimidazolium (Buzzeo *et al.* 2004; O'Mahony *et al.* 2008).

Turning to the anions, the common choice is to employ different WCAs. Yet, ILs based on perfluorinated anions such PF_6^- and BF_4^- are quite viscous due to their pseudo-spherical symmetry and strong H• • •F interactions via hydrogen bonding. Instead, anions such as TFSI and FSI have emerged as interesting candidates due to the low viscosity of their ILs. This is most likely related to the elongated shape and internal flexibility of these anions, resulting in plasticizing properties, together with a very delocalized negative charge (Johansson *et al.* 1998). Additionally, employing FSI leads to the formation of a stable SEI on reactive metals, such as Na, and some have even nick-named FSI the "magic anion" (Shkrob *et al.* 2014).

The addition of Na salt to the IL results in an increased viscosity of the electrolyte due to the formation of heavier and bigger ion pairs/aggregates (Mohd Noor *et al.* 2013; Forsyth *et al.* 2016). Also, it impacts negatively on the glass transition temperature (T_g), two clear signs of hindered ion mobility.

However, even highly viscous IL-based electrolytes have shown excellent sodium battery rate performance and stability as reported by Hagiwara's group (Ding *et al.* 2014). Despite the less than optimal physicochemical properties of the electrolyte, the transference number for the sodium cation ($t_{Na}{}^{+}$) increases with salt concentration (Matsumoto *et al.* 2015). An IL can dissolve up to 50 mol% of Na salt (ca. 3 M) to reach the most stable and high-rate cycling. This is clearly related to an altered sodium-ion conduction mechanism – not being vehicular, as this would be inversely proportional to the viscosity, but rather based on structural rearrangements, cooperativity or ion hopping (Forsyth *et al.* 2016).

In addition, working at higher temperatures will further facilitate the Na-ion transport, and this is indeed possible in IL-based electrolytes due to their thermal stability.

5.3.2. *Thermal stability*

The operation temperature of a sodium battery is to a large extent determined by the stability of the electrolyte employed versus decomposition and/or evaporation. As mentioned above, ILs and hence also IL-based electrolytes, most often present negligible vapor pressures due to the electrostatic interactions between the ions and therefore the operation temperature will normally be governed by the electrolyte decomposition.

Several studies have compiled the contributions of many different factors to determine the thermal stability of ILs; (i) the alkyl chain length, related to the formation of stable carbon cations or carbon radicals, (ii) alkyl functionalization and introduction of oxygen into the alkyl substituent and (iii) increasing the coordinating nature, nucleophilicity and hydrophilicity of the anion (Cao Mu 2014; Eshetu *et al.* 2017; MacFarlane *et al.* 2017b).

From the cation perspective, the general factors toward the decomposition temperature are the chemical nature of the cation and the anion, the functional groups, the cation alkyl chain length, and in general this results in temperature stabilities ranging from 100 to 400°C (Basile *et al.* 2018b). The general trend in IL cation chemical nature and stability is: phosphonium > pyrrolidinium > imidazolium > tetraalkyl ammonium (Kroon *et al.* 2007; MacFarlane *et al.* 2017b).

For the common fluorinated anions, TFSI, FSI and also bis(perfluoroethanesulfonyl)imide (BETI), the thermal decomposition of the ILs has been studied for a given cation, $[C_4mpyr]^+$, showing that the FSI anion is the least thermally stable anion (Eshetu *et al.* 2017), and similar results have also been observed for tetraalkyl ammonium based ILs, and this has been attributed to the FSO_2 group being more sensitive toward pyrolysis (Han *et al.* 2010). Still, the decomposition temperature of the IL $[C_4mpyr][FSI]$ is much higher (ca. 287°C) (Eshetu *et al.* 2017) than the boiling point of a traditional organic battery solvent such as DMC (92°C).

Similar trends have been reported for Na salt containing pyrrolidinium and phosphonium IL-based electrolytes, where the new interactions induced in the electrolyte also determine the decomposition temperature.

Moving the thermal stability one step further, a fire hazard analysis has also been made for a series of pyrrolidinium ILs and compared to conventional organic solvents (Monti *et al.* 2016; Eshetu *et al.* 2017), and the ignitability follows a similar trend to the decomposition temperature and thus increases with the cation chain length.

5.3.3. *Electrochemical stability*

A key criterion for selection of a solvent for electrochemical application is the electrochemical stability, and it is definitively a prerequisite for attaining high voltage batteries – such as sodium batteries – and not the least to be stable down to –2.71 V versus SHE – the sodium metal potential (and HC comes very close to this as already mentioned). The electrochemical stability is usually expressed by the width of the electrochemical stability window (ESW) i.e., the difference between the oxidation and reduction potentials.

An ESW of 5–6 V is not unusual for ILs (Buzzeo *et al.* 2004) and this clearly surpasses both aqueous and conventional organic liquid electrolytes (Buzzeo *et al.* 2006). Some caution is, however, warranted due to the sluggish kinetics of IL-based electrolytes, caused by their high viscosities – thus the effective ESWs might be narrower. Both the cation and the anion of the IL affect the ESW, a direct result of the oxidation and reduction reactions taking place at the anode and cathode, respectively, and the wide variety of IL chemistries opens for tuning the ESW.

In general, the ESW is affected by the IL cation core as: phosphonium > piperidinium > pyrrolidinium > tetraalkylammonium > imidazolium (De Vos *et al.* 2014), while the anion affects the ESW as: TFSI > FSI > PF_6^- > BF_4^- > halides (Jónsson Johansson 2015; MacFarlane *et al.* 2017a; Forero-Saboya *et al.* 2019; Hosseini-Bab-Anari *et al.* 2019). As a general rule, the anionic stability is directly related to the hydrophobicity of the anion. However, there are exceptions to this general trend depending on the cation–anion combination (MacFarlane *et al.* 2017a).

Impurities present in the electrolyte often have detrimental effects on the ESW. Water is the most common impurity and while it increases the fluidity of the electrolyte it decreases the ESW at both the cathodic and the anodic limits due to catalytic activity (O'Mahony *et al.* 2008). Additionally, IL anions such as PF_6^- and BF_4^- are prone to hydrolyze with subsequent generation of HF.

The ESWs also narrow as a function of temperature due to the electrochemical (side-)reactions occurring at a faster rate (O'Mahony *et al.* 2008).

5.4. Modeling IL-based electrolytes

Modeling and simulation of different sodium battery electrolyte materials is often employed to assist both the fundamental understanding of crucial phenomena such as Na-ion transport and the charge transfer in detail of existing systems, as well as to support the development of more performant systems. We will explain only a few select studies in more depth, that each target more general understanding of the special conditions and phenomena of IL-based electrolytes, and which can be extrapolated to cover many other Na conducting IL-based electrolytes.

For IL-based electrolytes, the starting point for modeling at the very local scale is the notion that there is a surplus of anions available for the Na^+ cation to coordinate and strictly no neutral solvent. In addition, large ionic aggregates do become more likely and hence also important to model.

The very nature of the strong coulombic interactions makes the use of accurate *ab initio* and DFT methods almost mandatory. By employing DFT, small systems like $[Na(TFSI)_n]^{1-n}$, n = 1–4, were modeled early on to study the Na^+ coordination and solvation. From only the computational data, the n = 2 species was suggested to be the preferred charge carrier, but with n = 3 as an almost equally probable candidate in terms of interaction energy. However, by also using experimental Raman spectra of NaTFSI in $[C_2mim][TFSI]$, the latter species was concluded to be the preferred local structure (Monti *et al.* 2014).

While indeed a very small and local model, this has huge implications for the fundamentals of the ion transport, as Na^+ is to be transported as a doubly negative anionic species in most (TFSI-based) IL-based electrolytes. This is a most profound difference, compared to an organic liquid electrolyte wherein the Na^+ can be surrounded by five to six solvent ligands and still be a positive species, which is compensated for by a less strongly solvated WCA.

In contrast, the IL-based electrolyte will be charged in neutral conditions and have at least three different species present: the $[Na(An)_3]^{2-}$ complex, the other anions (charge = –1) and the IL cations (charge = +1) (Figure 5.6). This can also be used to partially explain the low Na transference numbers (for systems of low Na salt content).

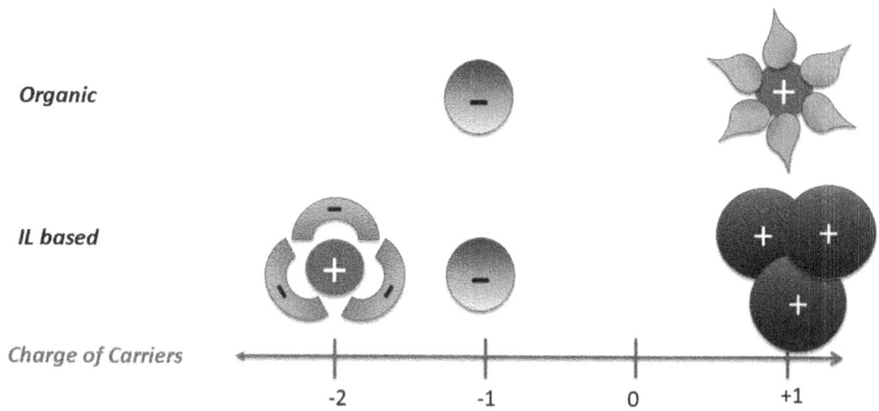

Figure 5.6. *Schematic of the main types of charge carriers and their charges present in organic liquid electrolytes and IL-based electrolytes. For a color version of this figure, see www.iste.co.uk/monconduit/batteries.zip*

Moving to a somewhat larger scale and systems, and with the intention to also include true dynamics, there are relatively many molecular dynamics (MD) studies on Na conducting IL-based electrolytes, especially with regard to elucidating the ion transport mechanism. Yet, as the interactions are special, the stance is to first validate any classical MD force-fields (FFs) by *ab initio* MD (AIMD). One such study found a simple Lennard–Jones potential in combination with care taken to the coulomb interactions to successfully reproduce the AIMD data for NaTFSI in [C_4mpyr][TFSI] (Vicent-Luna *et al.* 2016). Once the FF had been validated, several NaTFSI concentrations in the IL were studied and it was found that the ion conductivities decreased as a function of Na salt concentration – largely due to the formation of large aggregates. Overall, the diffusion was much slower for the Na^+ species than for the IL cations and the TFSI anions. Since the Na^+ transference numbers have been noted to increase as a function of salt concentration when HCEs are reached; this has also been investigated by MD simulations (Forsyth *et al.* 2016). One example is NaFSI in [C_3mpyr][FSI] – and at beyond 50 mol% of NaFSI, the FSI anions are shared between several sodium cations forming clusters, and the increase in transference number, reaching up to 0.25 – high for any IL-based electrolyte, is possibly due to an ion transport mechanism involving structural re-arrangements and Na^+ jumping between anionic coordination sites. The mean square displacement shows a faster Na^+ diffusion than for both the FSI anion and the IL cation for this IL-based HCE. Yet another MD study focused on the comparison between Na and Li by Na/LiFSI in [C_3mpyr][FSI] and also Na/LiPF_6 in [P_{1224}][PF_6] (Chen and Forsyth 2016). The effect of aggregate formation was more pronounced in the Na systems with both larger and more long-lasting structures, and increasing the salt concentration yielded larger aggregates. At least two different Na-ions could be discerned; some slow that were found to most often not be part of any large aggregate, while the more rapidly moving were mediated by rearrangements of the anions in the aggregates (Figure 5.7). Furthermore, the MD simulations indicate that the structural re-arrangements cause several Na^+ in close proximity to one another to move concerted. This is in line with later DFT results also for organic HCEs where Na^+ systems are prone to create diverse coordination environments, and are more easily affected by the salt concentration (Flores *et al.* 2017).

More recently, the behavior of these electrolytes at charged electrodes has also been investigated by MD simulations (Dmitrii *et al.* 2020). Here, the authors have shown that, while IL electrolytes tend to have nanostructured layering at a charged electrode as has been reported by several researchers for pure ILs and Li- and Na-based electrolytes, the HCE-based IL systems significantly disrupt the layering. Furthermore, at more negative electrode potentials the inner most layer appears to expel the pyrrolidiium cation and increases the concentration of Na^+ ions. This was hypothesized to lead to a better SEI layer and improved cycling of Na metal, which was also validated experimentally. This recent work opens up the possibility of applying electrode preconditioning together with tailoring electrolyte composition to improve electrochemical performance in Na devices. Moreover, the modeling work thus far demonstrates the power of its use not only to provide understanding of structure and dynamics in electrolytes but also to suggest potential composition and electrode treatment processes.

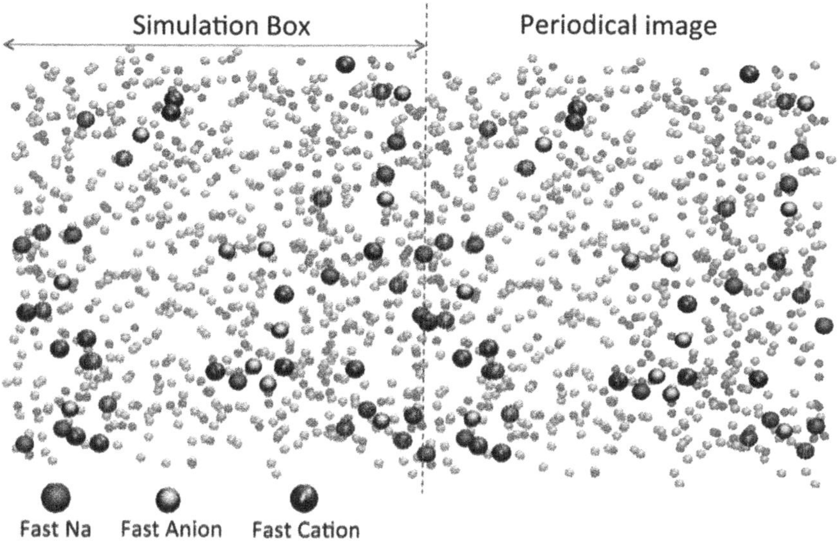

Figure 5.7. *Snap-shot from an MD trajectory showing fast ions to cluster. Reprinted with permission from Forsyth* et al. *(2016). Copyright (2016) American Chemical Society. For a color version of this figure, see www.iste.co.uk/monconduit/batteries.zip*

5.5. Conclusion and future perspectives

Today, the energy storage market for portable applications is totally dominated by LIBs. We also see them as totally instrumental for the current transition of the transportation sector to electrification and they are also emerging as contenders for large-scale storage for wind and solar power. However, the need for expensive Li-salts, the use of copper current collectors and the scarcity of some metals/materials necessary/preferred, such as cobalt, nickel and natural graphite, have triggered the need to research more sustainable, and still affordable and reliable energy storage technologies.

Sodium-based battery technologies are a logical alternative to LIBs due both to the many similarities to the present technology and to the availability of resources and projected low production costs. Sodium-based batteries should, however, ideally make use of a Na metal anode – as there is currently no obvious stronger candidate – and this puts the choice of electrolyte concept in focus.

ILs-based electrolytes present some unique properties that endows significant safety enhancements in comparison with conventional organic solvents, mostly related to higher decomposition temperatures. Furthermore, the intrinsic chemical composition of ILs allows for engineering the cations and the anions to fulfill specific target property values in terms of, e.g., mass transport, solubility, electrochemical and thermal stability. The solid-state analogues, OIPCs, can bring a new set of advantages such as no leakage and limiting dendritic growth. However, the unique and most wanted property of these electrolytes and their derivatives covered herein is the ability to work versus a Na metal anode – and thus keeping the promise of a competitive sodium-based battery technology.

However, several challenges must be addressed to move from fundamental and promises to a realization of sodium batteries using IL-based electrolytes:

– The ion transport during battery operation, especially at low temperatures, is limited by their high viscosities. This might also generate difficulties during the battery manufacturing, e.g. wetting of electrodes and separators. However, just as for LIB electrolytes, the optimal composition of these electrolytes is truly a multidimensional problem – and there is no reason to believe that we have exhausted our possibilities in neither choice of Na salt, IL or additional components (co-solvent for hybrids, additives). Here, the perspective of applying smarter and faster high-throughput screening techniques complemented with modeling utilizing the power of artificial intelligence approaches such a machine learning are key to creating a feasible optimization effort.

– The IL-based electrolytes must not only survive the harsh electrochemical conditions inside real sodium (metal) battery cells, using various electrode chemistries, but must do so with maintained capacity and power performance for several thousand cycles. Depending on the market/application targeted the sensitivity to operating temperature can be either an advantage (elevated) or a disadvantage (subambient). The control of purity and side-reactions are instrumental and as the quality of ILs and also Na salt have improved lately the perspective is that we will see advances in the real long-term performance of full cells – just as has been shown for organic liquid electrolyte based NIBs.

– The IL-based electrolytes must also show an economic viability in comparison with conventional organic liquid electrolytes. This can be made by smart up-scaling, by utilizing the unique prospects of recovering these electrolytes and by the use of less expensive cation and anion designs. It is, however, crucial to not look at production cost or purchasing cost of the IL-based electrolytes separately. If these electrolytes enable cells of Na metal to be more performant – in whatever targeted aspect, the cost of a single expensive component should be put in relation to the entire installation of an electrochemical storage solution – including aspects such as battery re-usage (second life) and re-cycling. A bit of a paradox is that as sodium-based batteries have a perspective of offering cells without any expensive or scarce metals, they will have less of monetary value in the re-cycling stage. Therefore, it is of utmost importance to create a framework for a circular economy for sodium-based batteries – even more than for LIBs.

5.6. Abbreviations

AIMD	*ab initio* MD
BC	Butylene carbonate
BETI	Bis(perfluoroethanesulfonyl)imide
CEI	Cathode electrolyte interphase
DEC	Di-ethyl carbonate
DFT	Density functional theory
Diglyme	Di-ethylene glycol di-methyl ether
DMC	Di-methyl carbonate
DME	Di-methoxyethane
EMC	Ethyl-methyl carbonate
EC	Ethylene carbonate
EMS	Ethyl methanesulfonate
ESW	Electrochemical stability window
FEC	Fluoroethylene carbonate

FF	Force-field
GPEs	Gel polymer electrolytes
HC	Hard carbon
HCE	Highly concentrated electrolyte
IL	Ionic liquid
MD	Molecular dynamics
NASICON	Sodium super ionic conductor
NVP	Sodium vanadium phosphate
OIPC	Organic ionic plastic crystal
PBA	Prussian blue analogue
PC	Propylene carbonate
PCL–PTMC	Poly(caprolactone)-poly(tri-methylene carbonate)
PEGDA	Poly(ethylene glycol) diacrylate
PEO	Polyethylene glycol
PTMC	Poly(tri-methylene carbonate)
PVdF-HFP	Poly(vinylidinefluoride-co-hexafluoropropylene)
SEI	Solid electrolyte interphase
SPE	Solid polymer electrolyte
SPI	Solid polymer interphase
SSEs	Solid-state electrolytes
Triglyme	Tri-ethylene glycol di-methyl ether
WCA	Weakly coordinating anion

5.7. References

Armand, M. (1983). Polymer solid electrolytes–An overview. *Solid State Ionics,* 9–10, 745–754.

Barnes, P., Smith, K., Parrish, R., Jones, C., Skinner, P., Storch, E., White, Q., Deng, C., Karsann, D., Lau, M.L., Dumais, J.J., Dufek, E.J., and Xiong, H. (2020). A non-aqueous sodium hexafluorophosphate-based electrolyte degradation study: Formation and mitigation of hydrofluoric acid. *Journal of Power Sources,* 447, 227363–227371.

Basile, A., Ferdousi, S.A., Makhlooghiazad, F., Yunis, R., Hilder, M., Forsyth, M., and Howlett, P.C. (2018a). Beneficial effect of added water on sodium metal cycling in super concentrated ionic liquid sodium electrolytes. *Journal of Power Sources,* 379, 344–349.

Basile, A., Hilder, M., Makhlooghiazad, F., Pozo-Gonzalo, C., Macfarlane, D.R., Howlett, P.C., and Forsyth, M. (2018b). Ionic liquids and organic ionic plastic crystals: Advanced electrolytes for safer high performance sodium energy storage technologies. *Advanced Energy Materials,* 8(17), 1703491.

Basile, A., Makhlooghiazad, F., Yunis, R., Macfarlane, D.R., Forsyth, M., and Howlett, P.C. (2017). Extensive sodium metal plating and stripping in a highly concentrated inorganic-organic ionic liquid electrolyte through surface pretreatment. *Chemelectrochem,* 4(5), 986–991.

Basile, A., Yoon, H., Macfarlane, D.R., Forsyth, M., and Howlett, P.C. (2016). Investigating non-fluorinated anions for sodium battery electrolytes based on ionic liquids. *Electrochemistry Communications,* 71, 48–51.

Boschin, A. and Johansson, P. (2015). Characterization of NaX (X: TFSI, FSI) – PEO based solid polymer electrolytes for sodium batteries. *Electrochimica Acta,* 175, 124–133.

Boschin, A. and Johansson, P. (2016). Plasticization of NaX-PEO solid polymer electrolytes by Pyr13X ionic liquids. *Electrochimica Acta,* 211, 1006–1015.

Buzzeo, M.C., Evans, R.G., and Compton, R.G. (2004). Non-haloaluminate room-temperature ionic liquids in electrochemistry—A review. *ChemPhysChem,* 5(8), 1106–1120.

Buzzeo, M.C., Hardacre, C., and Compton, R.G. (2006). Extended electrochemical windows made accessible by room temperature ionic liquid/organic solvent electrolyte systems. *ChemPhysChem,* 7(1), 176–180.

Cao, Y. and Mu, T. (2014). Comprehensive investigation on the thermal stability of 66 ionic liquids by thermogravimetric analysis. *Industrial & Engineering Chemistry Research,* 53(20), 8651–8664.

Chagas, L.G., Buchholz, D., Wu, L., Vortmann, B., and Passerini, S. (2014). Unexpected performance of layered sodium-ion cathode material in ionic liquid-based electrolyte. *Journal of Power Sources,* 247, 377–383.

Chen, F. and Forsyth, M. (2016). Elucidation of transport mechanism and enhanced alkali ion transference numbers in mixed alkali metal–organic ionic molten salts. *Physical Chemistry Chemical Physics,* 18(28), 19336–19344.

Chen, G., Bai, Y., Gao, Y., Wang, Z., Zhang, K., Ni, Q., Wu, F., Xu, H., and Wu, C. (2019). Inhibition of crystallization of poly(ethylene oxide) by ionic liquid: Insight into plasticizing mechanism and application for solid-state sodium ion batteries. *ACS Appl. Mater. Interfaces,* 11(46), 43252–43260.

Chimdi, T., Gunzelmann, D., Vongsvivut, J., and Forsyth, M. (2015). A study of phase behavior and conductivity of mixtures of the organic ionic plastic crystal N-methyl-N-methyl-pyrrolidinium dicyanamide with sodium dicyanamide. *Solid State Ionics,* 272, 74–83.

Colò, F., Bella, F., Nair, J.R., and Gerbaldi, C. (2017). Light-cured polymer electrolytes for safe, low-cost and sustainable sodium-ion batteries. *Journal of Power Sources,* 365, 293–302.

Croce, F., Appetecchi, G.B., Persi, L., and Scrosati, B. (1998). Nanocomposite polymer electrolytes for lithium batteries. *Letters to Nature,* 394, 456–458.

Croce, F., Persi, L., Scrosati, B., Serraino-Fiory, F., Plichta, E., and Hendrickson, M.A. (2001). Role of the ceramic fillers in enhancing the transport properties of composite polymer electrolytes. *Electrochimica Acta,* 46(16), 2457–2461.

Dahbi, M., Yabuuchi, N., Kubota, K., Tokiwa, K., and Komaba, S. (2014). Negative electrodes for Na-ion batteries. *Phys. Chem. Chem. Phys.,* 16(29), 15007–15028.

De Vos, N., Maton, C., and Stevens, C.V. (2014). Electrochemical stability of ionic liquids: General influences and degradation mechanisms. *ChemElectroChem,* 1(8), 1258–1270.

Deepak Kumar, S.A.H. (2010). Ionic liquid based sodium ion conducting gel polymer electrolytes. *Solid State Ionics,* 181, 416–423.

Ding, C., Nohira, T., Hagiwara, R., Matsumoto, K., Okamoto, Y., Fukunaga, A., Sakai, S., Nitta, K., and Inazawa, S. (2014). Na[FSA]-[C3C1pyrr][FSA] ionic liquids as electrolytes for sodium secondary batteries: Effects of Na ion concentration and operation temperature. *Journal of Power Sources,* 269, 124–128.

Ding, C., Nohira, T., Kuroda, K., Hagiwara, R., Fukunaga, A., Sakai, S., Nitta, K., and Inazawa, S. (2013). NaFSA–C1C3pyrFSA ionic liquids for sodium secondary battery operating over a wide temperature range. *Journal of Power Sources,* 238, 296–300.

Dmitrii A. Rakov, Fangfang Chen, Shammi A. Ferdousi, Hua Li, Thushan Pathirana, Alexandr Simonov, Patrick C. Howlett, Rob Atkin, Maria Forsyth. (2020). Engineering high energy density sodium battery anodes for improved cycling with superconcentrated ionic liquid electrolytes. *Nat. Mater.*, DOI: 10.1038/s41563-020-0673-0

Dunn, B., Kamath, H., and Tarascon, J.-M. (2011). Electrical energy storage for the grid: A battery of choices. *Science,* 334(6058), 928–935.

Eshetu, G.G., Jeong, S., Pandard, P., Lecocq, A., Marlair, G., and Passerini, S. (2017). Comprehensive insights into the thermal stability, biodegradability, and combustion chemistry of pyrrolidinium-based ionic liquids. *ChemSusChem,* 10(15), 3146–3159.

Fdz De Anastro, A., Lago, N., Berlanga, C., Galcerán, M., Hilder, M., Forsyth, M., and Mecerreyes, D. (2019a). Poly(ionic liquid) iongel membranes for all solid-state rechargeable sodium battery. *Journal of Membrane Science,* 582, 435–441.

Fdz De Anastro, A., Porcarelli, L., Hilder, M., Berlanga, C., Galceran, M., Howlett, P., Forsyth, M., and Mecerreyes, D. (2019b). UV-cross-linked ionogels for all-solid-state rechargeable sodium batteries. *ACS Applied Energy Materials,* 2(10), 6960–6966.

Ferdousi, S.A., Hilder, M., Basile, A., Zhu, H., O'dell, L.A., Saurel, D., Rojo, T., Armand, M., Forsyth, M., and Howlett, P.C. (2019). Water as an effective additive for high-energy-density Na metal batteries? Studies in a superconcentrated ionic liquid electrolyte. *ChemSusChem,* 12(8), 1700–1711.

Flores, E., Åvall, G., Jeschke, S., and Johansson, P. (2017). Solvation structure in dilute to highly concentrated electrolytes for lithium-ion and sodium-ion batteries. *Electrochimica Acta,* 233, 134–141.

Forero-Saboya, J., Hosseini-Bab-Anari, E., Abdelhamid, M.E., Moth-Poulsen, K., and Johansson, P. (2019). Solvent-free lithium and sodium containing electrolytes based on pseudo-delocalized anions. *Chemical Communications,* 55(5), 632–635.

Forsyth, M., Chimdi, T., Seeber, A., Gunzelmann, D., and Howlett, P.C. (2014). Structure and dynamics in an organic ionic plastic crystal, N-ethyl-N-methyl pyrrolidinium bis(trifluoromethanesulfonyl) amide, mixed with a sodium salt. *J. Mater. Chem. A,* 2(11), 3993–4003.

Forsyth, M., Yoon, H., Chen, F., Zhu, H., Macfarlane, D.R., Armand, M., and Howlett, P.C. (2016). Novel Na+ ion diffusion mechanism in mixed organic–inorganic ionic liquid electrolyte leading to high Na+ transference number and stable, high rate electrochemical cycling of sodium cells. *The Journal of Physical Chemistry C,* 120(8), 4276–4286.

Fukunaga, A., Nohira, T., Hagiwara, R., Numata, K., Itani, E., Sakai, S., and Nitta, K. (2016). Performance validation of sodium-ion batteries using an ionic liquid electrolyte. *Journal of Applied Electrochemistry,* 46(4), 487–496.

Fukunaga, A., Nohira, T., Hagiwara, R., Numata, K., Itani, E., Sakai, S., Nitta, K., and Inazawa, S. (2014). A safe and high-rate negative electrode for sodium-ion batteries: Hard carbon in NaFSA-C1C3pyrFSA ionic liquid at 363 K. *Journal of Power Sources,* 246, 387–391.

Girard, G.M.A., Hilder, M., Nucciarone, D., Whitbread, K., Zavorine, S., Moser, M., Forsyth, M., Macfarlane, D.R., and Howlett, P.C. (2017). Role of Li concentration and the SEI layer in enabling high performance Li metal electrodes using a phosphonium bis(fluorosulfonyl)imide ionic liquid. *The Journal of Physical Chemistry C,* 121(39), 21087–21095.

Guerfi, A., Dontigny, M., Charest, P., Petitclerc, M., Lagacé, M., Vijh, A., and Zaghib, K. (2010). Improved electrolytes for Li-ion batteries: Mixtures of ionic liquid and organic electrolyte with enhanced safety and electrochemical performance. *Journal of Power Sources,* 195(3), 845–852.

Han, H.-B., Liu, K., Feng, S.-W., Zhou, S.-S., Feng, W.-F., Nie, J., Li, H., Huang, X.-J., Matsumoto, H., Armand, M., and Zhou, Z.-B. (2010). Ionic liquid electrolytes based on multi-methoxyethyl substituted ammoniums and perfluorinated sulfonimides: Preparation, characterization, and properties. *Electrochimica Acta,* 55(23), 7134–7144.

Hilder, M., Howlett, P.C., Saurel, D., Gonzalo, E., Armand, M., Rojo, T., Macfarlane, D.R., and Forsyth, M. (2017). Small quaternary alkyl phosphonium bis(fluorosulfonyl)imide ionic liquid electrolytes for sodium-ion batteries with P2- and O3-Na 2/3 [Fe 2/3 Mn 1/3]O 2 cathode material. *Journal of Power Sources,* 349, 45–51.

Hosseini-Bab-Anari, E., Navarro-Suárez, A.M., Moth-Poulsen, K., and Johansson, P. (2019). Ionic liquid based battery electrolytes using lithium and sodium pseudo-delocalized pyridinium anion salts. *Physical Chemistry Chemical Physics,* 21(33), 18393–18399.

Hwang, J.-Y., Myung, S.-T., and Sun, Y.-K. (2017). Sodium-ion batteries: Present and future. *Chemical Society Reviews,* 46(12), 3529–3614.

Jache, B. and Adelhelm, P. (2014). Use of graphite as a highly reversible electrode with superior cycle life for sodium-ion batteries by making use of co-intercalation phenomena. *Angewandte Chemie International Edition,* 53(38), 10169–10173.

Jian, Z., Han, W., Lu, X., Yang, H., Hu, Y.-S., Zhou, J., Zhou, Z., Li, J., Chen, W., Chen, D., and Chen, L. (2013). Superior electrochemical performance and storage mechanism of Na3V2(PO_4)3cathode for room-temperature sodium-ion batteries. *Advanced Energy Materials,* 3(2), 156–160.

Johansson, P., Gejji, S.P., Tegenfeldt, J., and Lindgren, J. (1998). The imide ion: Potential energy surface and geometries. *Electrochimica Acta,* 43(10), 1375–1379.

Jónsson, E. and Johansson, P. (2015). Electrochemical oxidation stability of anions for modern battery electrolytes: A CBS and DFT study. *Physical Chemistry Chemical Physics,* 17(5), 3697–3703.

Kar, M., Winther-Jensen, B., Armand, M., Simons, T.J., Winther-Jensen, O., Forsyth, M., and Macfarlane, D.R. (2016). Stable zinc cycling in novel alkoxy-ammonium based ionic liquid electrolytes. *Electrochimica Acta,* 188, 461–471.

Kim, H., Hong, J., Park, Y.-U., Kim, J., Hwang, I., and Kang, K. (2015a). Sodium storage behavior in natural graphite using ether-based electrolyte systems. *Advanced Functional Materials,* 25(4), 534–541.

Kim, H., Hong, J., Yoon, G., Kim, H., Park, K.-Y., Park, M.-S., Yoon, W.-S., and Kang, K. (2015b). Sodium intercalation chemistry in graphite. *Energy & Environmental Science,* 8(10), 2963–2969.

Klongkan, S. and Pumchusak, J. (2015). Effects of nano alumina and plasticizers on morphology, ionic conductivity, thermal and mechanical properties of PEO-LiCF3SO3 solid polymer electrolyte. *Electrochimica Acta,* 161, 171–176.

Komaba, S., Murata, W., Ishikawa, T., Yabuuchi, N., Ozeki, T., Nakayama, T., Ogata, A., Gotoh, K., and Fujiwara, K. (2011). Electrochemical Na insertion and solid electrolyte interphase for hard-carbon electrodes and application to Na-ion batteries. *Advanced Functional Materials,* 21(20), 3859–3867.

Krause, L.J., Lamanna, W., Summerfield, J., Engle, M., Korba, G., Loch, R., and Atanasoski, R. (1997). Corrosion of aluminum at high voltages in non-aqueous electrolytes containing perfluoroalkylsulfonyl imides; new lithium salts for lithium-ion cells. *Journal of Power Sources,* 68(2), 320–325.

Kroon, M.C., Buijs, W., Peters, C.J., and Witkamp, G.-J. (2007). Quantum chemical aided prediction of the thermal decomposition mechanisms and temperatures of ionic liquids. *Thermochimica Acta,* 465(1), 40–47.

Kubota, K. and Komaba, S. (2015). Review—Practical issues and future perspective for Na-ion batteries. *Journal of The Electrochemical Society,* 162(14), A2538–A2550.

Kundu, D., Talaie, E., Duffort, V., and Nazar, L.F. (2015). The emerging chemistry of sodium ion batteries for electrochemical energy storage. *Angew Chem. Int. Ed. Engl.,* 54(11), 3431–3448.

Lei, D., He, Y.B., Huang, H., Yuan, Y., Zhong, G., Zhao, Q., Hao, X., Zhang, D., Lai, C., Zhang, S., Ma, J., Wei, Y., Yu, Q., Lv, W., Yu, Y., Li, B., Yang, Q.H., Yang, Y., Lu, J., and Kang, F. (2019). Cross-linked beta alumina nanowires with compact gel polymer electrolyte coating for ultra-stable sodium metal battery. *Nat. Commun.,* 10(1), 4244.

Lin, X., Kavian, R., Lu, Y., Hu, Q., Shao-Horn, Y., and Grinstaff, M.W. (2015). Thermally-responsive, nonflammable phosphonium ionic liquid electrolytes for lithium metal batteries: Operating at 100 degrees celsius. *Chem. Sci.,* 6(11), 6601–6606.

Liu, W., Liu, N., Sun, J., Hsu, P.-C., Li, Y., Lee, H.-W., and Cui, Y. (2015). Ionic conductivity enhancement of polymer electrolytes with ceramic nanowire fillers. *Nano Letters,* 15(4), 2740–2745.

Long, L., Wang, S., Xiao, M., and Meng, Y. (2016). Polymer electrolytes for lithium polymer batteries. *Journal of Materials Chemistry A,* 4(26), 10038–10069.

Macfarlane, D.R., Kar, M., and Pringle, J.M. (2017a). Electrochemistry of and in ionic liquids. In *Fundamentals of Ionic Liquids From Chemistry to Applications*, First Edition, Macfarlane, D.R., Kar, M., and Pringle, J.M. (eds), Wiley-VCH Verlag GmbH & Co. KGaA, 177–207.

Macfarlane, D.R., Kar, M., and Pringle, J.M. (2017b). Physical and thermal properties. In *Fundamentals of Ionic Liquids From Chemistry to Applications*, First Edition, Macfarlane, D.R., Kar, M., and Pringle, J.M. (eds), Wiley-VCH Verlag GmbH & Co. KGaA, 103–147.

Macfarlane, D.R., Tachikawa, N., Forsyth, M., Pringle, J.M., Howlett, P.C., Elliott, G.D., Davis, J.H., Watanabe, M., Simon, P., and Angell, C.A. (2014). Energy applications of ionic liquids. *Energy & Environmental Science,* 7(1), 232–250.

Makhlooghiazad, F., Guazzagaloppa, J., O'dell, L.A., Yunis, R., Basile, A., Howlett, P.C., and Forsyth, M. (2018). The influence of the size and symmetry of cations and anions on the physicochemical behavior of organic ionic plastic crystal electrolytes mixed with sodium salts. *Phys. Chem. Chem. Phys.,* 20(7), 4721–4731.

Makhlooghiazad, F., Gunzelmann, D., Hilder, M., Macfarlane, D.R., Armand, M., Howlett, P.C., and Forsyth, M. (2017a). Mixed phase solid-state plastic crystal electrolytes based on a phosphonium cation for sodium devices. *Advanced Energy Materials,* 7(2), 1601272– 1601272.

Makhlooghiazad, F., Howlett, P.C., Wang, X., Hilder, M., Macfarlane, D.R., Armand, M., and Forsyth, M. (2017b). Phosphonium plastic crystal salt alloyed with a sodium salt as a solid-state electrolyte for sodium devices: Phase behaviour and electrochemical performance. *Journal of Materials Chemistry A,* 5(12), 5770-5780.

Makhlooghiazad, F., Yunis, R., Mecerreyes, D., Armand, M., Howlett, P.C., and Forsyth, M. (2017c). Comparison of the physicochemical and electrochemical behaviour of mixed anion phosphonium based OIPCs electrolytes for sodium batteries. *Solid State Ionics,* 312, 44–52.

Manohar, C.V., Mendes, T.C., Kar, M., Wang, D., Xiao, C., Forsyth, M., Mitra, S., and Macfarlane, D.R. (2018a). Ionic liquid electrolytes supporting high energy density in sodium-ion batteries based on sodium vanadium phosphate composites. *Chem. Commun. (Camb.),* 54(28), 3500–3503.

Manohar, C.V., Raj K, A., Kar, M., Forsyth, M., Macfarlane, D.R., and Mitra, S. (2018b). Stability enhancing ionic liquid hybrid electrolyte for NVP@C cathode based sodium batteries. *Sustainable Energy & Fuels,* 2(3), 566–576.

Matsumoto, K., Okamoto, Y., Nohira, T., and Hagiwara, R. (2015). Thermal and transport properties of $Na[N(SO_2F)_2]$–[N-methyl-N-propylpyrrolidinium]$[N(SO_2F)_2]$ ionic liquids for Na secondary batteries. *The Journal of Physical Chemistry C,* 119(14), 7648–7655.

Mendes, T.C., Zhang, X., Wu, Y., Howlett, P.C., Forsyth, M., and Macfarlane, D.R. (2019). Supported ionic liquid gel membrane electrolytes for a safe and flexible sodium metal battery. *ACS Sustainable Chemistry & Engineering,* 7(4), 3722–3726.

Mohapatra, S.R., Thakur, A.K., and Choudhary, R.N.P. (2007). Studies on PEO-based sodium ion conducting composite polymer films. *Ionics,* 14(3), 255–262.

Mohd Noor, S.A., Howlett, P.C., Macfarlane, D.R., and Forsyth, M. (2013). Properties of sodium-based ionic liquid electrolytes for sodium secondary battery applications. *Electrochimica Acta,* 114, 766–771.

Monti, D., Jónsson, E., Palacín, M.R., and Johansson, P. (2014). Ionic liquid based electrolytes for sodium-ion batteries: Na+ solvation and ionic conductivity. *Journal of Power Sources,* 245, 630–636.

Monti, D., Ponrouch, A., Palacín, M.R., and Johansson, P. (2016). Towards safer sodium-ion batteries via organic solvent/ionic liquid based hybrid electrolytes. *Journal of Power Sources,* 324, 712–721.

Ni'mah, Y.L., Cheng, M.-Y., Cheng, J.H., Rick, J., and Hwang, B.-J. (2015). Solid-state polymer nanocomposite electrolyte of TiO_2/PEO/$NaClO_4$ for sodium ion batteries. *Journal of Power Sources,* 278, 375–381.

Nishi, Y. (2001). Lithium ion secondary batteries; past 10 years and the future. *Journal of Power Sources,* 100(1), 101–106.

O'mahony, A.M., Silvester, D.S., Aldous, L., Hardacre, C., and Compton, R.G. (2008). Effect of water on the electrochemical window and potential limits of room-temperature ionic liquids. *Journal of Chemical & Engineering Data,* 53(12), 2884–2891.

Oh, S.M., Myung, S.T., Yoon, C.S., Lu, J., Hassoun, J., Scrosati, B., Amine, K., and Sun, Y.K. (2014). Advanced Na[Ni0.25Fe0.5Mn0.25]O_2/C-Fe_3O_4 sodium-ion batteries using EMS electrolyte for energy storage. *Nano Lett.,* 14(3), 1620–1626.

Plylahan, N., Kerner, M., Lim, D.-H., Matic, A., and Johansson, P. (2016). Ionic liquid and hybrid ionic liquid/organic electrolytes for high temperature lithium-ion battery application. *Electrochimica Acta,* 216, 24–34.

Ponrouch, A., Dedryvère, R., Monti, D., Demet, A.E., Ateba Mba, J.M., Croguennec, L., Masquelier, C., Johansson, P., and Palacín, M.R. (2013). Towards high energy density sodium ion batteries through electrolyte optimization. *Energy & Environmental Science,* 6(8), 2361–2369.

Ponrouch, A., Marchante, E., Courty, M., Tarascon, J.-M., and Palacín, M.R. (2012). In search of an optimized electrolyte for Na-ion batteries. *Energy & Environmental Science,* 5(9), 8572–8583.

Ponrouch, A., Monti, D., Boschin, A., Steen, B., Johansson, P., and Palacín, M.R. (2015). Non-aqueous electrolytes for sodium-ion batteries. *Journal of Materials Chemistry A,* 3(1), 22–42.

Pringle, J.M., Howlett, P.C., Macfarlane, D.R., and Forsyth, M. (2010). Organic ionic plastic crystals: Recent advances. *Journal of Materials Chemistry,* 20(11), 2056–2062.

Qi, X., Ma, Q., Liu, L., Hu, Y.-S., Li, H., Zhou, Z., Huang, X., and Chen, L. (2016). Sodium bis(fluorosulfonyl)imide/poly(ethylene oxide) polymer electrolytes for sodium-ion batteries. *ChemElectroChem,* 3(11), 1741–1745.

Sångeland, C., Mogensen, R., Brandell, D., and Mindemark, J. (2019a). Stable cycling of sodium metal all-solid-state batteries with polycarbonate-based polymer electrolytes. *ACS Applied Polymer Materials,* 1(4), 825–832.

Sångeland, C., Younesi, R., Mindemark, J., and Brandell, D. (2019b). Towards room temperature operation of all-solid-state Na-ion batteries through polyester–polycarbonate-based polymer electrolytes. *Energy Storage Materials,* 19, 31–38.

Sato, T., Maruo, T., Marukane, S., and Takagi, K. (2004). Ionic liquids containing carbonate solvent as electrolytes for lithium ion cells. *Journal of Power Sources,* 138(1-2), 253–261.

Scheers, J., Lim, D.-H., Kim, J.-K., Paillard, E., Henderson, W.A., Johansson, P., Ahn, J.-H., and Jacobsson, P. (2014). All fluorine-free lithium battery electrolytes. *Journal of Power Sources,* 251, 451–458.

Serra Moreno, J., Armand, M., Berman, M.B., Greenbaum, S.G., Scrosati, B., and Panero, S. (2014). Composite PEOn:NaTFSI polymer electrolyte: Preparation, thermal and electrochemical characterization. *Journal of Power Sources,* 248, 695–702.

Shekibi, Y., Gray-Weale, A., Macfarlane, D.R., Hill, A.J., and Forsyth, M. (2007). Nanoparticle enhanced conductivity in organic ionic plastic crystals: Space charge versus strain induced defect mechanism. *The Journal of Physical Chemistry C,* 111(30), 11463–11468.

Shkrob, I.A., Marin, T.W., Zhu, Y., and Abraham, D.P. (2014). Why bis(fluorosulfonyl)imide is a "magic anion" for electrochemistry. *The Journal of Physical Chemistry C,* 118(34), 19661–19671.

Tarascon, J.M. and Armand, M. (2001). Issues and challenges facing rechargeable lithium batteries. *Nature,* 414(6861), 359–367.

Tsunashima, K. and Sugiya, M. (2007). Physical and electrochemical properties of low-viscosity phosphonium ionic liquids as potential electrolytes. *Electrochemistry Communications,* 9(9), 2353–2358.

Vicent-Luna, J.M., Ortiz-Roldan, J.M., Hamad, S., Tena-Zaera, R., Calero, S., and Anta, J.A. (2016). Quantum and classical molecular dynamics of ionic liquid electrolytes for Na/Li-based batteries: Molecular origins of the conductivity behavior. *ChemPhysChem,* 17(16), 2473–2481.

Welton, T. (2018). Ionic liquids: A brief history. *Biophysical Reviews,* 10(3), 691–706.

Weston, J.E. and Steele, B.C.H. (1982). Effects of inert fillers on the mechanical and electrochemical properties of lithium salt-poly(ethylene oxide) polymer electrolytes. *Solid State Ionics,* 7(1), 75–79.

Wright, P.V. (1998). Polymer electrolytes—The early days. *Electrochimica Acta,* 43(10), 1137–1143.

Xie, L., Xu, F., Qiu, F., Lu, H., and Yang, Y. (2007). Single-walled carbon nanotubes functionalized with high bonding density of polymer layers and enhanced mechanical properties of composites. *Macromolecules,* 40(9), 3296–3305.

Yang, Q., Zhang, Z., Sun, X.-G., Hu, Y.-S., Xing, H., and Dai, S. (2018). Ionic liquids and derived materials for lithium and sodium batteries. *Chemical Society Reviews,* 47(6), 2020–2064.

Yoon, H., Lane, G.H., Shekibi, Y., Howlett, P.C., Forsyth, M., Best, A.S., and Macfarlane, D.R. (2013). Lithium electrochemistry and cycling behaviour of ionic liquids using cyano based anions. *Energy & Environmental Science,* 6(3), 979–986.

Yoon, H., Zhu, H., Hervault, A., Armand, M., Macfarlane, D.R., and Forsyth, M. (2014a). Physicochemical properties of N-propyl-N-methylpyrrolidinium bis(fluorosulfonyl)imide for sodium metal battery applications. *Phys. Chem. Chem. Phys.,* 16(24), 12350–12355.

Yoon, H., Zhu, H., Hervault, A., Armand, M., Macfarlane, D.R., and Forsyth, M. (2014b). Physicochemical properties of N-propyl-N-methylpyrrolidinium bis(fluorosulfonyl)imide for sodium metal battery applications. *Physical Chemistry Chemical Physics,* 16(24), 12350–12355.

Zhou, Y., Wang, X., Zhu, H., Armand, M., Forsyth, M., Greene, G.W., Pringle, J.M., and Howlett, P.C. (2017a). N-ethyl-N-methylpyrrolidinium bis(fluorosulfonyl)imide-electrospun polyvinylidene fluoride composite electrolytes: Characterization and lithium cell studies. *Phys. Chem. Chem. Phys.,* 19(3), 2225–2234.

Zhou, Y., Wang, X., Zhu, H., Armand, M., Forsyth, M., Greene, G.W., Pringle, J.M., and Howlett, P.C. (2018). Ternary lithium-salt organic ionic plastic crystal polymer composite electrolytes for high voltage, all-solid-state batteries. *Energy Storage Materials,* 15, 407–414.

Zhou, Y., Wang, X., Zhu, H., Yoshizawa-Fujita, M., Miyachi, Y., Armand, M., Forsyth, M., Greene, G.W., Pringle, J.M., and Howlett, P.C. (2017b). Solid-state lithium conductors for lithium metal batteries based on electrospun nanofiber/plastic crystal composites. *ChemSusChem,* 10(15), 3135–3145.

Zhou, Z.-B. and Matsumoto, H. (2007). Lithium-doped, organic ionic plastic crystal electrolytes exhibiting high ambient-temperature conductivities. *Electrochemistry Communications,* 9(5), 1017–1022.

6

Solid Electrolyte Interphase in Na-ion batteries

Le Anh MA, Ronnie MOGENSEN, Andrew J. NAYLOR and Reza YOUNESI

Department of Chemistry - Ångström Laboratory, Uppsala University, Sweden

6.1. Introduction

6.1.1. *The solid electrolyte interphase*

The concept of solid electrolyte interphase (SEI) formation was first introduced by Peled in 1979 (Peled 1979) and is now a widely accepted phenomenon within the community. He proposed that non-aqueous alkali and alkaline earth batteries instantly may form a surface layer when the metal comes into contact with the electrolyte. This interphase layer between electrode and electrolyte has the characteristics of a solid electrolyte. The SEI consists of insoluble and soluble reduction products from electrolyte components (Figure 6.1). Ideally, the surface film acts as a passivation layer, which is electronically insulating and cationically conductive through Schottky defects, impermeable to solvent molecules, insoluble and inert to the chemical environment. Polar aprotic solvents used in non-aqueous electrolyte systems are thermodynamically unstable at the low electrode potential typical of the anode. However, the formation of such a passivation layer is crucial to make the battery kinetically stable in order to obtain a long cycle life (Peled and Menkin 2017).

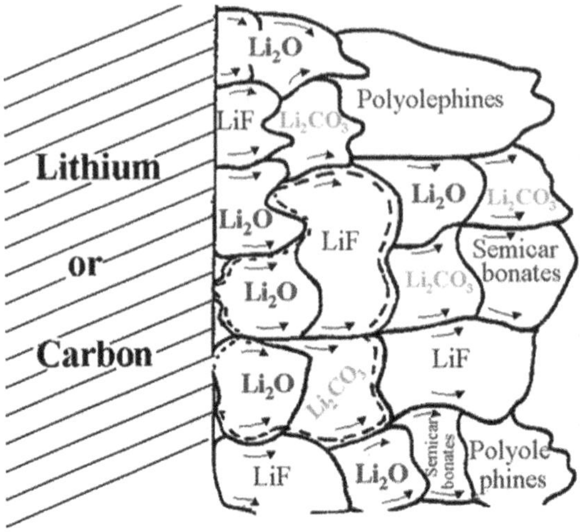

Figure 6.1. *Schematic illustration of SEI structures (*Peled and Menkin 2017*). For a color version of this figure, see www.iste.co.uk/monconduit/batteries.zip*

Before the introduction of the concept of SEI, it was thought that direct charge transfer between the electrode and the Li^+ ions in the electrolyte occurs according to the Butler–Volmer equation. The electron transfer was considered the rate-determining step. A passivation layer found on lithium metal anodes was determined to be a limitation on the cycle life of the cell (Brummer 1980; Newman 1980). It was concluded that, in order to obtain better electron transfer and hence a long cycle life and efficient battery performance, the anode surface must be free from a passivating layer. Peled's SEI concept later demonstrated the necessity of an SEI in order to obtain a kinetically stable anode. At low potentials, the electrolyte salt anions are often reduced to inorganic compounds such as MF and M_2O (M = Li, Na, or K) with low solubility in the electrolyte. Solvent components are decomposed to both inorganics, like carbonates, with low solubility and organic semicarbonates and polymers with higher solubility in the electrolyte solution.

6.1.2. *Characterization of the SEI*

Although the SEI in Li-based systems has been widely studied (Xu 2014), its poor performance is still one of the major ageing mechanisms in rechargeable batteries. SEI often does not effectively passivate the electrode surface or falls short of the properties expected for an "ideal" SEI. Therefore, understanding formation, composition and evolution of the SEI is still scientifically interesting. As the SEI is a very thin layer (nm range) made of inorganic and organic components, characterization of the SEI is challenging (Tripathi *et al.* 2018). Among others, the following techniques have been employed in order to analyze the physical and chemical characteristics of the SEI:

– X-ray photoelectron spectroscopy (XPS);

– electrochemical quartz-crystal microbalance (EQCM);

– scanning electron microscopy (SEM);

– transmission electron microscopy (TEM);

– electrochemical impedance spectroscopy (EIS);

– time-of-flight secondary ion mass spectrometry (ToF-SIMS);

– X-ray diffraction (XRD);

– surface-enhanced Raman spectroscopy (SERS);

– scanning tunneling microscopy (STM);

– energy-dispersive X-ray spectroscopy (EDS);

– Fourier transform infrared spectroscopy (FTIR);

– nuclear magnetic resonance (NMR);

– electron paramagnetic resonance (EPR);

– differential scanning calorimetry (DSC);

– thermogravimetric analysis (TGA);

– atomic force microscopy (AFM).

Due to the typically low thickness of the SEI layer, XPS is one of the most common techniques used to investigate the formation mechanisms and composition of the interphase. Such analysis is based on the electron ejection from the material near-surface by X-ray irradiation, and hence is a highly surface-sensitive method. Synchrotron-based XPS can, in addition, provide depth profiling of the composition of the SEI (see Figure 6.2). Low energy photoelectrons of 100–1,000 eV correspond to about 3–10 nm probing depth (Muñoz-Márquez *et al.* 2017) while high energy photoelectrons of 2–10 keV reveal composition at depths of 15–50 nm (Philippe *et al.* 2015).

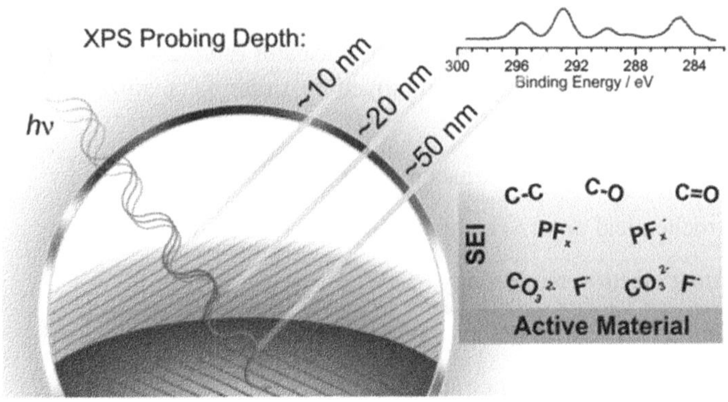

Figure 6.2. *A schematic diagram of depth profiling of the SEI using X-ray photoelectron spectroscopy with different photon energies. Adapted from Naylor* et al. *(2019). For a color version of this figure, see www.iste.co.uk/monconduit/batteries.zip*

ToF-SIMS is another surface-sensitive technique suitable for studying the SEI composition and structure. Charged ions are separated from the sample and are detected as mass/electron charge (m/z) signals. The analysis depth is 2 nm of the surface. However, ion sputtering such as Ar^+ can be used to form detailed depth profiles (Ota *et al.* 2003; Komaba *et al.* 2011b).

EIS can provide information about resistances within the cell for different electrode/electrolyte systems. However, one needs to be careful with the interpretation of the measurements due to the presence of multiple sources of resistances, which may not be related to the SEI (Bommier *et al.* 2016).

Although the concept of SEI formation and its characteristics in Na-ion batteries are largely similar to that for lithium-ion batteries, there exist differences between the SEI in Na-ion batteries compared to that in lithium-ion batteries, which influence the capacity loss and ageing mechanism (Song *et al.* 2018). This is due to the higher solubility of most sodium salts present in the SEI compared to their lithium counterparts (e.g. NaF vs. LiF). Also, the coordination structure of electrolyte salts with the same anions but alternative cations is different as demonstrated by Cresce *et al.* (2017). For example, $NaPF_6$ and $LiPF_6$ are dissolved in linear carbonates with different coordination numbers, which can influence the electrolyte conductivity, reduction potential and reduction products.

Further examples for the characterization of the SEI in sodium-ion cells and its comparison to lithium systems are discussed below. As SEI forms by decomposition of the electrolyte, its composition is directly influenced by the composition of the electrolyte solution. Several previous studies have characterized the composition of SEI formed from different aprotic electrolyte systems (Song *et al.* 2018; Huang *et al.* 2019; Sun *et al.* 2019). Therefore, the focus of this chapter is on differences between SEI in sodium-ion-based electrolytes compared to its counterpart in lithium-ion systems.

6.2. Physical properties of the Na-ion SEI

6.2.1. *Electrochemical stability*

Since SEI forms in expense of charge and electrolyte loss, its formation should ideally be completed during the initial cycles and shall remain stable over long-term cycling. Several studies have shown, through XPS measurements, that the SEI in a sodium-ion cell changes markedly in thickness and composition with the state of charge (Muñoz-Márquez *et al.* 2015; Mogensen *et al.* 2017). These changes even occur in systems where volume expansion of the active material is very limited, a strong indication of poor voltage stability for some of the deposited electrolyte decomposition products, which can lead to stripping of the SEI during the desodiation step (Carboni *et al.* 2019). More insight into this stripping process can be found in the work by Komaba *et al.* (2011b), where it was reported that sodium propyl carbonate is oxidized at potentials greater than 2.8 V versus Na/Na^+. In more recent work by Lee and Tang (2019), it was shown that oxidation of the reduction products in an SEI can occur at potentials as low as 1 V versus Na/Na^+ and thus within the typical operating voltage window of most anodes. This has important implications since, in many systems, the upper cut-off voltage for the anode can be a defining parameter for the capacity retention. This is particularly true in systems where surface catalytic properties play a role, such as tin and lead titanium oxides (Ellis *et al.* 2012, 2014; Muñoz-Márquez *et al.* 2015; Fukunishi *et al.* 2016). The electrochemical instability of the SEI on $Na_2Ti_3O_7$ anodes was demonstrated by Zarrabeitia *et al.* (2017) by analyzing the surface layer on $Na_2Ti_3O_7$ anodes by XPS at different states of charge in both half- and full-cells. The measurements, augmented by Auger parameter analysis, showed that it is mainly a surface layer consisting of alkylcarbonates and polyethylene oxide (PEO) type polymers that is stripped during charge. This effect persists in full-cells cycled with sodium iron phosphate, though the effect is much smaller. For the alloying anodes, such as tin

(Ellis *et al.* 2012; Fukunishi *et al.* 2016) and lead (Ellis *et al.* 2014), and conversion anodes, such as Fe_2O_3 (Valvo *et al.* 2014) and Sn_4P_3 (Wang *et al.* 2017; Mogensen *et al.* 2018), restricting the upper cut-off potential during cycling is imperative to achieving attractive long-term cycling performance. Even though the volume expansion in these types of materials is rather large, the impact of narrowing the cycling voltage window does not appear too detrimental toward the capacity for tin or tin phosphide. Rather, it suggests that the stripping of the SEI causes electrolyte fouling, or exacerbates agglomeration.

6.2.2. *Mechanical properties*

The mechanical properties of the SEI are dependent largely on the formation conditions, such as charge rate, composition of the electrolyte or temperature. The SEI structure and porosity are also influenced by the solubility or oxidation of the various species, as described previously.

In Weadock *et al.*'s (2013) work, AFM measurements were performed on SEI layers formed from a 1 M $NaPF_6$ in ethylene carbonate:diethyl carbonate (EC:DEC) electrolyte. It was found that the inner inorganic layer of the SEI had a much higher Young's modulus than the outer layer, consisting of "less reduced species" with a more organic character. Their measurements (Figure 6.3(a)) also show that the Young's modulus of the SEI varies by several orders of magnitude between 0.57 and 1796 MPa with an average of 304 MPa in experiments where three cyclic voltammetry (CV) scans between 0.05 and 2.1 V versus Na/Na^+ were performed. The corresponding values after 10 CV cycles changed the modulus values to between 3.91 and 756 MPa with an average of 197 MPa. For comparison, Young's modulus values for an analogous lithium-ion system were found to be in the range of 0.2–4.5 GPa, with even higher values after removing the outer organic layer (Shin *et al.* 2015). The cause of this decline in Young's modulus after 10 cycles is unknown but shows that, even at a low scan rate of 0.05 mV s^{-1}, the SEI constantly changes with further electrochemical cycling. A study by Carboni *et al.* (2019) combining TEM, XPS, and FTIR analysis of the SEI on hard carbon electrodes presented an SEI consisting of amorphous precipitates that grow in size and coverage over 20 cycles at slow rates. The SEI becomes rich in NaF and Na_2CO_3 species as cycling proceeds again corroborating that organic species are further reduced or dissolved with an increasing number of cycles.

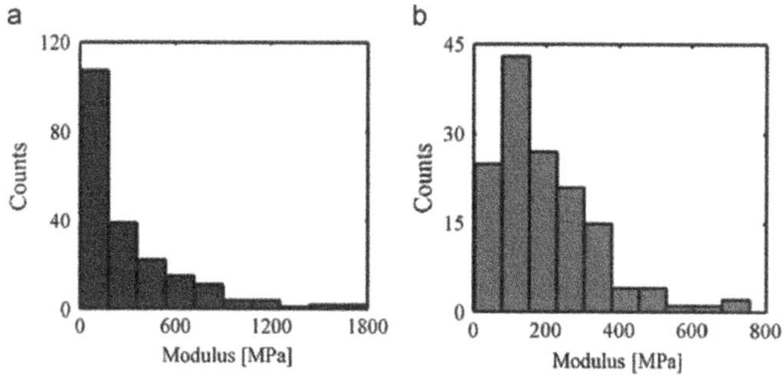

Figure 6.3. *(a and b) Histograms of Young's modulus for a Na-based SEI layer after 3 and 10 cycles (Weadock* et al. 2013*). For a color version of this figure, see www.iste.co.uk/monconduit/batteries.zip*

6.2.3. *Dissolution of SEI components*

Among different mechanisms leading to ageing in rechargeable batteries, SEI dissolution has rarely been studied in detail (Tasaki *et al.* 2009; Jones *et al.* 2011). As mentioned previously, the SEI formed in Na-ion batteries is expected to show higher solubility compared to its lithium analogues, due to the charge density and thus the dissociation energy in sodium salts being lower than their lithium counterparts. This effect can also be seen in the generally higher solubility of sodium salts compared to their lithium analogues. Although this argument is valid and true for aqueous electrolytes (Iermakova *et al.* 2015), the solubility of alkali salts in non-aqueous solvents can be radically different from aqueous solubility trends and also for different organic solvents.

Analysis of SEI dissolution can pose challenges and require different techniques than those usually employed to characterize SEI composition. The complication arises from the fact that common analytical techniques including spectrometry and imaging methods are performed on electrodes after cycling and often after drying/washing of samples. This means that some soluble species may be removed from the SEI during sample preparation. The other approach is to analyze electrolyte solutions where the soluble species reside. However, the challenge is to detect the composition of small concentrations of dissolved species in a bulk electrolyte solution.

Komaba *et al.* (2011b) showed self-discharge behavior for both lithium and sodium in a comparison of hard carbon electrodes cycled in propylene carbonate (PC). Although the difference was minor between sodium and lithium and the overall capacity loss was roughly 20% over a 30-day rest period, it showed that sodium lost more capacity. In further work, the discoloring of the PC-based electrolyte that occurs in both half- and full-cell tests of sodium ion batteries was discussed (Komaba *et al.* 2011a). It was shown, by analysis of the electrolyte from cycled cells using ATR FT-IR (Figure 6.4), that the color change is due to formation of sodium propyl carbonate. This type of analysis is very useful to detect soluble species and is underutilized when characterizing the practicality of electrolytes from a solubility perspective.

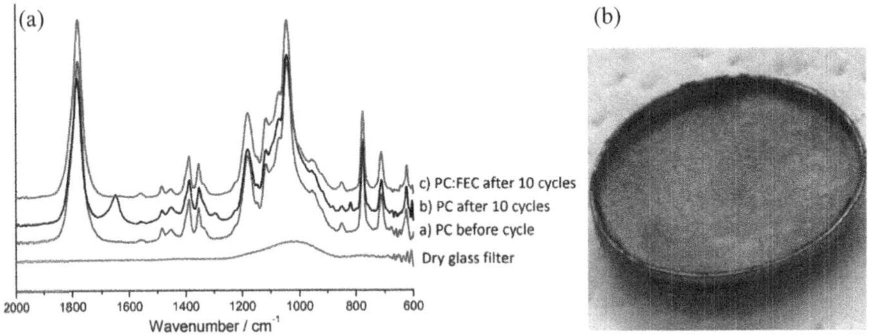

Figure 6.4. *(a) FTIR spectra of electrolytes from glass-fiber separators before and after cycling in hard carbon half-cells. (b) Photograph of the separator from a hard carbon half-cell after five cycles with 1 M $NaClO_4$ in PC electrolyte (Komaba* et al. *2011a). For a color version of this figure, see www.iste.co.uk/monconduit/batteries.zip*

Following on from this work, Mogensen *et al.* (2016) performed similar experiments for EC:DEC-based electrolytes and carbonaceous anodes using galvanostatic cycling with intermittent pauses. The results showed that capacity loss in sodium cells is higher than that in lithium cells, following the same cycling conditions and relaxation time in both cells (see Figure 6.5(a and b)). This measurement was combined with hard X-ray photoelectron spectroscopy (HAXPES) measurements of electrodes that were left to soak in electrolyte for varying amounts of time to compare self-discharge behaviour between lithium and sodium systems. Since such dissolution is a surface effect, a high surface area substrate consisting of Super P carbon was used to amplify the self-discharge rate. The thickness of the SEI was shown to decrease with pause or relaxation time as evidenced by the increasing substrate peak in the XPS spectra (Figure 6.5(c)). The study also showed that such capacity loss is dependent on the electrolyte composition and surface area of the electrode material (Figure 6.5(d)).

A contemporary study of hard carbon capacity loss by Bommier *et al.* (2016) using the same electrolyte mixture (EC:DEC 1 M $NaPF_6$) showed very similar results to the ATR FT-IR results from Komaba *et al.* (2011a). The study showed that EC:DEC mixtures also experience decomposition during cycling or spontaneous reaction with sodium metal to form $CH_3CH_2OCOONa$, with severe implications for capacity retention.

The decomposition species formed by carbonates-based electrolytes in Na-ion batteries can be divided into organic and inorganic species with the main constituents being NaF and Na_2CO_3 being most abundant, while Na_2O and NaOH mostly occur when oxygen and water contamination exists in the electrolyte (Philippe *et al.* 2014). Sodium alkyl carbonates ($ROCO_2Na$), PEO oligomers and similar reduction products from the solvent represent the organic part of the SEI, and it is these organic species that are the most soluble SEI components (Moshkovich *et al.* 2001; Komaba *et al.* 2011b; Muñoz-Márquez *et al.* 2015; Bommier *et al.* 2016; Zarrabeitia *et al.* 2017; Carboni *et al.* 2019). In fact, some of the organic species are so soluble that it can be debated if they are to be considered SEI components at all. This is suggested based on the evolution of the SEI during cycling and ageing at open circuit voltage (OCV) where inorganic species accumulate and constitute a larger part of the SEI with time as seen in XPS measurements and by the analysis of cycled electrolytes with FTIR.

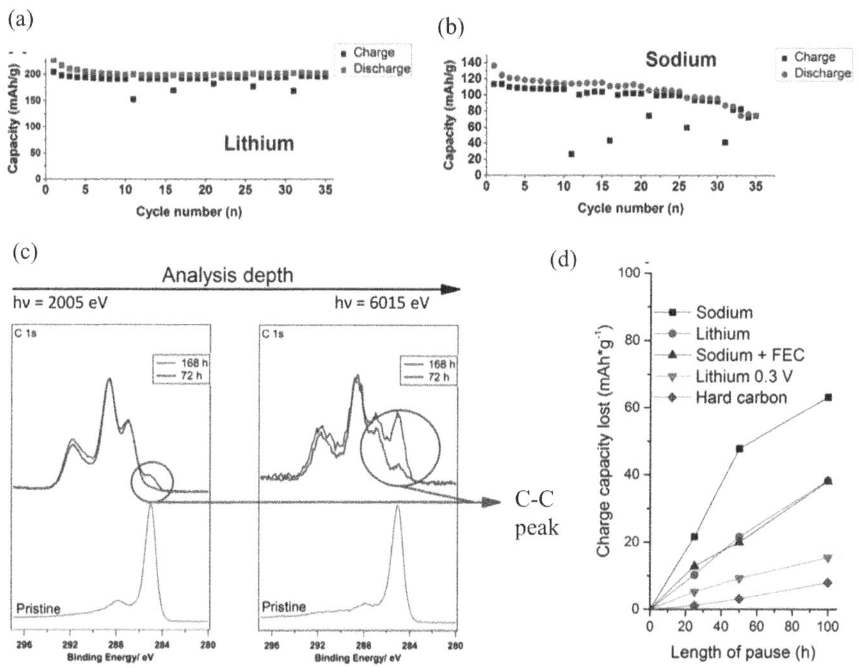

Figure 6.5. *(a and b) Galvanostatic cycling combined with extended relaxation time in lithium and sodium cells using the same electrolyte solvent (EC:DEC) and anion (PF_6). (c) C 1s HAXPES spectra (two photon energies) of carbon Super P electrodes sodiated in half-cells and then rested at OCV for 72 or 168 h. (d) Percentage charge capacity loss for different systems and pause lengths [modified from Mogensen* et al. *(2016)]. For a color version of this figure, see www.iste.co.uk/monconduit/batteries.zip*

Overall, the formation of the inorganic SEI species is usually described as a process where organic species that are first produced are reduced further. This would be supported by the descriptions of SEI layers as being mostly inorganic at the anode–SEI interface and more organic at the SEI–electrolyte interface. The reports of more inorganic species in the sodium SEI compared to the lithium SEI could, however, also be caused either by leaching of organic species in the cells or their dissolution during the washing of electrodes for XPS analysis.

6.3. Comparisons of SEI in sodium- and lithium-based electrolytes

6.3.1. *Formation and composition*

The composition and properties of the SEI are strongly correlated to the way in which the SEI forms, and hence the electrolyte employed as well as cell setup and testing parameters.

Philippe *et al.* (2014) investigated the SEI formation on Fe_2O_3 composite electrodes in Li and Na half-cells with $MClO_4$ (M = Li, Na) in EC:DEC (v/v 1:1) electrolyte. The SEI formation at different state of charges was studied for both Li and Na system via XPS analysis. From the C 1s spectra, the relative intensity of the carbon black peak compared to the total C 1s intensity can give an indication of the SEI thickness (Figure 6.6). In general, the SEI thickness in Na cells was greater when compared to corresponding Li systems. It was shown that during discharge, the intensity decreases, due to SEI formation on the electrode, while the SEI undergoes a slight dissolution/stripping in both cells when it is charged. Furthermore, it was demonstrated that the SEI in Li cells was based mostly on C-C and C-H components, indicative of organic compounds, whereas in Na cells the SEI contained more $-CO_3$ and -C=O environments indicating a more inorganic character (Figure 6.7). The study showed that the SEI structure in Li cells consisted of an inorganic-rich layer closer to the electrode interface and an outer layer containing organic decomposition products.

A similar phenomenon was observed in another study by Mogensen *et al.* (2018). It was demonstrated, via HAXPES, that the SEI growth was determined by the sodiation state of the electrode. When testing $NaPF_6$ and NaFSI electrolyte salts in EC:DEC on tin phosphide anodes, the carbon black peak intensity decreased significantly during sodiation and completely disappears when fully sodiated. Thus, the SEI layer was thick enough in order to block the signal from the anode substrate. However, when desodiating again, the peak reappeared, indicating a slight SEI stripping.

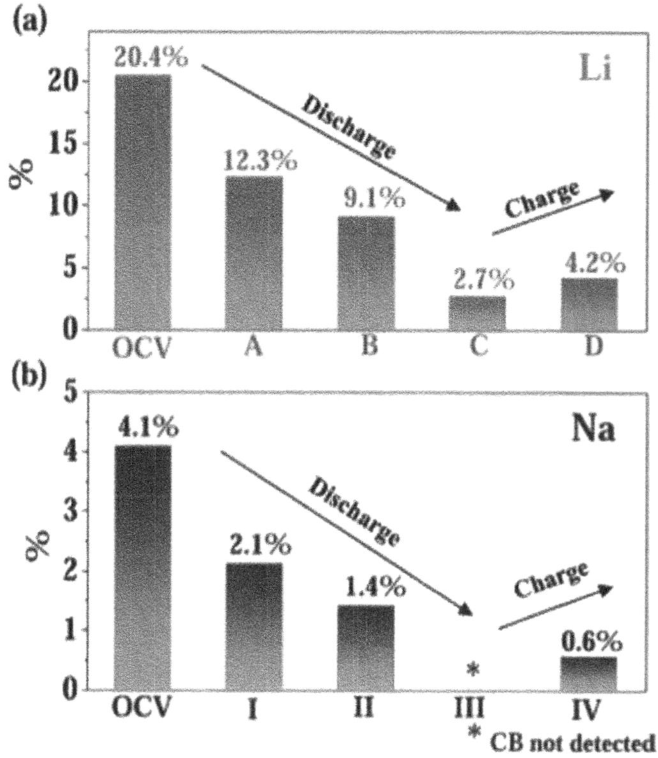

Figure 6.6. *Development of relative peak intensity for the carbon black peak measured in the C 1s XPS spectra during first discharge/charge cycle in Li and Na systems (Philippe* et al. *2014). For a color version of this figure, see www.iste.co.uk/monconduit/batteries.zip*

One important observation from half-cell studies is that a surface layer containing electrolyte decomposition products forms prior to cycling, while such an issue is less prominent in Li half-cells. This is indicative of cross-talk between the metallic counter electrode and the working electrode, which can lead to the formation of species that may mislead when characterizing surface composition on the working electrode (Iermakova *et al.* 2015). This is explained to be due to the high reactivity of sodium metal compared to lithium metal negative electrodes in contact with the aprotic carbonate-based electrolyte commonly used in batteries (Iermakova *et al.* 2015). Therefore, one must use either a stable counter electrode or an electrolyte solution that provides stable SEI formation on Na metal with high reversibility of plating-stripping processes.

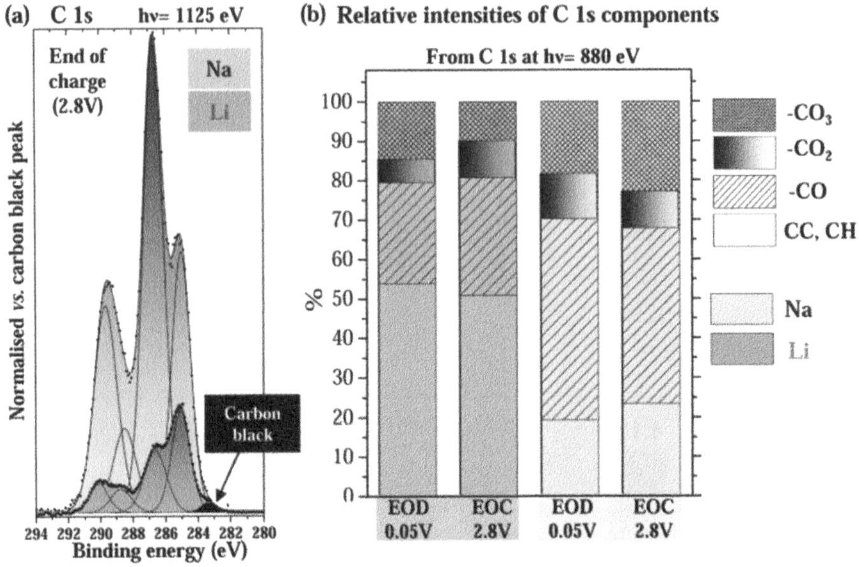

Figure 6.7. *(a) C 1s spectra of Fe_2O_3 electrodes after one CV cycle in a Li half-cell (grey) and a Na half-cell (blue). (b) Relative intensities of C 1s species after first discharge (EOD) and first charge (EOC) in Li and Na cells (Philippe* et al. *2014). For a color version of this figure, see www.iste.co.uk/monconduit/batteries.zip*

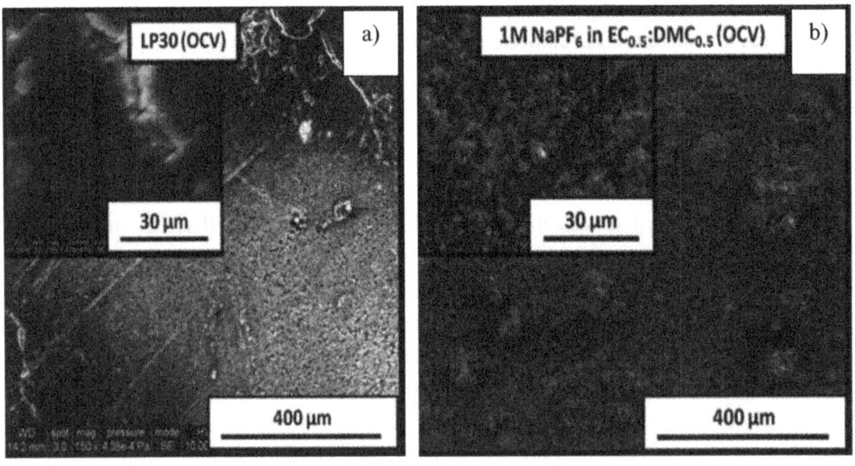

Figure 6.8. *SEM images of (a) Li metal surface after 24 h soaked in LP30 at OCV and (b) Na metal surface after 24 h soaked in 1 M $NaPF_6$ in EC: DMC (v:v, 1:1) at OCV (Iermakova* et al. *2015)*

The surface morphology of electrodes, studied by imaging techniques such as SEM, can give an indication of how the SEI forms. For example, Iermakova *et al.* (2015) demonstrated the difference between the surface morphologies of Li and Na metal anodes after immersion in electrolyte for 24 h (Figure 6.8). Their study revealed a smooth surface layer formation on the Li metal but a rough surface layer formation on the Na metal anode.

Due to the difference in Lewis acidities for Li and Na, the decomposition products from their respective electrolytes differ significantly in composition. Additionally, it is more difficult in Na cells to form a complete passivation layer since the solubility of most reduction productions are much higher than for Li components.

Moshkovich *et al.* (2001) conducted a comparative study of the electrochemical window and the processes on gold electrodes in $LiClO_4$, $NaClO_4$ and $KClO_4$ in PC. In all three cases, the reduction products on the gold surfaces behaved like SEI layers, conducting the corresponding alkali metal ions. According to FTIR and XPS, the main reduction product formed has been found to be the organic compound $CH_3CH(OCO_2M)CH_2OCO_2M$ (M = Li, Na, K). Residual oxygen and water in the electrolyte solution were typically reduced to oxides, peroxides and hydroxides. When comparing the SEI species in all three $MClO_4$ in PC solutions, Li and K salt solutions formed a more stable SEI than $NaClO_4$.

Due to the smaller cation size of Li ions compared to Na, Li forms stronger bonds in Li oxides, hydroxides, carbonates and other SEI species and is therefore largely insoluble in polar aprotic solvents such as PC. Na cations are smaller than K ions; hence, they have a higher charge density and therefore have stronger interactions with solvent molecules. As a result, Na salt products are generally more soluble in PC than K salts. $KClO_4$ has a maximum solubility in PC of 0.1 M, whereas $LiClO_4$ and $NaClO_4$ have a larger saturation concentration.

Another study on the surface layer formed from $MClO_4$ in PC solutions was conducted by Komaba *et al.* (2011b), utilizing XPS and TOF-SIMS. It was demonstrated that the surface films in Li cells contained more alkyl, alkylene groups and polymer species compared to Na cells. However, for Na cells more carbonates and alkoxides were present in the SEI. When the SEI surface was etched with Ar^+ sputtering, the XPS peak intensity for SEI species in Na cells decreased faster than that in Li cells, suggesting that the SEI thickness in Na cells was smaller than in Li cells. The difference in SEI composition for Li and Na cells was investigated via ToF-SIMS

(Figure 6.9). In Li cells, many fragment peaks with larger mass to charge ratio values (m/z), such as m/z = 133 corresponding to $C_3H_7O_3LiCl^+$ could be identified, whereas in Na cells, smaller fragment peaks, e.g. m/z = 62 and 85 indicating Na_2O^+ and Na_3O^+, were observed. Thus, it was concluded that the SEI layer in Li cells primarily consists of organic compounds, while in sodium it was mainly composed of inorganic species, which is consistent with former work from Moshkovich *et al.* (2001).

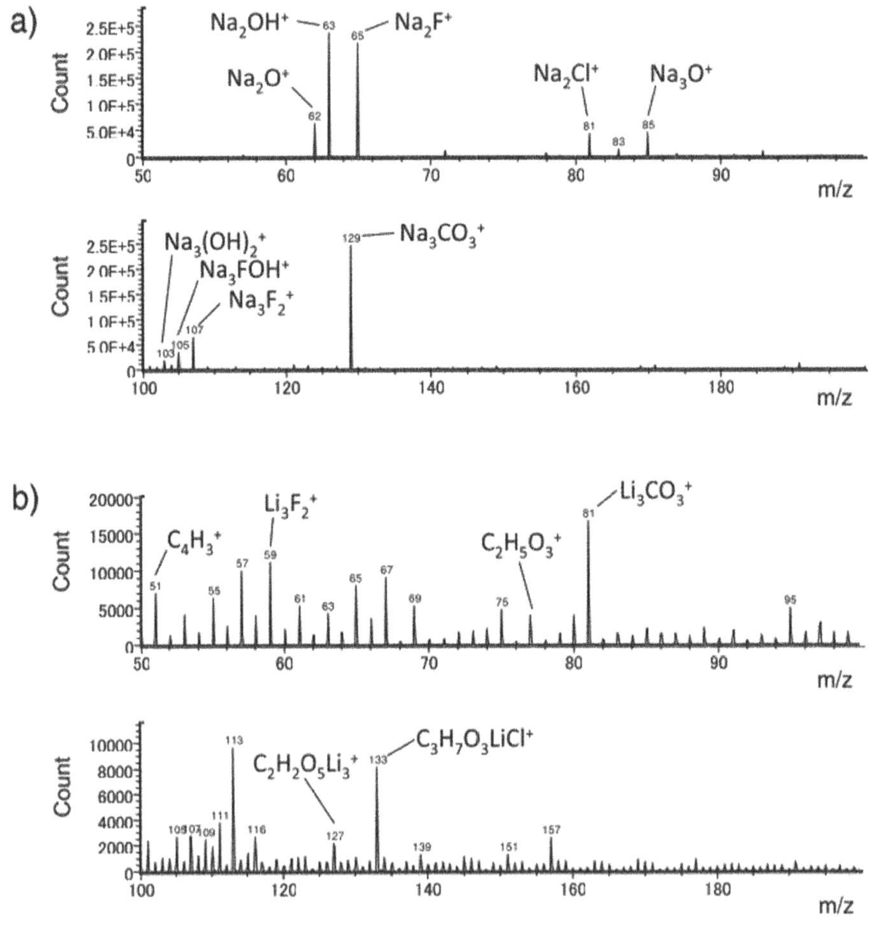

Figure 6.9. *ToF-SIMS spectra of hard carbon electrodes after the first cycle in (a) Na-cell and in (b) Li-cells (Komaba* et al. *2011b)*

Soto *et al.* (2017) also investigated SEI properties in Li and Na cells using a systematic experimental design. Initially, they cycled hard carbon electrodes in a Na half-cell. Afterwards, they replaced the Na counter electrode and electrolyte with Li equivalents. Finally, they switched the Li components back to Na components, hence cycling in Na/Li/Na environments. Then they performed the experiment through the opposite series, cycling in Li/Na/Li environment. The electrolyte used was $MClO_4$ (M = Li, Na) in EC/dimethylecarbonate (DMC) (weight ratio 1:2) with 10 wt.% fluoroethylenecarbonate (FEC). It was shown that, once a Li-based SEI was formed, Na^+ ion conductivity to the active material was considerable reduced. This was despite the Li SEI being formed after being in a Na environment, and indicated that the formed Li-SEI had a too low porosity to allow Na^+ ion conductivity. However, after the formation of a Na-based SEI, there were no consequences for the post-formation of the Li-SEI, implying that the Na-based SEI was more porous than Li-SEI, so Li-ions can pass through the Na-SEI (Figure 6.10).

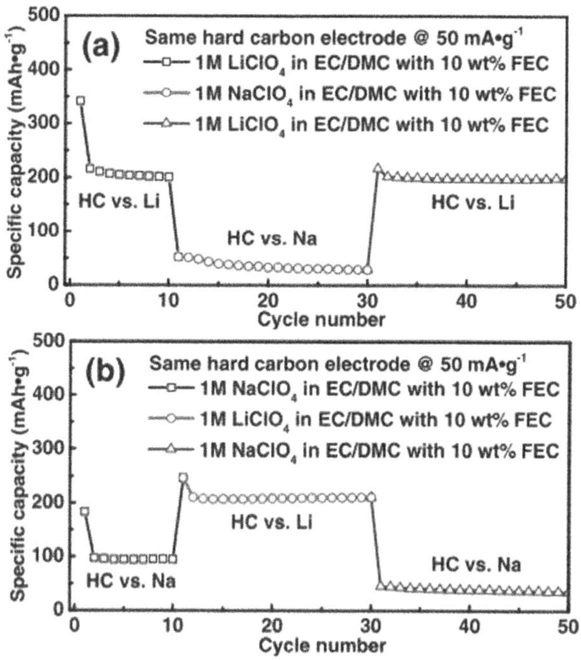

Figure 6.10. *Li- and Na-intercalation in hard carbon. (a) Cycling performance first in 1 M $LiClO_4$ EC:DMC with 10 wt.% FEC, afterwards cycled in 1 M $NaClO_4$ EC:DMC with 10 wt.% FEC, and then in 1 M $LiClO_4$ EC:DMC with 10 wt.% FEC. (b) Cycling performance first in 1 M $NaClO_4$ EC:DMC with 10 wt.% FEC, afterwards in 1 M $LiClO_4$ EC:DMC with 10 wt.% FEC, then again in 1 M $NaClO_4$ EC:DMC with 10 wt.% FEC (Soto* et al. *2017). For a color version of this figure, see www.iste.co.uk/monconduit/batteries.zip*

Kuo *et al.* (2017) studied the structure, stability and ionic conductivity of the SEI formed from the electrolyte $NaClO_4$ in EC:DEC on Sn anodes. Techniques included electrochemical methods, *in situ* surface-enhanced Raman scattering (SERS), supported by density functional theory (DFT) calculations. The decomposition products on the Sn anodes were found to be mainly Na_2O and Na_2CO_3. However, the DFT calculations showed that Na_2O is more stable in the applied voltage ranges than Na_2CO_3, and Na_2CO_3 favors further reaction to yield Na_2O under poor carbon and/or oxygen environments. In addition, when calculating the ionic conductivities, Na_2O shows a much larger ionic conductivity in a wider voltage range than Na_2CO_3. Thus, a SEI consisting mostly of Na_2O can be beneficial in terms of electrochemical window stability and ionic conductivity.

Eshetu *et al.* (2019) used XPS and Ar^+ sputtering to compare the composition of SEI on hard carbon in Na and Li electrolytes using different salt anions. Their results showed that the SEI in Na cells is richer in organic species compared to the SEI in Li cells. This was suggested to be due to the higher solubility of the inorganic Na salts and the Na-alkoxide intermediates compared to the analogous Li compounds.

6.3.2. *Resistance*

Iermakova *et al.* (2015) performed impedance spectroscopy measurements in order to compare the resistivity of the SEI formed in symmetrical Li/Li and Na/Na cells three different electrolytes: (1) $LiPF_6$ in 1:1 EC:DMC (LP30), (2) $NaPF_6$ in 1:1 EC:DMC and (3) $NaPF_6$ in 45:45:10 EC:PC:DMC. The measurements were conducted before and after CV cycling (Figure 6.11).

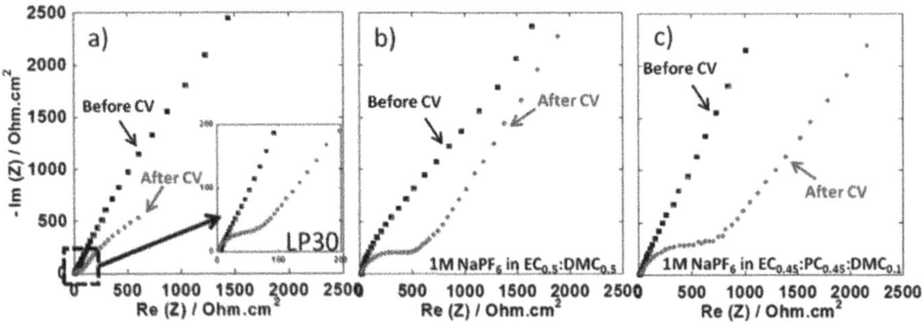

Figure 6.11. *Nyquist plots from impedance spectroscopy measurements before and after CV cycling in (a) LP30, (b) 1 M $NaPF_6$ EC:DMC (1:1) and (c) 1 M $NaPF_6$ EC:PC:DMC (45:45:10) (Iermakova* et al. *2015). For a color version of this figure, see www.iste.co.uk/monconduit/batteries.zip*

For all three electrolyte systems, the semi-circle, which corresponds to the charge transfer resistance, after CV cycling is more apparent than prior to CV, indicating the SEI formation. However, the semi-circle has a higher magnitude for Na cells (b and c) than for Li cells (a). In addition, an observed overpotential was not associated with ionic conductivity of the electrolyte, but rather linked with the interfacial resistance, hence the SEI.

Sodium metal is very reactive and therefore the SEI formed on sodium metal should be expected to be quite different in its nature. In the work by Iermakova *et al.* (2015), lithium and sodium versions of the electrolyte LP30 (EC:DMC 1:1, 1M (Li,Na)PF_6) were used to compare plating and stripping at Li or Na metal electrodes. The surface layer was investigated by FTIR and impedance spectroscopy. Sodium initially formed a highly resistive SEI layer as compared to lithium; they also showed that the resistance of the SEI evolves over time when lithium and sodium electrodes were kept immersed in the electrolyte without cycling. While the lithium electrode displayed a consistently increasing resistance, although with much lower resistance than sodium, the resistance evolution of the sodium was more complex. The resistance increased and then decreased to form a stable value after more than 72 h (Figure 6.12(a)). This behavior is supported by other works (Dugas *et al.* 2016), which showed that (1) carbonates made up the majority of the surface layers in sodium systems that had higher resistance than lithium, (2) the sodium SEI evolved more with cycling and/or time than the typical lithium SEI. Many studies attribute this to unstable SEI products that would either dissolve into the electrolyte or be stripped electrochemically (Moshkovich *et al.* 2001; Iermakova *et al.* 2015; Muñoz-Márquez *et al.* 2015; Zarrabeitia *et al.* 2017).

A similar evolution of the resistance was observed in half-cells, i.e. cells where SEI on electrodes other than sodium is studied, but where sodium metal is still present due to its use as the counter electrode. This is likely due to cross-talk effects involving the transport of decomposition products from the sodium metal counterelectrode to the working electrode, thus giving the impression that the SEI formed on the working electrode is very similar to the SEI formed on sodium metal (Bommier *et al.* 2016; Lee and Tang 2019). For example, the evolution of resistance during OCV conditions that occurs in symmetrical sodium-cells and half-cells is not evident in full-cells or in symmetrical cells with sodium metal absent (Figure 6.12(b) and (c)) (Iermakova *et al.* 2015).

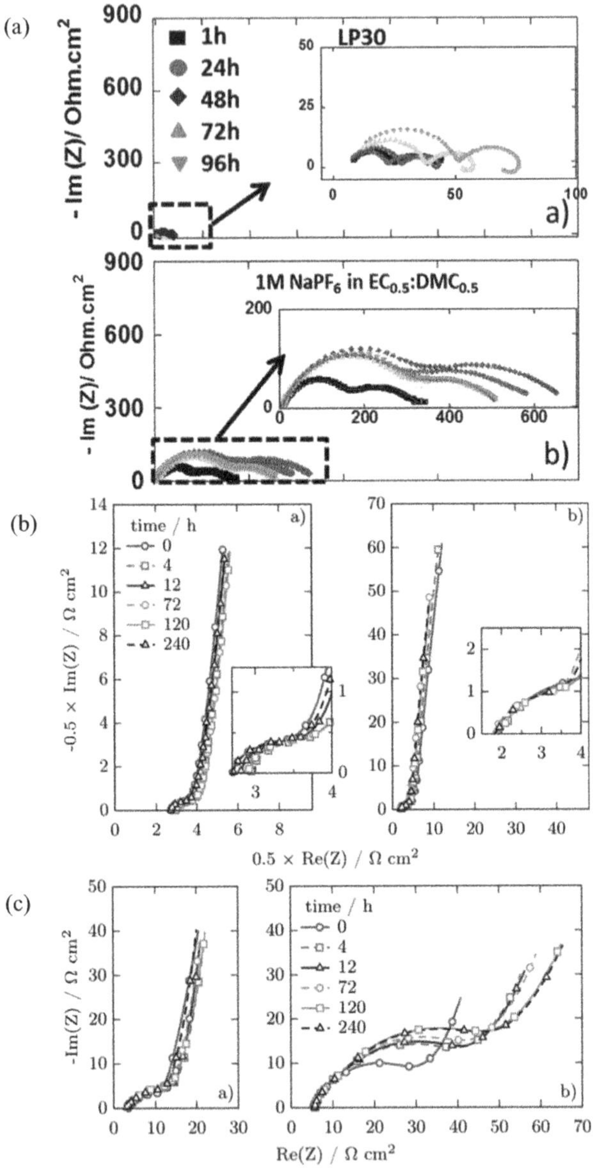

Figure 6.12. *(a) Nyquist plots from impedance spectroscopy measurements for Li (top) and Na (bottom) metal symmetric cells. (b) Impedance measurements of symmetric hard carbon cells with Li (left) or Na (right) electrolytes. (c) Impedance measurements for hard carbon cells with Li (left) or Na (right) metal counter electrode (Iermakova* et al. *2015). For a color version of this figure, see www.iste.co.uk/monconduit/batteries.zip*

Many sodium-ion battery researchers therefore warn against using half-cells in studies of the SEI. Both the unreliable potential of the sodium metal electrode and the SEI formation on sodium and subsequent cross-talk are far too prevalent to draw any conclusions about how full-cell systems will perform (Iermakova *et al.* 2015; Bommier *et al.* 2016; Zarrabeitia *et al.* 2017; Lee and Tang 2019). Although these issues are well known today, most of the information on SEI composition and behaviour found in literature is based on half-cell data since the studies still provide valuable insights and the sodium metal anode offers a convenient source of Na^+ ions.

6.4. Conclusion

Overall, formation of a passivating SEI is a necessity to realize high energy density Na-ion batteries based on non-aqueous electrolytes. This concept is similar to that in lithium-ion batteries; therefore, one can take experience from many previous studies on characteristics of the SEI in lithium-ion batteries. However, it should be noted that the SEI in Na-ion batteries has different features compared to that in lithium-ion batteries, mainly originating from the lower charge density of Na^+ compared to that of Li^+.

This difference in charge density for the cations influences solvation structure in electrolytes, formation and composition of the SEI, physical properties including solubility and resistance of the SEI, as well as, mechanical properties of SEI species. Therefore, electrolyte systems with the same solvent and anion but different cation may provide extremely dissimilar properties and thereby distinct electrochemical performance.

Therefore, the aforementioned differences need to be taken into account when designing new electrolyte solutions for Na-ion batteries. A long-life and viable sodium-ion battery requires a truly stable SEI that remains intact over long-term cell cycling.

6.5. References

Bommier, C., Leonard, D., Jian, Z., Stickle, W.F., Greaney, P.A., and Ji, X. (2016). New paradigms on the nature of solid electrolyte interphase formation and capacity fading of hard carbon anodes in Na-ion batteries. *Advanced Materials Interfaces*, 3(19), 1–10.

Brummer, S.B. (1980). No Title. In *Proceedings of the Workshop on Lithium Non-Aqueous Battery Electrochemistry*, Yeager, E.G. *et al.* (eds). The Electrochemical Society, Princeton, NJ, 80–87.

Carboni, M., Manzi, J., Armstrong, A.R., Billaud, J., Brutti, S., and Younesi, R. (2019). Analysis of the solid electrolyte interphase on hard carbon electrodes in sodium-ion batteries. *ChemElectroChem*, 6(6), 1745–1753.

Cresce, A. V., Russell, S.M., Borodin, O., Allen, J.A., Schroeder, M.A., Dai, M., Peng, J., Gobet, M.P., Greenbaum, S.G., Rogers, R.E., and Xu, K. (2017). Solvation behavior of carbonate-based electrolytes in sodium ion batteries. *Physical Chemistry Chemical Physics*, 19(1), 574–586.

Dugas, R., Ponrouch, A., Gachot, G., David, R., Palacin, M.R., and Tarascon, J.M. (2016). Na reactivity toward carbonate-based electrolytes: The effect of FEC as additive. *Journal of The Electrochemical Society*, 163(10), A2333–A2339.

Ellis, L.D., Hatchard, T.D., and Obrovac, M.N. (2012). Reversible insertion of sodium in tin. *Journal of The Electrochemical Society*, 159(11), A1801–A1805.

Ellis, L.D., Wilkes, B.N., Hatchard, T.D., and Obrovac, M.N. (2014). In situ XRD study of silicon, lead and bismuth negative electrodes in nonaqueous sodium cells. *Journal of The Electrochemical Society*, 161(3), A416–A421.

Eshetu, G.G., Diemant, T., Hekmatfar, M., Grugeon, S., Behm, R.J., Laruelle, S., Armand, M., and Passerini, S. (2019). Impact of the electrolyte salt anion on the solid electrolyte interphase formation in sodium ion batteries. *Nano Energy*, 55, 327–340.

Fukunishi, M., Yabuuchi, N., Dahbi, M., Son, J.Y., Cui, Y., Oji, H., and Komaba, S. (2016). Impact of the cut-off voltage on cyclability and passive interphase of Sn-polyacrylate composite electrodes for sodium-ion batteries. *Journal of Physical Chemistry C*, 120(28), 15017–15026.

Huang, Y., Zhao, L., Li, L., Xie, M., Wu, F., and Chen, R. (2019). Electrolytes and electrolyte/electrode interfaces in sodium-ion batteries: From scientific research to practical application. *Advanced Materials*, 31(21), 1808393.

Iermakova, D.I., Dugas, R., Palacín, M.R., and Ponrouch, A. (2015). On the comparative stability of Li and Na metal anode interfaces in conventional alkyl carbonate electrolytes. *Journal of The Electrochemical Society*, 162(13), A7060–A7066.

Jones, J., Anouti, M., Caillon-Caravanier, M., Willmann, P., Sizaret, P.-Y., and Lemordant, D. (2011). Solubilization of SEI lithium salts in alkylcarbonate solvents. *Fluid Phase Equilibria*, 305(2), 121–126.

Komaba, S., Ishikawa, T., Yabuuchi, N., Murata, W., Ito, A., and Ohsawa, Y. (2011a). Fluorinated ethylene carbonate as electrolyte additive for rechargeable Na batteries. *ACS Applied Materials and Interfaces*, 3(11), 4165–4168.

Komaba, S., Murata, W., Ishikawa, T., Yabuuchi, N., Ozeki, T., Nakayama, T., Ogata, A., Gotoh, K., and Fujiwara, K. (2011b). Electrochemical Na insertion and solid electrolyte interphase for hard-carbon electrodes and application to Na-ion batteries. *Advanced Functional Materials*, 21(20), 3859–3867.

Kuo, L.-Y., Moradabadi, A., Huang, H.-F., Hwang, B.-J., and Kaghazchi, P. (2017). Structure and ionic conductivity of the solid electrolyte interphase layer on tin anodes in Na-ion batteries. *Journal of Power Sources*, 341, 107–113.

Lee, S.E. and Tang, M.H. (2019). Electroactive decomposition products cause erroneous intercalation signals in sodium-ion batteries. *Electrochemistry Communications*, 100(January), 70–73.

Mogensen, R., Brandell, D., and Younesi, R. (2016). Solubility of the solid electrolyte interphase (SEI) in sodium ion batteries. *ACS Energy Letters*, 1(6), 1173–1178.

Mogensen, R., Maibach, J., Naylor, A.J., and Younesi, R. (2018). Capacity fading mechanism of tin phosphide anodes in sodium-ion batteries. *Dalton Transactions*, 47(31), 10752–10758.

Mogensen, R., Maibach, J., Brant, W.R., Brandell, D., and Younesi, R. (2017). Evolution of the solid electrolyte interphase on tin phosphide anodes in sodium ion batteries probed by hard x-ray photoelectron spectroscopy. *Electrochimica Acta*, 245, 696–704.

Moshkovich, M., Gofer, Y., and Aurbach, D. (2001). Investigation of the electrochemical windows of aprotic alkali metal (Li, Na, K) salt solutions. *Journal of The Electrochemical Society*, 148(4), E155.

Muñoz-Márquez, M.Á., Saurel, D., Gómez-Cámer, J.L., Casas-Cabanas, M., Castillo-Martínez, E., and Rojo, T. (2017). Na-ion batteries for large scale applications: A review on anode materials and solid electrolyte interphase formation. *Advanced Energy Materials*, 7(20), 1700463.

Muñoz-Márquez, M.A., Zarrabeitia, M., Castillo-Martínez, E., Eguía-Barrio, A., Rojo, T., and Casas-Cabanas, M. (2015). Composition and evolution of the solid-electrolyte interphase in $Na_2Ti_3O_7$ electrodes for Na-Ion batteries: XPS and auger parameter analysis. *ACS Applied Materials and Interfaces*, 7(14), 7801–7808.

Naylor, A.J., Carboni, M., Valvo, M., and Younesi, R. (2019). Interfacial reaction mechanisms on graphite anodes for K-ion batteries. *ACS Applied Materials & Interfaces*, 11(49), 45636–45645.

Newman, G.H. (1980). No Title. In *Proceedings of the Workshop on Lithium Non-Aqueous Battery Electrochemistry*, Vol. 80–87, Yeager, E.G. *et al.* (eds). The Electrochemical Society, Princeton, New Jersey.

Ota, H., Akai, T., Namita, H., Yamaguchi, S., and Nomura, M. (2003). XAFS and TOF-SIMS analysis of SEI layers on electrodes. *Journal of Power Sources*, 119–121, 567–571.

Peled, E. (1979). The electrochemical behavior of alkali and alkaline earth metals in nonaqueous battery systems–The solid electrolyte interphase model. *Journal of The Electrochemical Society*, 126(12), 2047.

Peled, E. and Menkin, S. (2017). Review–SEI: Past, present and future. *Journal of The Electrochemical Society*, 164(7), A1703–A1719.

Philippe, B., Valvo, M., Lindgren, F., Rensmo, H., and Edström, K. (2014). Investigation of the electrode/electrolyte interface of Fe_2O_3 composite electrodes: Li vs Na batteries. *Chemistry of Materials*, 26(17), 5028–5041.

Philippe, B., Hahlin, M., Edström, K., Gustafsson, T., Siegbahn, H., and Rensmo, H. (2015). Photoelectron spectroscopy for lithium battery interface studies. *Journal of the Electrochemical Society*, 163(2), A178–A191.

Shin, H., Park, J., Han, S., Marie, A., and Lu, W. (2015). Component-/structure-dependent elasticity of solid electrolyte interphase layer in Li-ion batteries: Experimental and computational studies. *Journal of Power Sources*, 277, 169–179.

Song, J., Xiao, B., Lin, Y., Xu, K., and Li, X. (2018). Interphases in sodium-ion batteries. *Advanced Energy Materials*, 8(17), 1703082.

Soto, F.A., Yan, P., Engelhard, M.H., Marzouk, A., Wang, C., Xu, G., Chen, Z., Amine, K., Liu, J., Sprenkle, V.L., El-Mellouhi, F., Balbuena, P.B., and Li, X. (2017). Tuning the solid electrolyte interphase for selective Li- and Na-ion storage in hard carbon. *Advanced Materials*, 29(18), 1606860.

Sun, Y., Shi, P., Xiang, H., Liang, X., and Yu, Y. (2019). High-safety nonaqueous electrolytes and interphases for sodium-ion batteries. *Small*, 15(14), 1805479.

Tasaki, K., Goldberg, A., Lian, J.-J., Walker, M., Timmons, A., and Harris, S.J. (2009). Solubility of lithium salts formed on the lithium-ion battery negative electrode surface in organic solvents. *Journal of The Electrochemical Society*, 156(12), A1019.

Tripathi, A.M., Su, W.N., and Hwang, B.J. (2018). In situ analytical techniques for battery interface analysis. *Chemical Society Reviews*, 47(3), 736–751.

Valvo, M., Lindgren, F., Lafont, U., Björefors, F., and Edström, K. (2014). Towards more sustainable negative electrodes in Na-ion batteries via nanostructured iron oxide. *Journal of Power Sources*, 245, 967–978.

Wang, W., Zhang, J., Yu, D.Y.W., and Li, Q. (2017). Improving the cycling stability of Sn_4P_3 anode for sodium-ion battery. *Journal of Power Sources*, 364, 420–425.

Weadock, N., Varongchayakul, N., Wan, J., Lee, S., Seog, J., and Hu, L. (2013). Determination of mechanical properties of the SEI in sodium ion batteries via colloidal probe microscopy. *Nano Energy*, 2(5), 713–719.

Xu, K. (2014). Electrolytes and interphases in Li-ion batteries and beyond. *Chemical Reviews*, 114(23), 11503–11618.

Zarrabeitia, M., Muñoz-Márquez, M., Nobili, F., Rojo, T., and Casas-Cabanas, M. (2017). Influence of using metallic Na on the interfacial and transport properties of Na-ion batteries. *Batteries*, 3(4), 16.

7

Batteries Containing Prussian Blue Analogue Electrodes

Colin D. WESSELLS
Natron-Energy, Santa Clara, USA

7.1. Introduction

7.1.1. *Chapter introduction*

The purpose of this chapter is to provide a practical introduction to the use of Prussian blue and its analogues as electrodes in sodium-ion cells. It describes the relevant physical and electrochemical properties of these materials, provides illustrative examples of their use in batteries and summarizes the current commercial state-of-the-art. However, the purpose of this chapter is not to provide an exhaustive review of the extraordinary variety of Prussian blue batteries that have been reported over the past decade. Readers seeking a thorough review of recent literature should consult (Hurlbutt *et al.* 2018; Wang *et al.* 2018b).

7.1.2. *History of Prussian blue*

The discovery of ferric ferrocyanide, more commonly known as Prussian blue, predates its use in batteries by nearly three centuries. It was discovered in Berlin in the early 18th Century during a failed attempt to synthesize red cochineal pigment. Prussian blue was the first synthetic blue pigment, which made it an inexpensive alternative to other blue pigments of that time. Its relatively low cost led to its rapid

Na-ion Batteries,
coordinated by Laure MONCONDUIT and Laurence CROGUENNEC. © ISTE Ltd 2020.

adoption by the pigment and dyestuff industries for applications ranging from fine art to machining.

The early scientific study of Prussian blue during the 19th and early 20th Centuries focused on its chemical and structural properties, and it was not until the 1960s that its crystal structure, including the ordering transition metal valences and spins was well understood. In 1978, Vernon Neff discovered that thin films of Prussian blue undergo reversible electrochromic cycling (Neff 1978). Soon afterwards, the first Prussian blue battery was built with lithium metal as an anode (Messina and Perichon 1980). Throughout the 1980s and 1990s, various research groups experimented with half and full battery cells containing thin films of Prussian blue or its analogues, but these early efforts were rarely successful in replicating the long cycle life achieved in electrochromic cells.

In 2011, researchers reported the ability to charge and discharge a bulk composite Prussian blue analogue (PBA) electrode for tens of thousands of deep discharge cycles (Wessells *et al.* 2011). These results arrived at an opportune time at which the battery research community began to diversify its efforts to include new materials systems for stationary energy storage applications. The discovery of such long cycle life in a cell perceived as scalable renewed interest in PBAs as a class of battery electrodes. For the rest of the decade, an increasing number of research groups have investigated the use of these materials as anodes and cathodes in cells containing both aqueous and organic electrolytes. There have even been at least two attempts to commercialize battery products containing cells based on PBA electrodes: Natron Energy, which uses two different PBAs for its anode and cathode, and Sharp Laboratories of America, which paired a PBA cathode with a hard carbon anode in a high-voltage cell.

7.1.3. *Physical characteristics: structure, composition and morphology*

An understanding of Prussian blue batteries begins with an understanding of the composition, crystal structure and morphology of these materials. PBAs are complex transition metal cyanides, which means that transition metal cations are bonded to both the carbon and nitrogen ends of cyanide ligands. The general chemical formula for PBAs is: $A_xP_y[R(CN)_6]_z(H_2O)_n$, where A is a mobile alkali or alkaline earth cation and P and R are transition metal cations. The stoichiometry x:y:z must fulfill charge neutrality and is influenced by synthesis conditions. The bonding structure of PBAs is a three-dimensional framework of octahedral $R(CN)_6^{n-}$ complexes that are linked together by P^{m+} transition metal cations that coordinate to the nitrogen ends of the cyanide groups. The resulting crystal structure is most commonly a Perovskite-type cubic structure in which both the P site and R site transition metal cations are octahedrally coordinated and charge neutrality is maintained by mobile interstitial A cations. The mobile A site species is most commonly Na^+ or K^+. The PBA structure accommodates a wide variety of transition metals in the P sites. The R sites are most commonly occupied by Fe, Mn, or Cr.

The unit cell of a defect-free cubic PBA lattice is shown in Figure 7.1. Each unit cell contains four formula units at full site occupancy (x:y:z = 2:1:1). The lattice parameter a_0 depends on the P and R species, and is typically between 10.0 and 10.7 Å. The combination of a relatively low sixfold coordination number of the P and R sites with a long cyanide bridging ligand results in unusually large interstices and channels through the crystal structure. The interstitial "A sites" have a hard sphere radius of approximately 2.5 Å, while the channels between them along the <100>-type directions have a hard sphere radius of approximately 1.6 Å. The A sites and the channels between them are therefore significantly larger than the alkali cations that move in and out of the structure during electrochemical cycling (e.g. Na^+, 1.16 Å). These large interstices and channels are the differentiating characteristics between the "open framework" structure of PBAs and the structures of oxide and poly-anion-based ceramic electrode materials.

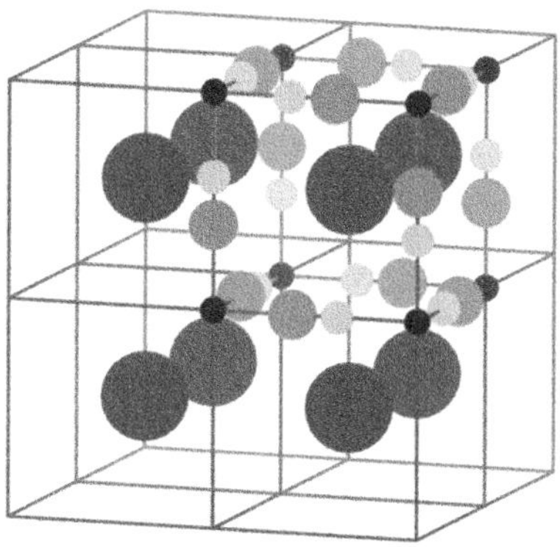

Figure 7.1. *The unit cell of a defect free, cubic Prussian blue structure. The large spheres represent the interstitial A sites that may contain mobile cations such as Na^+. For clarity, one-eighth of the unit cell is fully populated with the P, R, C and N atoms. For a color version of this figure, see www.iste.co.uk/monconduit/batteries.zip*

The occupancy of the A sites varies during electrochemical cycling as mobile cations are inserted and removed from the structure. Under most circumstances, PBAs have a stoichiometry including $x \leq 2$, which means that the occupancy of each A site does not exceed one. Under these conditions, the lattice strain observed during cycling is no greater than the strain associated with the change in the R-C or P-N bond length as the valence state of the R or P transition metal changes, typically <1%. For instance, in the case of reduction of a hexacyanoferrate-based cathode, a negative strain of up to about −0.1 Å may be observed, which is fully accounted for by the 0.05 Å decrease in bond length of the two Fe-C bonds along the <100>-type directions (Wessells *et al.* 2011). For this reason, PBAs are often referred to as "zero strain materials" though in fact they undergo a non-zero, albeit small strain during cycling. The observation of this strain is actually an indication of the stability of the PBA framework during cycling: the bond length of the $R(CN)_6$ groups is able to vary freely to maintain a stable state as cations are inserted and removed from the A sites.

The PB structure also contains a significant quantity of water. The total amount of water present in a PBA may exceed 20% of its total mass, which may be expressed as n = 4 or greater in the general PBA chemical formula. The water concentration in PBAs is orders of magnitude higher than the water impurities found in other electrode systems and commercial Li-ion cells. The water content of PBAs is typically measured using thermogravimetric analysis or Karl–Fisher analysis. Four types of water may be present in a PBA: surface water, zeolitic or "free" interstitial water, interstitial water that is coordinated to mobile A site cations, and water that is bound to the transition metal cyanide framework (Herren *et al.* 1980; Wardecki *et al.* 2017).

The surface and zeolitic interstitial water rapidly equilibrate with the surrounding environment and the quantities present depend on the relative humidity and temperature. Therefore, the concentration of mobile water in the PBA structure is not only quite high, but also dependent on process conditions. This introduces unique engineering challenges for electrode fabrication and cell operation. For instance, electrode slurries based on common organic solvents and fluoridated polymer binders have rheological properties that depend on the amount of water present. To minimize process variability, PBA electrode materials must be dried to a well-defined residual water content before electrode processing and cell assembly. The surface water and some interstitial water (likely zeolitic) can be removed via vacuum drying at modest temperatures without damaging the material. However, full dehydration has been found to correlate with reduced crystallinity. In some PBAs, full dehydration enhances electrochemical performance, while in others, it may degrade it (Song *et al.* 2015; Yang *et al.* 2016). For this reason, a tolerance to residual water must be considered during the selection of materials systems for processing PBA electrodes and for cells containing them. The possibility that water from the PBA structure might be released into the cell poses risks to the stability of other species typically found in nominally anhydrous cells, such as aluminum current collectors, fluoridated electrolyte salt anions and some separators that contain ceramic particles. Water release from the PBA structure into the cell also poses a risk for the formation of a stable solid electrolyte interphase on carbon anodes. As of 2019, no PBA/carbon cell has been successfully commercialized and the interactions between PBAs, the water they release, and the anode SEI are not well understood.

The electrochemical properties of PBAs also depend on the concentration of point defects the structure contains. Among these, the concentration of $R(CN)_6$ vacancies is of primary importance. The absence of a $R(CN)_6$ group from the framework leaves behind a vacant site in the octahedral coordination of each of the six adjacent P cations. These vacant sites are filled by water that coordinates directly to the P cation (Figure 7.2). This "lattice bound" water is not easily removed from the structure. The presence of $R(CN)_6$ vacancies is sometimes explicitly denoted in the chemical formula of a PBA, as in: $A_xP[R(CN)_6]_z(Vac)_{1-z}(H_2O)_n$, where the formula is normalized to full site occupancy of the framework. Depending on the synthesis conditions, (1-z) may be as great as 0.5, meaning that up to half of the $R(CN)_6$ sites are vacant. At a concentration of high $R(CN)_6$ vacancies, the vacancies may form a continuous network of micropores having a radius of a few Å. It is generally believed that charge storage of monovalent alkali cations occurs at the A sites, but in some cases storage of multivalent species occurs in the $R(CN)_6$ vacancies (Wang *et al.* 2015a). The lattice parameter of the PBA remains practically unchanged as the concentration of electrochemical active $R(CN)_6$ groups is varied. The volumetric energy density of a cell containing a PBA electrode therefore depends on the concentration of $R(CN)_6$ vacancies, and in general, the vacancy concentration should be minimized.

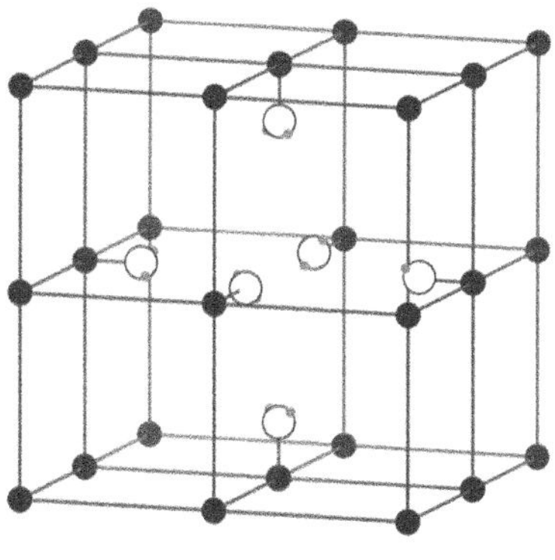

Figure 7.2. *The unit cell of a cubic Prussian blue structure containing a central $R(CN)_6$ vacancy. Six lattice-coordinated water molecules complete the coordination shells of the adjacent P-site cations. C, N and A site atoms omitted for clarity. For a color version of this figure, see www.iste.co.uk/monconduit/batteries.zip*

The electrochemical properties of PBAs also depend on their morphology. Two morphological length scales are relevant: the size of the primary grains, and the size of the secondary agglomerates. Many PBAs have solubilities of 10 ppm or less, orders of magnitude less than their commercially viable synthesis concentrations of 10% or more. This extreme supersaturation results in rapid precipitation of nanoparticulate grains. For PBA materials relevant to batteries, primary grain size ranges from about 20 nm up to about 1 μm. Unless these grains are agglomerated into larger particles, they are challenging to convert into high-performance electrodes using conventional slurry processing methods. The greater the number of discrete active material particles in the slurry, the more challenging it is to fully disperse them with the nanocarbon black that is typically used as a conductive additive. Agglomeration of the primary PBA grains into secondary structures of 10 μm or more therefore facilitates successful dispersion of the PBA and carbon particles, resulting in a high-capacity, low-impedance electrode. However, these agglomerates should not become too large, or their charge transport properties will limit the cell's rate capability (Shrivastava and Smith 2018).

7.1.4. *Synthetic methods*

This chapter considers methods for bulk synthesis of PBAs, and not the synthesis of thin films, which are impractical for most battery applications. There are three general routes for bulk synthesis of PBAs:

1) reaction of a transition metal salt and a hexacyanometallate salt;

2) reaction of a transition metal salt with a simple cyanide;

3) decomposition of a hexacynometallate salt.

Route 1 is the most common because it allows for the widest variety of PBA compositions, including multiple P and R species. The general reaction for Route 1 is represented by equation [7.1]:

$$\mathrm{x}A^{+} + \mathrm{y}P^{m+} + \mathrm{z}[R(CN)_6]^{n-} \rightarrow A_xP_y[R(CN)_6]_z(H_2O)_n \qquad [7.1]$$

where stoichiometry x, y, z fulfills charge neutrality and n depends on the identity of species A, the ratio of y/z (or equivalently, the concentration of $R(CN)_6$ vacancies) and the extent to which the product is dried after synthesis.

Depending on the solubility of the PBA, equation [7.1] may proceed very quickly, resulting in a nanoparticulate product. This is the case for pigment-grade Prussian blue and common PBA cathodes such as copper hexacyanoferrate (CuHCF) and nickel hexacyanoferrate (NiHCF), each of which has solubilities of less than 10 ppm in water at room temperature. Those nanoparticulate PBAs generally have a high $R(CN)_6$ vacancy content ($1\text{-}z > 0.25$), which results in low specific and volumetric capacity and a high water content. Numerous modifications to Route 1 have been developed to slow down the rate of nucleation to favor the growth of larger particles containing a lower defect concentration. These include chelation of P^{m+} with carboxylic acids, the addition of concentrated polymers, the use of high-temperature, high-pressure solvothermal conditions, and even sonication (Wu *et al.* 2006; Ji *et al.* 2016; Su *et al.* 2017).

Route 1 is the most commonly used method for the production of the mixed-metal hexacyanoferrates that are used as cathodes. In a typical synthesis, a precursor solution containing dissolved salts of multiple transition metal cations P is reacted with a solution of hexacyanoferrate to form the mixed-metal hexacyanoferrate product. Many of these mixed-metal hexacyanoferrates are effectively solid solutions, in which the ratio of the various P species may be varied continuously while maintaining a single phase. The single-phase nature of these materials has been confirmed using both physical characterization techniques such as X-ray diffraction, which show a continuous variation in lattice parameter with composition, and by electrochemical experiments that show a continuous variation in reaction potential with composition (Wessells *et al.* 2012a). The ability to form a solid solution at the P site introduces the opportunity to fine tune a PBA to optimize for multiple functional properties such as solubility, morphology or even reaction potential and cycle life for a particular insertion species. During synthesis of a mixed-metal PBA, care must be taken to account for the difference in the solubilities of the various pure phase of each P species. Synthesis of a mixed-metal PBA may result in a composition that is richer in the P species for which the pure PBA is least soluble.

Route 1 is also the most commonly used method for production of core-shell materials in which a core particle of the first PBA is coated with a shell of a second PBA that has a different composition from the core. The core–shell architecture is attractive because conformal, pinhole free epitaxial growth can be achieved for core and shell material pairs having a small mismatch in lattice parameter (Catala *et al.* 2009). This allows for effective chemical protection of a PBA that has desirable electrochemical properties but poor chemical stability. For instance, CuHCF, which has a high reaction potential and unacceptably high solubility in aqueous electrolytes may be protected by an epitaxial shell of less soluble NiHCF (Li *et al.* 2015). Unfortunately, the core-shell PBA materials demonstrated to date have been synthesized at concentrations too low (<1%) to be economically attractive.

Finally, Route 1 is the synthesis method of choice for the materials used in all publicly documented efforts to commercialize PBA batteries. In addition to the flexibility in PBA composition that Route 1 offers, it can be performed at very high concentrations. This allows relatively small reactors to be used and keeps the synthesis and purification waste streams to an acceptable volume. For example, as of 2019, Natron Energy uses Route 1 to regularly produce batches of a mixed metal hexacyanoferrate cathode at a scale of hundreds of kilograms per batch at a concentration of about 12% PBA weight in solution.

In Route 2, a PBA product is synthesized by the direct reaction of a transition metal salt with cyanide anions (equation [7.2]):

$$\mathrm{x}A^{+} + (\mathrm{y}+\mathrm{z})P^{m+} + 6\mathrm{z}(CN)^{-} \rightarrow A_xP_y[P(CN)_6]_z(H_2O)_n \qquad [7.2]$$

The variety of PBA compositions produced by Route 2 is limited by the presence of just one transition metal P in solution. This makes Route 2 appropriate for the synthesis of Prussian blue, manganese hexacyanomanganate, and other PBAs that contain a single transition metal. Route 2 is unlikely to be practical for PBAs that contain multiple transition metals because each of them may form simple or complex cyanides having different solubilities. Despite this limitation, Route 2 also offers advantages. For instance, simple salts of hexacyanomanganate are not commercially available in bulk, so Route 2 is the most economic method for production of manganese hexacyanomanganate (equation [7.3]):

$$\mathrm{x}A^{+} + (\mathrm{y}+\mathrm{z})Mn^{2+} + 6\mathrm{z}(CN)^{-} \rightarrow A_xMn_y[Mn(CN)_6]_z(H_2O)_n \qquad [7.3]$$

Direct synthesis from Mn^{2+} and CN^{-} as in equation [7.3] results in a material having an initial valence state of hexacyanomanganate(II), allowing the first charge of a cell in which that material is configured as the anode to proceed immediately with hexacyanomanganate(II/I) reduction at low potential and avoiding the high potential hexacyanomanganate(III/II) reaction. That high potential plateau is unavoidable during syntheses of hexacyanomanganate-based PBAs from the $K_3Mn^{III}(CN)_6$ salt that is commercially available in small quantities, introducing a considerable challenge for production and operation of a practical cell.

In Route 3, a PBA is formed from the products of decomposition of a hexacyanometallate salt (equation [7.4]):

$$xA^+ + (y + z)[R(CN)_6]^{n-} \rightarrow A_xR_y[R(CN)_6]_z(H_2O)_n \quad [7.4]$$

The kinetics of Route 3 is generally slow for stable hexacyanometallates such as hexacyanoferrate. Syntheses following Route 3 generally rely on high temperature or a pH at which the hexacyanometallate anions are subject to rapid ligand exchange. For instance, heating dilute hexacyanoferrate in hydrochloric acid for several hours results in the formation of Prussian blue (Su *et al.* 2017; Wang *et al.* 2018b). Route 3 is favored for the synthesis of PB having high crystallinity and a large primary crystal grain size of up to 100 μm or more. Materials produced via Route 3 are well suited for crystallographic characterization, mass transport measurements and as model substrates for core shell PBAs. Route 3 cannot be scaled industrially because a stoichiometric quantity of HCN is released as a by-product, in contrast to Routes 1 and 2 that have high yields on the cyanide-containing precursors.

7.2. Electrochemistry of PBAs

7.2.1. *Mechanism and resulting characteristics*

PBAs undergo reversible electrochemical reactions in which mobile cations such as Na^+ are inserted and removed from the interstitial A sites. During these reactions, the $PR(CN)_6$ framework structure undergoes little or no changes. As Na^+ is inserted or removed from the framework, the valence state of at least one of the P or R site transition metals varies to maintain charge neutrality. In most cases, insertion or removal of Na^+ from a PBA proceeds by a single-phase process in which the concentration of Na^+ in the structure and the average valence of the electrochemically active transition metal site varies continuously. The general equation for the reaction of Na^+ with a PBA in which the R site is electrochemically active is:

$$Na_xP_y[R^{m+}(CN)_6]_z + \delta\left(Na^+ + e^-\right) \rightarrow$$
$$Na_{x+\delta}P_y[R^{m+}(CN)_6]_{z-\delta}[R^{m-1}(CN)_6]_\delta \quad [7.5]$$

During the electrochemical cycling of PBAs, the reaction potential varies continuously with the state of charge. In the most common case of a simple solid solution reaction, the potential profile has a smooth S-shaped curve, the shape of which is defined primarily by the entropy associated with the varying occupancy of Na^+ in the interstitial A sites (Figure 7.3). Deviations of PBA reaction potential profiles from the theoretical solid solution profile are most commonly attributed to an enthalpy of mixing that is asymmetric about 50% state of charge (McCargar and Neff 1988; Shrivastava and Smith 2018). Further evidence for the solid solution reaction mechanism of most PBAs is the linear evolution in the lattice parameter with the extent of reaction.

The low-strain solid solution mechanism by which PBAs are charged and discharged is highly reversible. Thin films of PB used in electrochromic applications have been shown to survive hundreds of thousands of full cycles (Itaya and Shibayama 1982). Bulk PBA electrodes relevant to batteries survive for tens of thousands of cycles (Wessells *et al.* 2011). This cycle life is unique among all classes of battery electrode materials, most of which last 1,000–10,000 deep discharge cycles (Figure 7.4). To achieve such a long cycle life, the PBA electrode must be chemically stable in the cell across the entire state-of-charge range over which it is cycled. In effect, the PBA must be insoluble and otherwise inert to the electrolyte and any other chemical species it contacts. Under those conditions, PBA electrodes can be cycled almost indefinitely regardless of the electrochemically active P or R species. This makes fabrication of cells containing PBAs for both the cathode and anode active materials attractive, as it introduces the opportunity for a cell life of tens of thousands of cycles (see section 7.3).

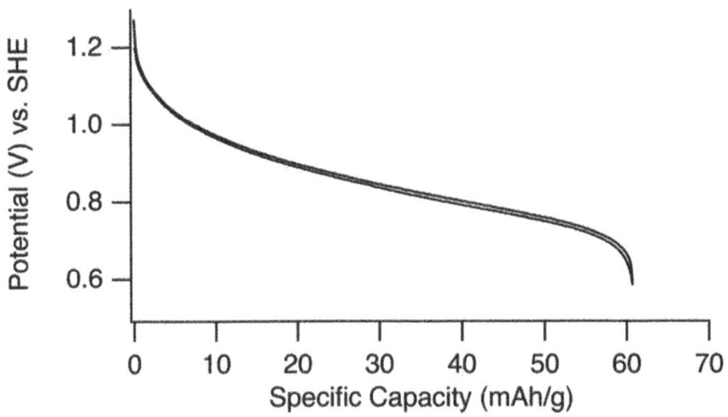

Figure 7.3. *The charge and discharge reaction potential profiles of copper hexacyanoferrate in an aqueous 1 M Na^+ electrolyte during galvanostatic cycling at C/6 (Courtesy Natron Energy)*

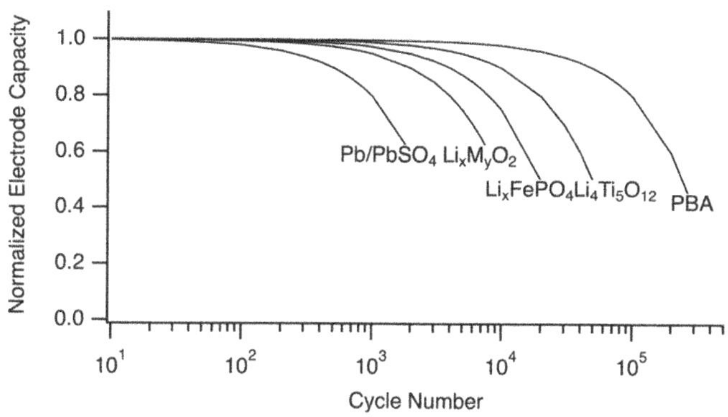

Figure 7.4. *Semiquantitative comparison of the typical deep discharge cycle life of lead acid, lithium-ion and PBA electrode materials*

The reaction of insertion species such as Na^+ with PBA materials proceeds with unusually fast kinetics. Thin films of PB can be fully charged or discharged in just seconds (Itaya and Shibayama 1982). Bulk PBA electrodes relevant to battery cells have been shown to sustain rates of 60 C, equivalent to a full charge discharge in 1 min (see section 7.3.1). In comparison, commercial lead acid and Li-ion cells rarely operate at a sustained charge or discharge rate greater than 10 C. During cycling of a PBA electrode, the rate of reaction is not limited by either the transport of charge carriers across the surface interfaces of the PBA particles or by bulk transport of Na^+ through the PBA's open framework structure. Rather, the reaction rate of a PBA is typically limited by electronic conductivity throughout the bulk of the PBA particle (Shrivastava and Smith 2018). This places great importance on control of the morphology of the PBA particles: too large, and the rate of reaction may be undesirably slow, but too small, and those particles may not be readily processable into composite electrodes in which all of the particles are in good electronic contact with the conductive additives.

7.2.2. *Reaction potentials*

A PBA's reaction potential depends most strongly on the electrochemically active transition metal species and on which site is electrochemically active. A schematic of typical potential ranges for electrochemically active species in the PBA structure is shown in Figure 7.5. PBA cathodes are commonly based on the $Fe^{III/II}(CN)_6$ redox couple, which is typically centered between 0.7 and 1.1 V with respect to the standard hydrogen electrode (SHE), or equivalently, between 3.4 and 3.8 V versus Na^+/Na. The lowest known reversible insertion reactions for the PBA structure are $Mn^{II/I}(CN)_6$ and $Cr^{III/II}(CN)_6$, which occur at potentials as low as about −1.0 V versus SHE, or equivalently, 1.7 V versus Na^+/Na. To maximize full cell potential, PBA anodes are usually selected for the use of one of those two reactions. There are a handful of reports of cycling PBA materials at even lower potentials via displacement reactions to form P or R metals (Nie *et al.* 2014; Deng *et al.* 2018).

Some PBAs undergo multiple electrochemical reactions. Both the P and R sites may be active, or one of those sites may undergo multiple reactions. This is of great practical interest because of the low specific capacity of PBAs. A practical specific capacity of only 60–80 mAh/g per reaction is achieved, so accessing multiple reactions is essential to achieving a competitive energy density for footprint-constrained applications. The most studied example of multielectron redox in the PBA structure is PB itself. In aqueous electrolytes, reversible cycling of P site $Fe^{III/II}$ occurs at a center potential of about 0.3 V versus SHE, while $Fe^{III/II}(CN)_6$ cycles at about 1.0 V. Other hexacyanoferrate-based cathodes such as manganese hexacyanoferrate and cobalt hexacyanoferrate are also electrochemically active at both the P and R sites. Each of these "two electron transfer" cathodes has a specific capacity of 140 mAh/g or more. For this reason, these three PBAs are the most commonly used cathodes in high-voltage, high-energy cells.

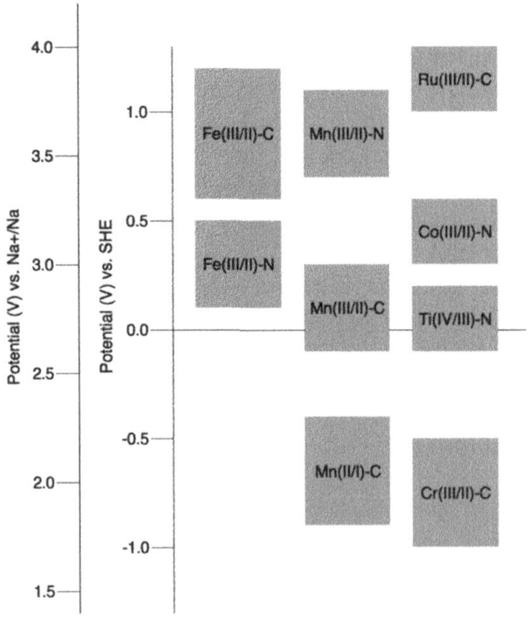

Figure 7.5. *The ranges of the typical reaction potentials of electrochemically active groups in the PBA structure*

The reaction potential of a site in the PBA structure depends on the composition of the framework around that site, the insertion species and the extent of hydration of the structure, among other factors. Scholz and Dostal (1995) performed the first systematic evaluation of the influence of framework composition on electrochemically active groups in the PBA structure. They found that the reaction potential of hexacyanoferrate-based PBAs varied by nearly 400 mV depending on the P site cation. Furthermore, regardless of the particular R site species, the potential of the R site is higher in the presence of a smaller, more highly charged P site species. In effect, more highly polarizing P site transition metals exert a stronger inductive effect on the $R(CN)_6$ complex, raising its reaction potential. This is possible because of strong hybridization of the R site's 3*d* electrons with the molecular orbitals of the surrounding CN ligands (Firouzi 2018). Delocalization of the redox-active electron across the entire $R(CN)_6$ complex renders the reaction potential of that complex sensitive to the surrounding environment. In a sense, it is not the R site transition metal that undergoes electrochemical cycling, but rather, the entire $R(CN)_6$ group.

The PBA structure accommodates a wide variety of P site transition metals that can form effective solid solutions at the P sites. This introduces the opportunity to fine tune the reaction potential of the R site by blending multiple transition metals at the P site. For example, the reaction potential of the copper-nickel hexacyanoferrate system varies continuously and approximately linearly with the ratio of P site Cu/Ni in the structure (Wessells *et al.* 2012a). This effect may also be used to adjust the reaction potential of the R site such that it is closer to the reaction potential of the P site, as is commonly done in manganese-metal hexacyanoferrates. The effect of the R site species on the reaction potential of the P site transition metal has not yet been systematically studied.

The choice of insertion species also affects a PBA's reaction potential. In both aqueous and organic electrolytes, PBAs have higher R-site potentials during reactions with larger, less strongly solvated insertion species. In aqueous electrolytes, the reaction of cathodes such as copper hexacyanoferrate and nickel hexacyanoferrate with Na^+ have center potentials as much as 200 mV lower than their reactions with K^+ (Wessells *et al.* 2012b; Erinmwingbovo *et al.* 2017). Similarly, in anhydrous, carbonate-based electrolytes, R-site cycling of PB with K^+ occurs at a potential about 400 mV higher than the potential for cycling with Na^+ (Wang *et al.* 2015b; He and Nazar 2017). This effect is not limited to hexacyanoferrates, as shown by the lower reaction potential of manganese hexacyanomangnate with Na^+ as compared to K^+. The insertion species also has a secondary effect on the slope of a PBA's reaction potential profile. Flatter potential profiles are generally observed for larger monovalent insertion species (McCargar and Neff 1988; Wessells *et al.* 2012b).

Finally, the quantity of interstitial water in a PBA's structure affects is reaction potential. The magnitude of this effect depends on the P and R species. It may also depend on the concentration of $R(CN)_6$ vacancies present in the framework. The most studied example of the dependence of PBA reaction potential on hydration is manganese hexacyanoferrate (Song *et al.* 2015). When interstitial water is present, this material undergoes two distinct reactions: $Fe^{III/II}(CN)_6$ at low potential, followed by $Mn^{III/II}$ at higher potential. However, full dehydration of the structure results in a downward shift in the potential of the latter reaction by over 100 mV, resulting in a single reaction plateau that has a specific capacity consistent with two electron transfer. In fully dehydrated manganese hexacyanoferrate, both the R and P sites are still active, but because their potentials are so close together, the two reactions proceed concurrently. The underlying reason behind the effect of PBA hydration on reaction potential is not understood. As of 2019, systematic studies of electrochemical properties as a function of hydration have not been reported.

The initial composition of a PBA determines the amount of electrochemical capacity that can be reversibly cycled. For instance, consider the manganese hexacyanoferrate cathode synthesized in a reduced Mn^{II}/Fe^{II} state. In the case of a defect-free framework having the stoichiometry $Na_2Mn[Fe(CN)_6]$, there are two Na^+ per formula unit, and both the Fe and Mn may be fully oxidized during charging as Na^+ is removed from the structure. On the other hand, if there are hexacyanoferrate vacancies in the structure, then there will be a corresponding decrease in the amount of Na^+ that is initially present so that charge neutrality is maintained. The charge ratio of $Fe^{II}(CN)_6^{4-}$ to Na^+ is 4:1, so every hexacyanoferrate vacancy results in the loss of four Na^+. Thus, the accessible specific capacity depends strongly on the initial concentration of Na^+, and therefore, the stoichiometry of the PBA framework. As the lattice parameter of PBAs is effectively constant with A site occupancy for the same P and R valences, the charge and energy densities of the electrode also scale proportionally with quantity of Na^+ initially present in the structure.

7.2.3. *PBA cathodes*

PBA cathodes are most commonly based on the electrochemical cycling of hexacyanoferrate at high potential. These PBAs can be divided into two classes: materials in which only the hexacyanoferrate groups are active, and materials in which the P site transition metal is also active. The latter class of materials is relevant to practical batteries because of the high specific capacity associated with multielectron transfer reactions. The best characterized of these materials are PB, manganese hexacyanoferrate and cobalt hexacyanoferrate.

The ability of PB to undergo two electrochemical reactions was first documented by Neff in his seminal publication of the discovery of electrochromism of PB films (Neff 1978). The reaction at low potential, during which P site Fe is active, received the vast majority of subsequent attention because it is associated with the electrochromic transition of opaque PB to colorless, transparent Prussian white (PW, also known as Everitt's salt). By the early 1980s, other researchers had successfully built displays that could be cycled on and off hundreds of thousands of times as P site Fe was oxidized and reduced (Itaya and Shibayama 1982). The upper reaction, during which hexacyanoferrate is active, received less attention because it resulted in a less interesting electrochromic transition: conversion of PB to Berlin Green upon oxidation. Both reactions have been utilized during operation of Prussian blue electrodes as cathodes in battery cells.

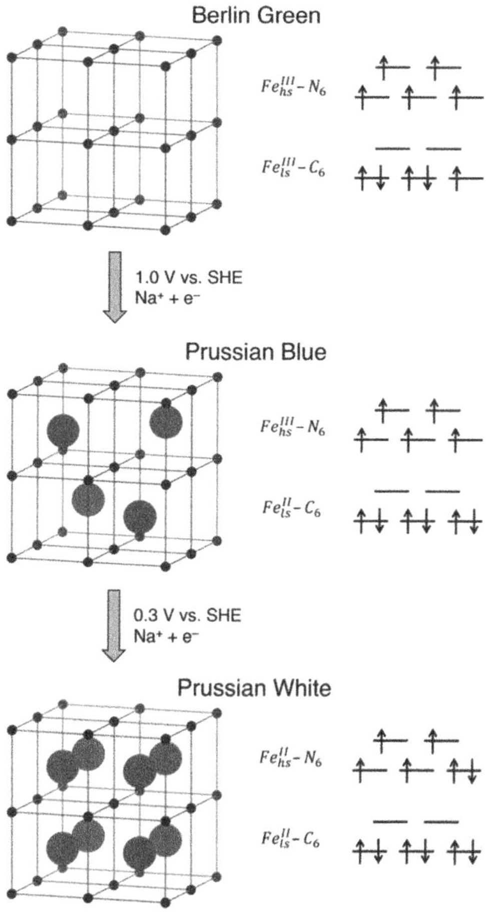

Figure 7.6. *Schematic of the electrochemical reactions of Prussian blue and the corresponding changes in the electronic states of the P and R sites. For a color version of this figure, see www.iste.co.uk/monconduit/batteries.zip*

The chemistry of PB poses substantial challenges for the development of a high-performance electrode. Prussian blue is extremely insoluble, which results in rapid precipitation of nanoparticles during synthesis in aqueous solutions. These nanoparticles may be as small as 10–20 nm and may not readily flocculate into larger agglomerates. While this makes PB attractive as a pigment that can be readily dispersed in other media, it also makes isolation and purification of PB powder challenging at high volume. Syntheses employing aids that boost particle size, such as chelating agents or solvothermal conditions, may be used to increase particle size and crystallinity to address this difficulty (see section 7.1.3).

A second challenge is synthesis with the correct initial composition. Prussian blue is the mixed valence phase that contains oxidized P site Fe^{III} and reduced R site $Fe^{II}(CN)_6$. In effect, PB has an initial state of charge of 50%. This is impractical for battery cells, which are typically assembled at a state of full discharge. Therefore, the synthesis should be performed from precursors and under conditions that result in the fully reduced PW state: $Na_xFe^{II}_y[Fe^{II}(CN)_6]_z$. Initial charge of this material proceeds through both of the two reactions, as is desirable for practical cells. While synthesis of PW from a Fe^{II} salt and $Fe^{II}(CN)_6$ is not difficult, the fully reduced material has an open circuit potential that is low enough to be unstable in air (Rudola *et al.* 2017). To avoid partial oxidation, all postsynthesis processing of PW must therefore be performed under inert gas, which introduces considerable cost and complexity at high volume. Alternatively, the material may be allowed to oxidize partially, resulting in a non-zero initial state of charge and a sodium hydroxide byproduct. This may be preferable to the additional cost of inert gas processing if the cell design can accept a basic impurity.

In addition to these difficulties in synthesis and processing, the electrochemical characteristics of PB pose challenges to practical cell design. In most PB samples, the two reaction plateaus have a difference in potential of about 0.65 V. While such a large difference in reaction potential for the same transition metal may seem counterintuitive, the P site Fe is in a high spin state and weakly bonded to nitrogen, while the R site Fe is in a low spin state and strongly bonded to carbon (Figure 7.6). The different electronic states of the P and R site Fe results in the wide gap between their reaction potentials. Accessing the full capacity of PB requires cycling the electrode over a potential range of over 1.0 V (Itaya and Shibayama 1982; Neff 1985). In many applications batteries should maintain a minimum voltage of at least half of their maximum voltage to avoid the use of a prohibitively expensive power conversion system. Using this rule of thumb, any anode paired with a PB cathode should operate entirely at potentials at least 1 V lower than the bottom-of-discharge potential of PB, or equivalently, below about −1 V versus SHE. One exception is a rhombohedral phase of PW, in which cycling of R site $Fe(CN)_6$ occurs 300 mV lower than in typical PB samples (Wang 2015b). However, that material undergoes a rapid phase change to a cubic structure on contact with humid air, with unknown effects on its electrochemical properties (Rudola *et al.* 2017).

Alternatively, just one of PB's two reaction plateaus could be cycled, allowing for cells containing higher potential anodes to operate in an acceptable voltage range. In this case, the potentials of the individual reactions are of great interest. The potentials of these reactions are somewhat dependent on synthesis methods and initial composition and structure (Table 7.1). For most samples of PB and PW, the lower reaction of P site Fe is centered at about 0.3 V versus SHE. In most samples of PB, the upper reaction of R site $Fe(CN)_6$ is centered at 1.0 V, although for samples of PW its potential varies with synthesis route and initial composition. Thus, the lower reaction is in the middle of the electrochemical stability range of aqueous electrolytes, while the upper reaction has such a high potential that oxidation of water would be practically unavoidable. The potentials of PB's two reactions, as well as the large difference between them, make PB a plausible candidate for practical cells only in the presence of an anhydrous electrolyte and an anode with a very low reaction potential, such as hard carbon.

Reference	Material/ synthesis route	Structure/composition	P and R site potentials versus SHE (electrolyte)
Hu *et al.* (2019)	PW/Route 1	Cubic $Na_{1.02}Fe[Fe(CN)_6]_{0.95}(H_2O)_n$	0.2/0.7 V (Na^+, anhydrous)
He and Nazar (2017)	PB/Route 3	Monoclinic $K_{1.7}Fe[Fe(CN)_6]_{0.9}(H_2O)_{0.4}$	0.4/1.0 V (K^+, anhydrous carbonate)
Rudola *et al.* (2017)	PW/Route 3	Monoclinic $Na_2Fe[Fe(CN)_6](H_2O)_2$	0.3/1.3 V (Na^+, anhydrous glyme)
Wang *et al.* (2015b)	PW/Route 3	Rhombohedral $Na_{1.92}Fe[Fe(CN)_6]\ (H_2O)_{0.08}$	0.3/ 0.6 V (Na^+, anhydrous carbonate)
Wu *et al.* (2015a)	PB/Route 1	Cubic $Na_{1.33}Fe[Fe(CN)_6]_{0.82}(H_2O)_n$	0.3/ 1.0 V (Na^+ aqueous)
You *et al.* (2014)	PB/Route 3	Cubic $Na_{0.61}Fe[Fe(CN)_6]_{0.94}(H_2O)_n$	0.2/ 0.8 V (Na^+, anhydrous carbonate)

Table 7.1. *Syntheses, compositions, structures, and reaction potentials of PB and PW*

After PB, manganese hexacyanoferrate is the best-studied PBA cathode in which both the P and R sites are electrochemically active. Manganese hexacyanoferrate offers advantages over PB both in ease of synthesis and processing as well as cell operation. Synthesis of manganese hexacyanoferrate is straightforward. Its solubility in water is much higher than most hexacyanoferrate-based PBAs, at up to about 250 ppm at neutral pH (for comparison, the solubility of PB is less than 10 ppm). Its relatively high solubility makes formation of micrometer-scale particles from high-concentration precursors possible even in the absence of aids such as chelating agents. The result is a material that can be easily purified and processed into electrodes. In addition, its open circuit potential in the fully discharged $Na_xMn^{II}_y[Fe^{II}(CN)_6]_z$ state is high enough that it is not readily oxidized in air.

The main consideration for synthesis of manganese hexacyanoferrate is therefore to obtain a sodium-rich initial composition that is as close to $Na_2Mn^{II}[Fe^{II}(CN)_6]$ as possible so that a full two electrons of electrochemical charge capacity can be accessed. The simplest approach to obtaining this composition is to perform the synthesis between equimolar Mn^{2+} and $Fe(CN)_6^{4-}$ precursors in the presence of a large excess (>10:1) of Na^+. The crystal structure of manganese hexacyanoferrate depends on the concentration of $Fe(CN)_6$ vacancies, interstitial water and A site occupancy. Samples having a high vacancy concentration or a low A site occupancy (typically at intermediate state of charge) are cubic. However, at full discharge, low-vacancy manganese hexacyanoferrate has a monoclinic structure in which the CN bridging ligands are tilted a few degrees away from the <100>-type directions between the P and R sites. In these samples, there is a reversible phase change between a monoclinic structure at low state of charge and a cubic structure at high state of charge (Sottman *et al.* 2016). It is unknown if this phase change decreases the reversibility of cycling. Full dehydration of the monoclinic phase results in significant additional distortion of the framework structure and the formation of a rhombohedral phase in which the CN ligands are rotated off-axis by 20° or more (Song *et al.* 2015).

Starting from full discharge, oxidation of the P and R sites in manganese hexacyanoferrate proceeds in an order reversed from what is observed in PB: first $Fe^{II}(CN)_6$ is oxidized at a low potential, and then Mn^{II} is oxidized to Mn^{III} at a higher potential (Figure 7.7). The potentials of these reactions are strongly dependent on the hydration of the structure. In electrodes that have not been fully dehydrated before cycling, $Fe^{II}(CN)_6$ cycles at a center potential of about 0.8 V versus SHE, or equivalently, about 3.5 V versus Na^+/Na. This reaction is followed by $Mn^{III/II}$ cycling at a center potential of approximately 1.0 V versus SHE, or 3.7 V versus Na^+/Na. The two reactions are much closer together than the two reactions of PB, which makes two-electron cycling of manganese hexacyanoferrate more practical. In fact, the two reactions are so close together in potential that they are not entirely sequential. Rather, oxidation of P site Mn begins before the $Fe(CN)_6$ groups are fully oxidized. In hydrous manganese hexacyanoferrate, at 50% state of charge about 80%

of the $Fe(CN)_6$ and 20% of the Mn is oxidized (Wu *et al.* 2017). With increasing dehydration, the two reaction potentials move closer to one another. In the case of fully dehydrated manganese hexacyanoferrate, the two reactions occur at virtually the same potential, and one large reaction plateau having a capacity corresponding to two-electron cycling is observed. While these shifts in potential occur with a change in crystal structure from monoclinic to cubic, it is not generally true that the reaction potential of a PBA depends strongly on its structure, as is observed for the structure-independent lower reaction potential of PB.

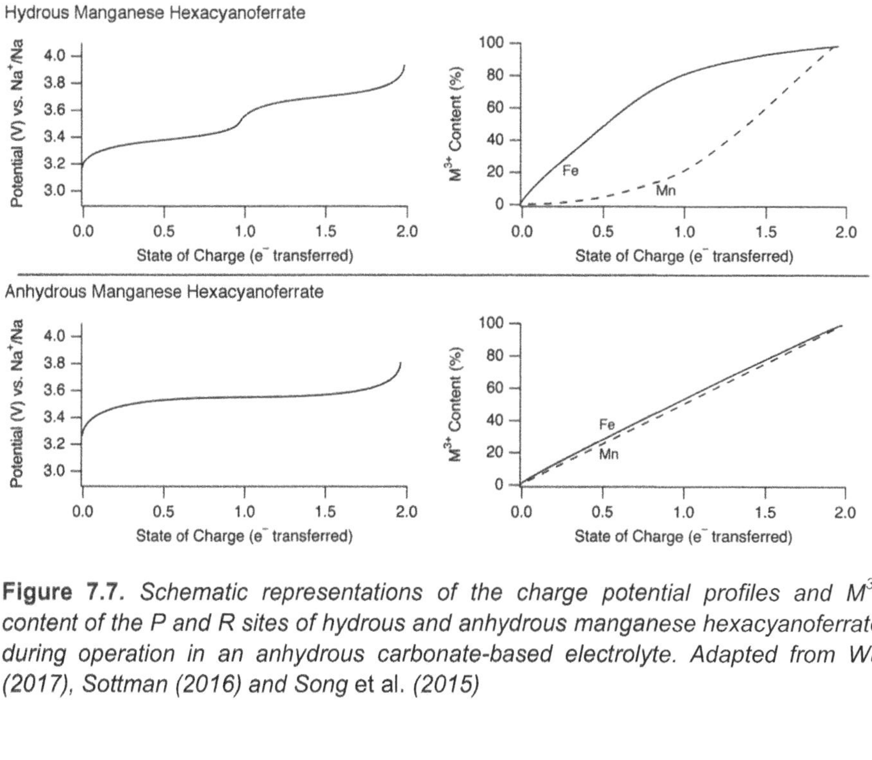

Figure 7.7. *Schematic representations of the charge potential profiles and M^{3+} content of the P and R sites of hydrous and anhydrous manganese hexacyanoferrate during operation in an anhydrous carbonate-based electrolyte. Adapted from Wu (2017), Sottman (2016) and Song* et al. *(2015)*

The chemical stability of manganese hexacyanoferrate poses a major challenge for its use in practical cells. Even in anhydrous electrolyte, two-electron cycling of fully dehydrated manganese hexacyanoferrate has been demonstrated for only 500 cycles, orders of magnitude less than other hexacyanoferrate-based cathodes (Song 2015). The degradation mechanism during cycling is not yet understood, but manganese hexacyanomanganate appears to be less chemically stable than other hexacyanoferrate-based PBAs. The high solubility that makes its synthesis straightforward also introduces a risk of dissolution in the cell's electrolyte. This effect is strongest in aqueous electrolytes that are similar in composition to the solution in which the material is synthesized. Furthermore, even partial oxidation of manganese hexacyanoferrate results in formation of Mn^{III}, which is well known to undergo disproportionation and reaction with water to form MnO_2 and Mn^{2+}. Operation of manganese hexacyanofererate at a high state of charge therefore introduces the risk that manganese may be lost from the structure in a reaction with an aqueous electrolyte, or even in a reaction with interstitial water.

One approach to stabilize manganese hexacyanoferrate is to partially substitute the manganese with other transition metals to form a mixed-metal hexacyanoferrate phase. The choice of the substituting species is typically made on a basis of the chemical stability or reaction potential of its pure hexacyanoferrate. For instance, partial substitution of Mn with either Fe or Ni has been found to decrease the rate of capacity loss of manganese hexacyanoferrate during cycling in a propylene carbonate electrolyte by approximately 50% (Morimoto *et al.* 2016). Further study of the stability and degradation mechanisms of manganese hexacyanoferrate with and without partial substitution with other P site transition metals is essential for the development of a practical cell.

Cobalt hexacyanoferrate has also been reported to undergo two-electron cycling at high potential. As in the case of PB, the P site reaction of $Co^{III/II}$ occurs at a lower potential than that of the R site Fe. The two reaction potentials appear to be dependent on synthesis conditions and the choice of an aqueous or anhydrous electrolyte, and occur at approximately 0.6 and 1.1 V versus SHE, respectively (Pasta *et al.* 2016; Wu *et al.* 2016). The relatively wide gap between the reaction potentials of cobalt hexacyanoferrate is similar to the one observed for PB, making this material suitable for high-voltage cells. The reaction potentials of cobalt hexacyanoferrate are higher than those of PB and manganese hexacyanoferrate and introduce the opportunity for high-energy density. However, the high, volatile cost of cobalt makes this material unattractive in comparison to cheaper alternatives.

PBA cathodes that undergo only single-electron transfer cycling are generally less attractive than two-electron transfer materials for commercial applications due to their low specific capacity and energy. They are used primarily as systems for understanding the properties of PBA electrodes. The most commonly studied single electron transfer PBA cathodes are the hexacyanoferrates of copper, nickel and zinc.

In 2011, copper hexacyanoferrate was found to be capable of cycling K^+ for tens of thousands of cycles, a service life comparable to that of electrochromic PB films (Wessells *et al.* 2011). This discovery triggered the modern field of PBA batteries including Na^+ batteries. Aside from its low specific capacity of about 60–65 mAh/g, copper hexacyanoferrate offers an attractive combination of electrochemical properties including a relatively high reaction potential (0.85 V vs. SHE), excellent kinetics and rate capability, and a smoothly varying potential profile that is representative of a classic solid solution reaction (Figure 7.3). It also undergoes reversible reactions with a wide variety of alkali, alkaline earth and transition metal cations (Wang *et al.* 2015). Synthesis of copper hexacyanoferrate in its discharged state from a $Fe^{II}(CN)_6^{4-}$ precursor results in fine nanoparticles that are poorly crystalline and extremely challenging to filter. For this reason, copper hexacyanoferrate is typically synthesized in its oxidized state using $Fe^{III}(CN)_6^{3-}$ as a precursor. Doing so results in somewhat larger particles that can be isolated using common filter media. However, it also results in a material that has an impractical, high initial charge state. This can be addressed through exposure of copper hexacyanoferrate to a mild reducing agent such as sodium thiosulfate. Chemical reduction of copper hexacyanoferrate has been found to result in the same phase as is obtained through electrochemical reduction (Pasta *et al.* 2012).

Nickel hexacyanoferrate is frequently used as a model system for PBA cathodes. It has one of the lowest reaction potentials of any hexacyanoferrate-based cathode at 0.6 V versus SHE. This mild potential, in combination with its extremely low solubility allows it to be used for long-term experiments evaluating cycle life as a function of electrolyte composition. Its low solubility also makes it attractive for use as a protective shell deposited on the surface of other PBAs. As is the case for copper hexacyanoferrate, it affords great flexibility in the choice of insertion species of any PBA cathode, as it has been found to reversibly cycle a wide variety of monovalent and divalent cations (Wang *et al.* 2013).

Of all single electron transfer PBA cathodes, zinc hexacyanoferrate has perhaps the most complex chemistry and electrochemistry. It can be synthesized in both a cubic phase in which octahedrally coordinated Zn^{2+} occupies the P sites, as well as in a hexagonal phase in which Zn^{2+} is tetrahedrally coordinated to the $Fe(CN)_6^{4-}$ octahedra. The latter phase is stable under conditions relevant to materials processing and cell operation (Rodríguez-Hernández *et al.* 2007). It has a very high reaction potential, centered at about 1.1 V versus SHE. Zinc hexacyanoferrate is unique among PBAs in that its P site transition metal may also act as a metallic anode at a practical potential. This has led to the construction of cells containing a zinc hexacyanoferrate cathode and a zinc metal anode that operates at an average voltage of 1.7 V (Zhang *et al.* 2014). Saturation of the electrolyte with Zn^{2+} limits the solubility of the cathode, extending the life of the cell. This strategy is practical only in the case of zinc hexacyanoferrate, as the other common P site transition metals such as Fe, Cu or Ni each undergo electroplating reactions at much higher potentials than Zn.

7.2.4. *PBA anodes*

Unlike PBA cathodes that rely at least in part on the electrochemical activity of hexacyanoferrate, PBA anodes generally operate by the reversible cycling of hexacyanomanganate or hexacyanochromate. The ability for those complexes to undergo reversible electrochemical reactions has been known for decades. Electrochemical activity of free $Mn(CN)_6^{4-}$ anions in aqueous solution was first reported in 1941 (Treadwell 1941). However, PBAs containing hexacyanomanganate or hexacyanochromate received little study in the following decades, perhaps because they do not offer the compelling electrochromic properties of PB. Finally, in 1995 the electrochemical activity of both hexacyanomanganate- or hexacyanochromate- based PBAs was reported by Scholz and Dostal (1995). These materials were found to have reaction potentials that are much lower than those of the hexacyanoferrate-based cathodes, between −1 and 0 V versus SHE (Table 7.2).

Unlike the hexacyanoferrate-based cathodes that have a seemingly endless variety of compositions, there are few chemically stable hexacyanomanganate- or hexacyanochromate-based PBAs. The relative strengths of the transition metal carbon bond in hexacyanometallates is Fe $\approx$ Co > Cr > Mn, as determined by increasing lability toward radiocyanide (Adamson *et al.* 1951). Both simple salts and PBAs containing hexacyanoferrate and hexacyanocobaltate are chemically stable at room temperature in the presence of water and oxygen. On the other hand, the relatively weak Mn-C or Cr-C bonds in hexacyanomanganate or hexacyanochromate compounds make them prone to unwanted decomposition in water and air, resulting in the formation of metal oxide and hydroxide phases.

The relative strengths of the metal-cyanide bonds in hexacyanometallate complexes may also be elucidated from the spontaneous decomposition of some PBAs by a linkage isomerization process (Shriver *et al.* 1965). During linkage isomerization, the cyanide bridging ligands flip directions, resulting in the conversion of a $P[R(CN)_6]$ framework into a $R[P(CN)_6]$ framework. Common examples of PBA linkage isomerization include iron hexacyanomanganate and iron hexacyanochromate, which spontaneously decompose to form manganese hexacyanoferrate and chromium hexacyanoferrate, respectively (Brown *et al.* 1968; Buschmann *et al.* 1999). The general rule of thumb for PBA anodes is that the P and R site transition metals should be selected such that $R(CN)_6$ is more stable than $P(CN)_6$ to avoid linkage isomerization and a loss of the desired electrochemical properties. Among the hexacyanomanganates of Fe, Co, Cr, Mn and Zn, only the last two are stable against linkage isomerization. Chromium hexacyanomanganate is challenging to produce with high purity due to the tendency of both its Cr^{2+} and $Mn(CN)_6^{3-}$ precursors to decompose during its synthesis. Hexacyanochromate-based PBAs of Mn, Cr and Co have each been shown to cycle reversibly.

Material	Potential vs. SHE (electrolyte)	Reference
$Na_xMn^{II}[Mn^{II/I}(CN)_6]$	−0.72 V (aqueous 1 M Na^+)	Firouzi *et al.* (2018)
$Na_xZn^{II}[Mn^{II/I}(CN)_6]$	−0.55 V (aqueous 1 M Na^+)	Natron Energy, unpublished data
$Na_xCr^{II}[Mn^{II/I}(CN)_6]$	−0.32 V (aqueous 10 M Na^+)	Pasta *et al.* (2014)
Chromium hexacyanochromate (III/II)	−0.55 V (aqueous 0.1 M K^+)	Eftekhari (2003)
Manganese hexacyanochromate (III/II)	−0.86 V (aqueous 1 M Na^+)	Nakamoto *et al.* (2018); Wheeler *et al.* (2019)
Cobalt hexacyanochromate (III/II)	−1.0 V (aqueous 0.1 M K^+)	Bagkar *et al.* (2006)

Table 7.2. *Reaction potentials of PBA anodes*

PBA anodes undergo reactions with sodium at relatively high potentials centered at 1.7 V versus Na^+/Na or more. These potentials are much higher than those observed for the reaction of carbon anodes with sodium, which typically occur below 1 V. This makes PBA anodes impractical for cells requiring a high energy density. Rather, PBA anodes are more practically used in combination with a PBA cathode that offers similar rate capability and reversibility, albeit with low energy density. In some cases, hexacyanomanganate-based PBAs have been described as cathodes in half cells containing sodium metal anodes (Lee *et al.* 2014). While it is true that these materials could be used as cathodes in combination with a low-potential anode, their reaction potentials are much lower than those of hexacyanoferrate-based PBAs that there is no practical justification for doing so. Rather, the practical application of hexacyanomanganate- and hexacyanochromate-based PBAs is limited to their use as anodes in cells for which high energy density is not required.

The most studied PBA anode is manganese hexacyanomanganate. This material was originally described in the first publication on the electrochemistry of the free hexacyanomanganate anion (Treadwell *et al.* 1941). Hexacyanomanganate-based PBAs received little attention in the subsequent decades, with determination of their crystal structures not arriving until 1964 (Ferrari *et al.* 1964), and the first reports on their electrochemical properties in 1995 (Scholz and Dostal 1995). At that time, manganese hexacyanomanganate was found to undergo three reversible reactions: P site $Mn^{III/II}$ at 0.7 V versus SHE, $Mn^{III/II}(CN)_6$ at 0 V, and $Mn^{II/I}(CN)_6$ at –0.72 V. The Mn^{I} valence state is unusual and is not observed in other manganese-based battery electrode systems such as lithium manganese oxides. In ceramics and in liquid solutions, manganese is typically found in a high spin state. Reduction of high spin Mn^{2+} proceeds directly to Mn^0. However, just as is the case for PB, in manganese hexacyanomanganate the P site Mn has a high spin state, while the R site Mn has a low spin state. Low spin $Mn^{I}(CN)_6$ is a stable configuration that is isoelectric with $Fe^{II}(CN)_6$. The low spin state of $Mn^{I}(CN)_6$ has been directly observed using advanced X-ray spectroscopy and scattering techniques (Firouzi *et al.* 2018).

Synthesis of manganese hexacyanomanganate may be performed by either Route 1 or 2. The latter is more common. Route 1 synthesis involves the reaction of dissolved Mn^{2+} with $Mn^{III}(CN)_6^{3-}$, which is commercially available as $K_3Mn(CN)_6$ at gram-scale. This results in precipitation of manganese hexacyanomanganate(III). In this reaction, there is no isomerization of valence analogous to the one that occurs during synthesis of PB from Fe^{2+} and $Fe^{III}(CN)_6^{4-}$ (Robin 1962). Production of a high-purity product by this route is challenging because $Mn^{III}(CN)_6^{3-}$ rapidly hydrolyzes to MnO_2 unless a large excess of CN^- is present. However, if that excess of CN^- is present during a Route 1 synthesis, it will undergo a competing reaction with free Mn^{2+} to form $Mn^{II}(CN)_6^{4-}$, and eventually manganese hexacyanomanganate(II). Synthesis via Route 1 results in an undesirable initial state of charge. The first charge of the material must proceed through the entire $Mn^{III/II}(CN)_6$ reaction plateau at 0 V versus SHE before the lower $Mn^{II/I}(CN)_6$ plateau may be cycled. Together, these difficulties make Route 1 synthesis of manganese hexacyanomanganate unattractive for practical cells.

Route 2 synthesis of manganese hexacyanomanganate involves the direct reaction of dissolved aqueous Mn^{2+} with CN^- under inert atmosphere. The reaction proceeds by simultaneous conversion of the Mn^{2+} aqua ion to hexacyanomanganate and condensation of those anions into a PBA structure:

$$xNa^+ + (y + z)Mn^{2+}(H_2O)_6 + 6z(CN)^- \rightarrow$$

$$xNa^+ + (y + z)Mn^{2+}(H_2O)_{6-p}(CN)_p + \big(6z - p(y + z)\big)(CN)^- \rightarrow$$

$$Na_xMn_y[Mn(CN)_6]_z(H_2O)_n \qquad [7.6]$$

No thorough study of this reaction mechanism has been performed, and the mechanism by which the intermediate manganese cyano-aqua ions condense is unknown. As these complexes oligomerize, some of the bridging CN ligands must undergo linkage isomerization to form $Mn(CN)_6$ groups. Route 2 synthesis of manganese hexacyanomanganate provides advantages over Route 1. Both the Mn^{2+} and CN^- precursors are inexpensive commodities, allowing for production via Route 2 to be done at effectively unlimited volume. Route 2 also results in formation of manganese hexacyanomanganate(II), which is in the correct initial state of charge to immediately begin charging and cycling of the $Mn^{II/I}(CN)_6$ reaction.

The composition, crystal structure, morphology and electrochemical properties of manganese hexacyanomanganate synthesized by Route 2 depend strongly on reaction conditions such as concentration and the rate at which the precursor solutions are combined. For example, the reaction of Mn^{2+} with a 5% excess of NaCN at a low concentration of 6.5% weight per volume results in a face centered cubic phase composed of 2–3 μm particles. This material has a composition of $Na_{1.24}Mn[Mn(CN)_6]_{0.81}$ and reacts with Na^+ by a simple solid-solution reaction (Figure 7.8) in which the A site occupancy does not exceed one (Firouzi *et al.* 2018). In effect, this grade of manganese hexacyanoferrate is a model PBA material both in terms of structure and electrochemical behavior.

In contrast, the reaction of Mn^{2+} with a 5% excess of NaCN at a high concentration of 11% weight per volume produces 1–2 μm particles of a monoclinic phase in which the CN ligands are tilted slightly away from the <100>-type directions. This material has a composition of $Na_{1.75}Mn[Mn(CN)_6]_{0.94}$ (neglecting its water content) and undergoes electrochemical reduction to $Na_{2.69}Mn[Mn(CN)_6]_{0.94}$. The potential profile shows a clear deviation from theoretical solid solution behavior beginning at a state of charge of about 35%, when the A site occupancy exceeds one. That deviation from the expected electrochemical behavior corresponds to a reversible phase change to a cubic structure. Other high-sodium grades of manganese hexacyanomanganate have also been shown to undergo phase changes when the A site occupancy exceeds one (Lee *et al.* 2014). This illustrates that PBA anodes may undergo unexpected phase transformations depending on their composition and the maximum A site occupancy reached during Na^+ insertion.

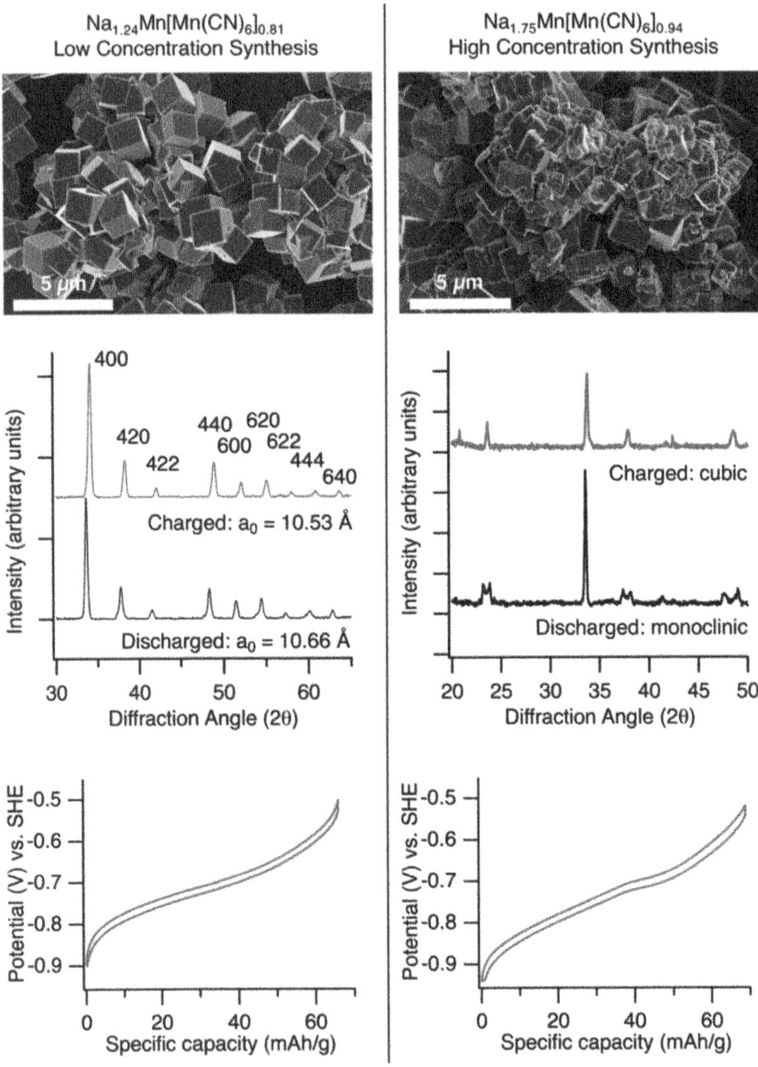

Figure 7.8. *Scanning electron micrographs, X-ray diffraction spectra, and galvanostatic profiles (1C-1C) of manganese hexacyanomanganate synthesized at low or high concentration. Low-concentration data adapted from Firouzi (2018) under the Creative Commons license; http://creativecommons.org/licenses/by/4.0/. High-concentration data: courtesy Natron Energy. For a color version of this figure, see www.iste.co.uk/monconduit/batteries.zip*

Zinc hexacyanomanganate is the one other hexacyanomanganate-based PBA anode that has been produced with high purity and crystallinity. It is synthesized via a Route 1 reaction of aqueous Zn^{2+} and $Mn^{III}(CN)_6^{3-}$, resulting in a material having an initial composition of $K_xZn_y[Mn(CN)_6]_z$. Charging to negative potential in Na^+ or K^+ electrolyte results in two reactions, which may be attributed to reduction to $Mn^{II}(CN)_6$ and $Mn^{I}(CN)_6$, respectively. The potential of the $Na_xZn_y[Mn^{II/I}(CN)_6]_z$ reaction is centered at −0.55 V versus SHE, about 0.17 V higher than the corresponding reaction in manganese hexacyanomanganate. Its relatively high reaction potential makes zinc hexacyanomanganate a less attractive anode for practical cells. However, zinc hexacyanomanganate is useful as a model system for understanding the electrochemistry of the $Mn(CN)_6$ complex. Comparison of the reaction potentials of manganese, zinc and chromium hexacyanomanganate shows that regardless of the P site transition metal, the $Mn^{III/II}(CN)_6$ and $Mn^{II/I}(CN)_6$ reactions are separated by about 0.75 V (Figure 7.9). Furthermore, lower reaction potentials are observed for increasing lattice parameter, or effectively, a larger, less electron-withdrawing P site species. This trend closely matches the behavior of hexacyanoferrate-based cathodes (Scholz and Dostal 1995). The wide gap in potential between the $Mn^{III/II}(CN)_6$ and $Mn^{II/I}(CN)_6$ reactions prevents them both from being used in a practical cell. When paired with any known Na^+ cathode, only the lower $Mn^{II/I}(CN)_6$ reaction provides an acceptable full cell voltage.

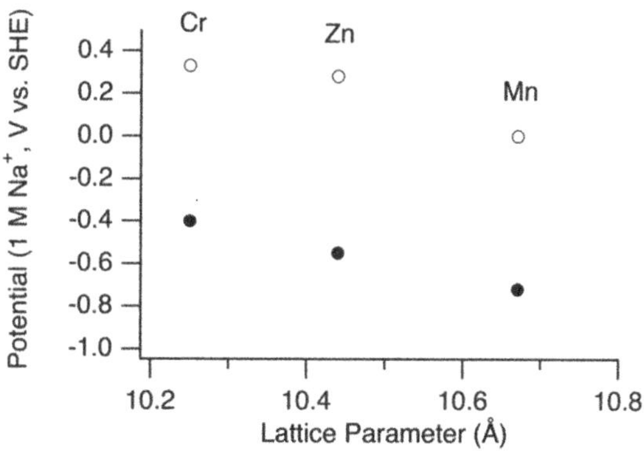

Figure 7.9. *The effect of lattice parameter on the reaction potentials of hexacyanomanganates of chromium, zinc and manganese*

Hexacyanochromate-based anodes also have low reaction potentials that make them interesting as anode materials. These materials undergo reversible cycling of $Cr^{III/II}(CN)_6$. The most promising of them is manganese hexacyanochromate, which was recently reported to have the lowest known reaction potential for alkali cation insertion in any PBA at potentials as low as −1.1 V versus SHE. Synthesis of manganese hexacyanochromate is performed via a Route 1 reaction of aqueous Mn^{2+} with aqueous $Cr^{III}(CN)_6^{3-}$, which is commercially available at gram-scale as a potassium salt, or which may be readily synthesized. The manganese hexacyanochromate samples reported to date have relatively high R site vacancy concentrations and the classic cubic PBA structure. It is unknown whether a low-vacancy manganese hexacyanochromate has the same lattice distortions observed in low-vacancy manganese hexacyanoferrate and manganese hexacyanomanganate. In 2018, Nakamoto *et al.* (2018) reported reversible cycling of $Na_xMn^{II}[Cr^{III/II}(CN)_6]$ at a center potential of –0.9 V versus SHE in saturated (17 M) aqueous sodium perchlorate. Soon afterward, Wheeler *et al.* (2019) reported a reaction potential of –0.86 V in saturated (37 M) aqueous sodium bis(trifluorosulfonimide), and a much lower potential of –1.1 V in an anhydrous electrolyte of 1 M Na^+ in propylene carbonate. This apparent 240 mV variation in reaction potential with electrolyte composition is among the largest reported for any PBA. Additional study is required to understand the source of this variation in potential, which may be caused by differences in electrolyte composition, hydration of the structure or merely a difference in reference electrodes. Another critical area of future investigation is the long-term electrochemical stability of hexacyanochromate-based PBAs, which has not yet been studied.

7.3. Prussian blue batteries

7.3.1. *Cells containing two PBA electrodes*

PBA electrodes are most commonly used in cells in which both the electrodes rely on PBA active materials. This choice has obvious advantages: both the electrodes have high rate capability, long cycle life and are chemically stable and nonvolatile. In the absence of other chemical reactions that degrade the electrodes or otherwise imbalance the cell, the cell should have long cycle life. These performance advantages come at a cost of relatively low energy density. On an active material mass basis, the maximum theoretical specific energy for a cell containing a PBA cathode and anode is about 75 Wh/kg for two electron transfer in a manganese hexacyanoferrate cathode and single electron transfer in a manganese hexacyanochromate anode. This specific energy is competitive with lead acid cells, but far below the 200–300 Wh/kg specific energy of modern lithium-ion cells. Cells containing two PBA electrodes are therefore best suited for applications that value power or service life, but not high energy.

Most of the early investigations of PBAs in battery cells involved the use of two PBA electrodes. In a logical extension of research into electrochromic displays based on PB, in 1985 Neff reported a cell containing two PB electrodes and an aqueous K^+ electrolyte (Neff 1985). The anode operated by reduction to PW, while the cathode operated by oxidation to Berlin Green, with an average full cell voltage of 0.6 V. This proof of concept was reproduced by several other academic groups. Soon afterwards, researchers began to explore the use of other PBAs in battery cells. These included a cell containing a copper hexacyanoferrate cathode and a PB anode (Grabner and Kalwellis-Mohn 1987). In 1989, the first cell containing PBAs relying on the electrochemical activity of different R site transition metals was demonstrated when a nickel hexacyanoruthenate cathode was paired with nickel hexacyanoferrate anode to produce a 0.6 V Na^+ cell (Kalwellis-Mohn and Grabner 1989). This cell was unique in that electrochemical activity of $Fe(CN)_6$ was used in the cell's negative electrode. The reaction potential of nickel hexacyanoruthenate with Na^+ is centered above 1.1 V versus SHE, the highest of any known PBA. Unfortunately, ruthenium is prohibitively expensive and cannot be used as the active species in a commercially viable cell. A more practical combination of electrochemically active R site transition metals is the pairing of chromium hexacyanoferrate cathode with a chromium hexacyanochromate anode to produce a 1.5 V cell (Eftekhari 2003).

More recently, a copper hexacyanoferrate cathode was combined with a manganese hexacyanomanganate(III/II) anode in aqueous 10 M Na^+ to produce a 1 V cell (Pasta *et al.* 2012). The bulk composite electrodes used in this cell also contained a binder and conductive carbon additives frequently found in lithium-ion electrodes and were prepared using processes and equipment common to the lithium-ion industry. The resulting cell was capable of sustained charge and discharge at rates of up to 50 C, and had no capacity or energy loss after over 1,000 deep discharge cycles. This proved that a PBA cell containing electrodes processed via scalable methods could achieve high performance. However, that cell was flooded with a large excess of electrolyte and was not representative of a realistic commercial design.

The first cell containing two PBA electrodes and having a practical design was reported in 2018 (Firouzi *et al.* 2018). In that cell, a copper hexacyanoferrate cathode was paired with a manganese hexacyanomanganate (II/I) anode in an organic-aqueous co-solvent Na^+ electrolyte to achieve an average discharge voltage of 1.5 V (Figure 7.10). The electrodes in that cell were prepared using industry-standard inactive materials and processes. The cell itself was a pouch cell containing commodity components including a thin porous separator, metal foil current collectors and tabs, and aluminum/polymer laminate packaging. This cell was shown to be capable of sustaining continuous ±12 C cycling over its full voltage range and was operated for over 1,000 cycles. The unusual majority organic, minority aqueous electrolyte was selected to suppress the solubility of PBAs that often limits their service life in purely aqueous electrolytes, while still retaining the high ionic conductivity and non-volatility of aqueous electrolytes.

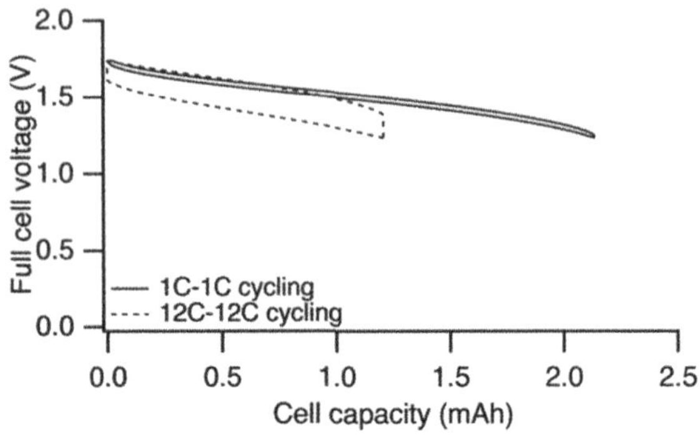

Figure 7.10. *The voltage profiles of cell containing a copper hexacyanoferrate cathode and a manganese hexacyanomanganate(II/I) anode during galvanostatic cycling. Adapted from Firouzi (2018) under the Creative Commons license; http://creativecommons.org/licenses/by/4.0/*

The most recent research into cells containing two PBA electrodes has focused on improving the energy density by increasing voltage and specific capacity. The highest voltage and energy cells reported to date combine two electron transfer cycling of a manganese hexacyanoferrate cathode with single electron transfer of a manganese hexacyanochromate anode in concentrated aqueous Na^+ electrolyte (Nakamoto *et al.* 2018; Wheeler *et al.* 2019). These cells operate at voltages of up to 2.5 V at top of charge, with two reaction plateaus at approximately 2.0 and 1.5 V in discharge. Unfortunately, only one of the two studies on this cell chemistry reported cycle life, observing capacity loss at a rate of up to 40% after just 100 cycles. This suggests that one or both of these PBAs is electrochemically unstable during cycling or is chemically unstable in aqueous electrolyte. More investigation of the failure modes in this high-energy PBA cell is required before it can be considered as a viable technology.

Natron Energy began the first effort to develop a commercial cell containing two PBA electrodes in 2012. As of early 2019, Natron manufactures 4.1 Ah cells for use in industrial and grid storage applications (Figure 7.11). Natron's cell contains a mixed metal hexacyanoferrate cathode in with a blend of manganese and other inexpensive transition metals form a solid solution at the P sites. This cathode is paired with a manganese hexacyanomanganate (II/I) anode in an organic Na^+ electrolyte to achieve an average cell voltage of 1.57 V. The cell operates between

1.85 and 1.0 V, with over 95% of capacity accessed above 1.25 V during discharge at a 1 C rate (Figure 7.12). All other cell components are borrowed from commodity lithium-ion pouch cells. The slurry-processed composite electrodes also contain nanoparticulate carbon black and an adhesive polymer binder. Those electrodes are coated onto aluminum foil electrodes that are stacked with a porous thin film separator, bonded to metal tabs and sealed in laminate packaging.

Figure 7.11. *A commercial PBA cell produced by Natron Energy. This 1.57 V, 4.1 Ah cell has approximate planar dimensions of 19 x 22 cm. Courtesy Natron Energy. For a color version of this figure, see www.iste.co.uk/monconduit/batteries.zip*

The materials in Natron's cell were selected to maximize rate capability and cycle life at a practical cost. The P site of the mixed metal hexacyanoferrate cathode contains a majority of manganese and a minority of other inexpensive transition metals such as iron. Partial substitution of manganese with other species raises the reaction potential of R site Fe, moving it closer to the reaction potential of P site Mn and allowing the cell to operate in a higher, narrower voltage range. Manganese hexacyanomanganate (II/I) is favored over hexacyanochromate-based anodes because of its scalable Route 2 synthesis from inexpensive precursors. Natron has reportedly developed a wide variety of anhydrous organic and aqueous electrolytes. Its commercial cells contain an anhydrous electrolyte in which acetonitrile is used as the primary solvent to maximize Na^+ conductivity. The low viscosity of acetonitrile results in an ionic conductivity of about 40 mS/cm for 1 M Na^+ salt, which is five to ten times greater than the ionic conductivity of the electrolytes found in commercial lithium-ion cells. The choice of an anhydrous electrolyte also extends the life of the cell by eliminating the risk of dissolution or hydrolysis of the PBAs, two commonly observed failure modes in aqueous electrolytes.

These materials design decisions result in unusually low impedance and fast kinetics. The Natron cell may be discharged at rates of up to 60 C, and recharged at rates of up to

20 C (Figure 7.12). These high rates correspond to power densities and specific powers of over 1200 W/L and 800 W/kg, respectively, or 40 times greater than the cell's energy density and specific power of 30 Wh/L and 22 Wh/kg. At 50% state of charge, the cell's impedance is 2.3 mΩ, which equates to an impedance per energy stored of 0.33 mΩ/Wh. In contrast, commercially available high-power lithium-ion cells containing a $LiFePO_4$ electrode have an impedance of 6 mΩ for an energy of 8.25 Wh, or equivalently, 0.73 mΩ/Wh (A123 2011). This comparison illustrates the fundamental tradeoff for cells containing two PBA electrodes: high intrinsic power capability deriving from a low impedance, but at the cost of a relatively low energy density.

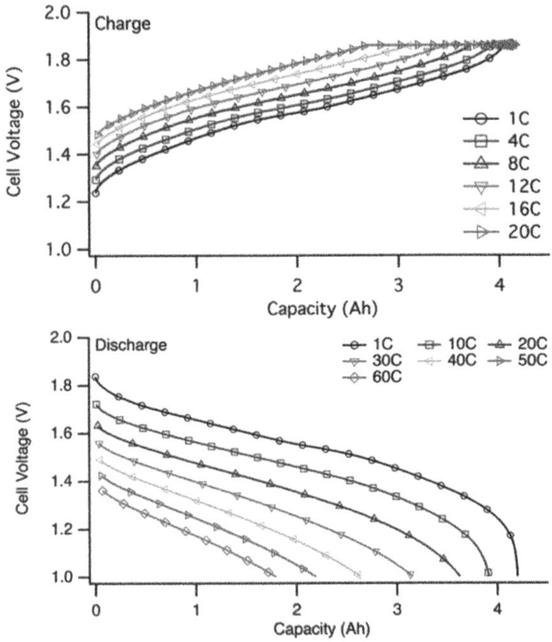

Figure 7.12. *Galvanostatic charge (top panel) and discharge (bottom panel) profiles of a 4.1 Ah cell containing two PBA electrodes and an anhydrous Na^+ electrolyte. Courtesy Natron Energy. For a color version of this figure, see www.iste.co.uk/monconduit/batteries.zip*

Natron's cell also provides unusually long cycle life. By pairing two PBA electrodes with one another, neither electrode limits the cycle life of the cell. For example, no change in capacity is observed after 4,000 full cycles of 1 C–1 C charging at 25°C (Figure 7.13). The coulombic and energy efficiencies of the cells are stable during this test, indicating that the PBA electrodes do not lose capacity or gain in impedance. Furthermore, to achieve such stable cycling, there must not be a significant slip in the relative states of charge of the two PBA electrodes. That is possible because of their relatively mild reaction potentials, which are within the electrochemical stability windows of acetonitrile and other solvents used in organic battery electrolytes. Some advanced lithium-ion cells can also operate for several thousand cycles, albeit at a lower fractional capacity utilization and with some capacity loss. What makes Natron's all-PBA cell unique is how stable the cell truly is.

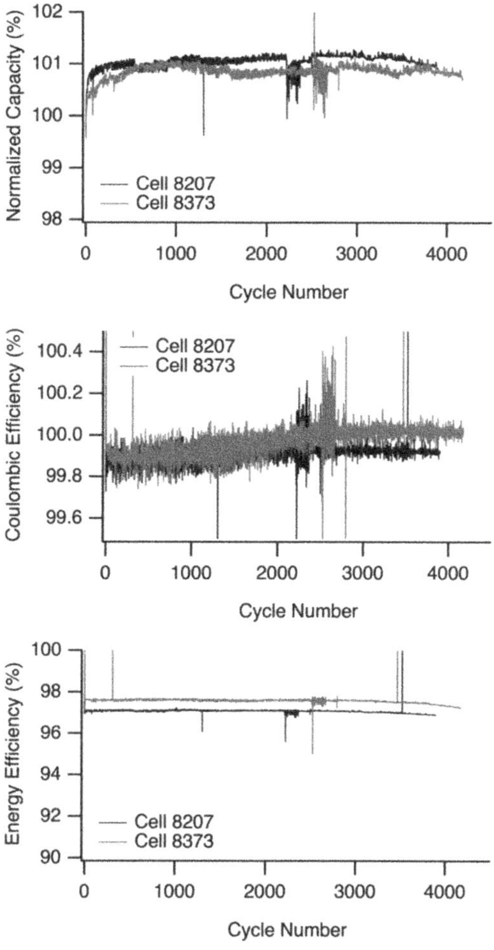

Figure 7.13. *The normalized capacity, coulombic efficiency and energy efficiency of 4.1 Ah cells containing two PBA electrodes and an anhydrous Na^+ electrolyte during continuous, full capacity utilization 1 C–1 C cycling at 25°. Courtesy Natron Energy. For a color version of this figure, see www.iste.co.uk/monconduit/batteries.zip*

Long-term continuous cycling testing of the Natron cell has been performed at rates of up to 12 C–12 C (Figure 7.14). For this test, cells were continuously cycled over their full voltage range at a 12 C rate, with no constant voltage step at top or charge or any rest periods. Approximately 75% of the cell's total capacity was accessed under this duty cycle. Every 100 cycles, three 1 C–1 C cycles were performed to determine the total capacity and state of health of the cells. Under this test condition, the cell stack temperature equilibrates at 10–15°C above ambient temperature, or at 35–40°C during tests performed at a nominal temperature of 25°C. After 25,000 tested 12 C–12 C cycles over a period of 6 months, a capacity loss of about 5% is observed. These results validate the premise that the low strain cycling of the open framework PBA structure can be applied to build a practical cell having a significantly longer cycle life than other cell technologies.

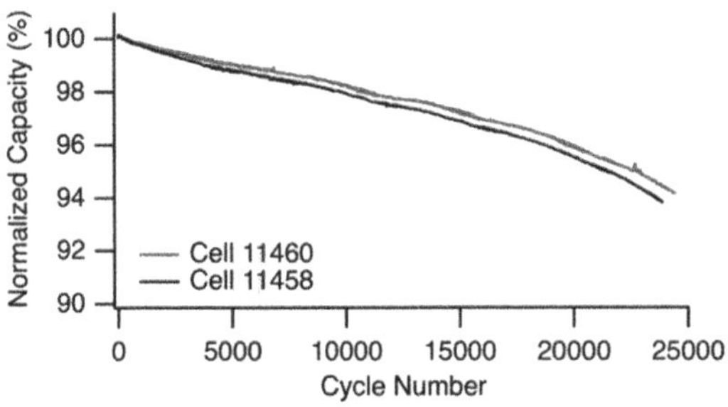

Figure 7.14. *Normalized capacities of 4.1 Ah cells containing two PBA electrodes and an anhydrous Na^+ electrolyte (1 C–1 C state of health checks during continuous 12 C–12 C cycling). Courtesy Natron Energy. For a color version of this figure, see www.iste.co.uk/monconduit/batteries.zip*

In addition to offering high-rate capability and long cycle life, cells containing two PBA electrodes have attractive safety characteristics. Those containing aqueous electrolytes are non-flammable. In the case that an anhydrous organic electrolyte is used, cells containing two PBA electrodes still offer safety advantages in comparison to commercial lithium-ion cells. Even at top of charge, the reaction potentials of PBA anodes are at least 1.5 V above the potential of sodium metal plating, which dramatically reduces the risk that metallic dendrites will form during overcharge. Furthermore, the relatively low specific energy of cells containing PBA electrodes makes them more thermally stable during a short circuit than other cell technologies. For instance, the maximum temperature increase in Natron's cell during a hard short circuit (<3 mΩ) is less than 30°C above ambient temperature, despite a peak current of over 1,000 A, or 250 C (Figure 7.15). In comparison to a typical lithium-ion cell, cells containing two PBA electrodes have five to ten times greater thermal mass per energy stored, making them much more stable during abuse or failure. Continued operation of cells containing two PBA electrodes after a short circuit is also possible. Such a short circuit results in overdischarge of the manganese hexacyanomanganate anode into the $Mn^{III/II}(CN)_6$ reaction plateau at a potential of 0 V versus SHE. PBA cathodes are fully discharged but chemically stable at that potential, so the two electrodes may be shorted without damaging the cell.

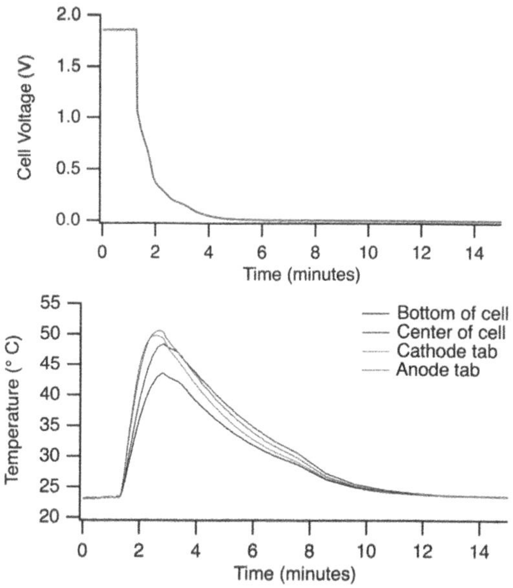

Figure 7.15. *Cell voltage (top panel) and temperatures at key locations (bottom panel) during a short circuit test (3 mΩ circuit) of Natron's production pouch cell. Courtesy Natron Energy. For a color version of this figure, see www.iste.co.uk/monconduit/batteries.zip*

Natron Energy presently markets and sells battery pack products based on those cells for use in uninterruptable power supply (UPS) systems. In the UPS application, a battery provides backup power during a grid power outage until grid power is restored or until the load is transferred to a generator. In the latter case, the battery typically carries the load for a no more than 2–3 min. For such a short run time, a high sustained C rate, or equivalently, a high power-to-energy ratio allows a smaller, lower capacity battery to be used. For instance, Natron's BlueTray4000 battery pack provides up to 8 kW of sustained discharge power despite having a total energy content of 210 Wh (Figure 7.16). More generally, products based on cells containing two PBA electrodes will be used in applications requiring at least one of the following: high sustained discharge power, fast recharge or frequent cycling. These applications include not only UPS, but also materials handling (opportunity fast charging), kinetic energy recovery (frequent fast charge and discharge), behind-the-meter energy storage for sites that have volatile loads (high cycle life) and application stacking of grid storage applications (high cycle life).

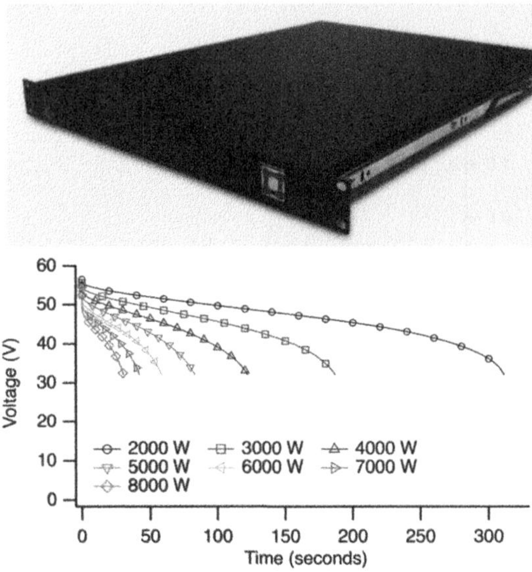

Figure 7.16. *The BlueTray4000 UPS battery tray presently sold by Natron Energy. The tray provides up to 8 kW of continuous power from 210 Wh of total energy. Courtesy Natron Energy. For a color version of this figure, see www.iste.co.uk/monconduit/batteries.zip*

7.3.2. *Cells containing one PBA electrode*

The second general class of PBA batteries contains one PBA electrode and a second electrode in which a different type of active material stores energy. A vast assortment of electrode materials has been paired with PBA cathodes. Both lithium and sodium metal are frequently used as counter electrodes during research of PBA cathodes. These cells are generally used as research tools, and not for practical devices, so they are not discussed further. Other anode materials that have been paired with PBA cathodes include metals such as Zn, ceramics such as $NaTi_2(PO_4)_3$, various polymers and carbons. Some of these cells contained aqueous electrolytes, while others contained anhydrous organic electrolytes. Table 7.3 lists a selection of cells containing one PBA electrode.

Cathode	Anode	Electrolyte	Voltage (V)	Reference
Prussian blue	Zn	Aq. NH_4Cl	1.6	Itaya and Shibayama (1983)
Copper hexacyanoferrate	Zn	Aq. $ZnSO_4$	1.7	Trocoli and La Mantia (2015)
Zinc hexacyanoferrate	Zn	Aq. $ZnSO_4$	1.7	Zhang *et al.* (2014)
Nickel hexacyanoferrate	Zn	$Zn(ClO_4)_2$/acetonitrile	1.3	Chae *et al.* (2016)
Prussian blue	Bi/BiOCl	$NaClO_4$/tetraglyme	2.2	Dutta *et al.* (2019)
Indium hexacyanoferrate	WO_3	K^+-polymer	1.0	Tung *et al.* (2003)
Nickel hexacyanoferrate	$NaTi_2(PO_4)_3$	Aq. Na_2SO_4	1.27	Wu *et al.* (2013)
Cobalt hexacyanoferrate	$NaTi_2(PO_4)_3$	Aq. Na_2SO_4	1.5	Wu *et al.* (2015b)
Manganese hexacyanoferrate	$NaTi_2(PO_4)_3$	Aq. Na_2SO_4	1.5	Nakamoto *et al.* (2017)
Copper hexacyanoferrate	Charcoal/polypyrrole	Aq. $NaClO_4$	1.1	Pasta *et al.* (2012)
Cobalt-copper hexacyanoferrate	SNDI (polymer)	Aq. Na_2SO_4	1.1	Kim *et al.* (2014)
Manganese hexacyanoferrate	Hard carbon	$NaClO_4$/EC/DEC/FEC	3.2	Song *et al.* (2015)
Nickel hexacyanoferrate	Graphite	KPF_6/$LiPF_6$/EC/DMC	3.4	Zheng *et al.* (2018)

Table 7.3. *List of cells containing one PBA electrode and a second electrode of a different type of active material*

Metal anodes have been paired with PBA cathodes since at least 1982, when Itaya described a cell in which a PB cathode was paired with a zinc anode in aqueous electrolyte (Itaya 1983). The result was a 1.6 V cell in which PB cycled NH_4^+ and zinc was reversibly plated and stripped at the anode current collector. More recently, other research groups have paired various hexacyanoferrate-based cathodes with zinc metal anodes in a variety of aqueous and organic electrolytes (Zhang *et al.* 2014; Trocoli *et al.* 2015; Chae *et al.* 2016). In some of these cells, the

PBA cathode reversibly cycles Zn^{2+}, rather than a monovalent alkali cation as in other PBA batteries. This illustrates the breadth of intercalation species the open framework PBA structure can accommodate. To date, PBA-zinc cells have used only cathodes that undergo single electron transfer. An obvious area of investigation for high energy density PBA cells is the combination of a two-electron transfer cathode with a zinc anode. A major open question for these cells is the practical cycle life of the zinc anode, which is typically orders of magnitude shorter than the cycle life of common PBA cathodes. Either Fe or Cu could be substituted in place of Zn, albeit with a lower cell voltage. A higher voltage PBA-metal cell was recently reported by Dutta *et al.* (2019), who paired a PB cathode with a bismuth oxychloride (BiOCl) anode in anhydrous organic electrolyte. In this cell, BiOCl undergoes a reversible displacement reaction to form metallic Bi at low potential. The result is cell voltage of 2.2 V, much higher than those of cells containing zinc anodes.

Ceramics have also been used as anodes in combination with PBA cathodes. In general, these ceramics have crystal structures in which large channels allow rapid, highly reversible insertion of Na^+ or K^+. For instance, tungsten trioxide (WO_3) has been paired with an indium hexacyanoferrate cathode and a K^+-saturated polymer electrolyte to produce an electrochromic battery cell operating at 1 V (Tung *et al.* 2003). More recently, several research groups have combined various PBA cathodes with a sodium titanium phosphate ($NaTi_2(PO_4)_3$) anode (Wu *et al.* 2013, 2015b; Nakamoto *et al.* 2017). Both the tungsten bronze WO_3 structure and the NaSICON $NaTi_2(PO_4)_3$ structure contain large channels through which Na^+ is readily transported, allowing them to operate by low strain insertion mechanisms analogous to that of Na^+ insertion into the PBA structure. The reaction potential of $NaTi_2(PO_4)_3$ is –0.7 V versus SHE, near the lower limit of the electrochemical stability window of aqueous electrolytes. When paired with a two-electron transfer PBA cathode such as manganese hexacyanoferrate, the resulting cell operates at an average voltage of up to about 1.5 V, quite high for an aqueous electrolyte cell. The cost structure and energy density of PBA versus $NaTi_2(PO_4)_3$ cells are attractive. However, the cycle life of these cells is just a few hundred cycles, much less than the several thousand cycles observed for both PBA cathodes and $NaTi_2(PO_4)_3$ in half cells. Further investigation of the emergent failure modes of PBA versus $NaTi_2(PO_4)_3$ cells is required to understand if they are commercially viable.

Polymers and small molecule organics have also been paired with PBA cathodes in aqueous electrolytes. Two recent examples include a hybrid activated charcoal/polypyrrole anode (Pasta *et al.* 2012) and a naphthalene diimide anode (Kim *et al.* 2014). The activated charcoal/polypyrrole anode is notable because it has a cycle life comparable to those of PBA cathodes, introducing the opportunity for a full cell that lasts for tens of thousands of cycles. However, each of these anodes has a relatively high reaction potential of about –0.2 V versus SHE, which limits full cell voltage to about 1.1 V.

Finally, PBA cathodes have been paired with carbon anodes in anhydrous electrolytes to produce high energy density cells that operate at voltages of over 3 V. In these cells, the PBA cathode acts as a substitute for the conventional oxide or phosphate-based electrodes found in other high voltage Na^+ and Li^+ cells (Figure 7.17). In 2015, Song et al. described a cell combining a two-electron transfer manganese hexacyanomanganate cathode and a hard carbon anode in a carbonate-based electrolyte (Song *et al.* 2015). In this study, Song *et al.* (2015) meticulously vacuum dried the manganese hexacyanoferrate cathode to fully dehydrate it. Doing so was critical to achieving acceptable cell stability. Without this dehydration step, zeolitic water may leach out of a PBA cathode and undergo unwanted reactions with a low potential anode. A significant effort to commercialize a cell containing a manganese hexacyanoferrate cathode and a hard carbon anode was made by Sharp Laboratories of America beginning in 2013. While this effort was eventually halted, the high energy and low materials cost of this cell chemistry make it an attractive target for future commercialization efforts. Analogous cells in which a PBA cathode is paired with a graphite anode that cycles Li^+ have also been demonstrated (Zheng *et al.* 2018).

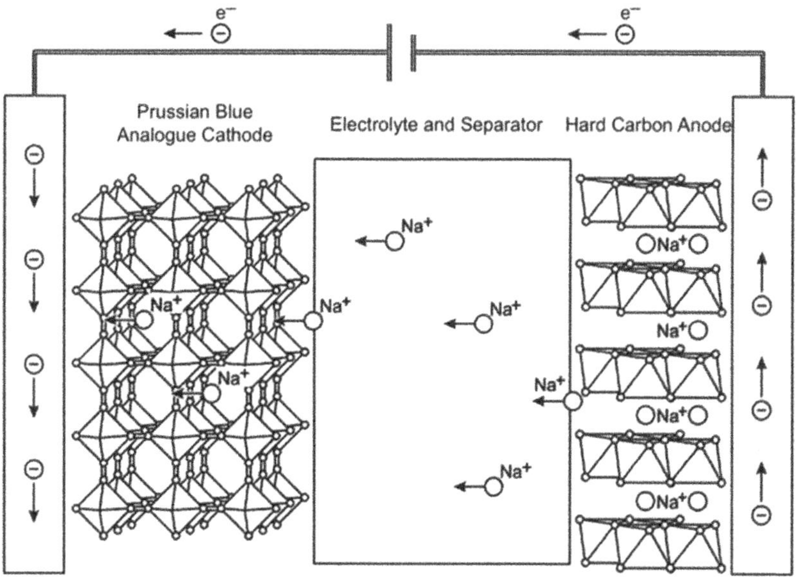

Figure 7.17. *Schematic of a high-voltage PBA versus hard carbon cell as developed by Sharp Laboratories of America. Adapted from Lu (2017)*

7.3.3. *Challenges for PBA batteries*

Successful research, development, and commercialization of any emerging battery technology require understanding and overcoming challenges in areas such as performance, safety, and scalability of manufacturing. PBAs have chemical and electrochemical properties that result in unique challenges, three of which are highlighted here: state-of-charge slip, stability of two electron cycling, and materials safety.

An imbalance in the coulombic efficiencies of the two electrodes in a cell will result in the overcharge of the more efficient electrode, and the undercharge of the less efficient electrode. This imbalance typically increases with the duration of operation (cycling, float, etc.) and is known as state-of-charge slip. Even if each of the two electrodes is perfectly stable, state-of-charge slip places a limit on the service life of a cell before an unacceptable amount of capacity and energy is lost (Burns *et al.* 2011; Smith *et al.* 2012). In most lithium-ion cells, at least one of the two electrodes operates with a flat potential profile (e.g. a graphite anode or $LiFePO_4$ cathode). In this case, the slope of the full cell voltage profile as a function of the state of charge (dV/dQ) is determined primarily by the slope of the potential profile of one electrode. However, in a cell containing two PBA electrodes, both electrodes have sloping potential profiles. The cell's dV/dQ is steeper than those of lithium-ion cells, so a smaller SOC slip is required to cause a comparable loss in capacity. Effectively, a cell containing two PBA electrodes must have a more precise balancing of its half cell coulombic efficiencies than a lithium-ion cell to achieve the same service life. For aqueous electrolyte cells, this will require nearly perfect balancing water splitting and recombination. In anhydrous organic electrolytes, water from the PBA electrodes themselves may undergo unbalanced reactions that results in unacceptable state-of-charge slip. Virtually no research has been performed on the relationship between composition properties such as water content and the coulombic efficiency of PBA electrodes. A greater understanding of these phenomena is a pre-requisite for successful development of practical cells based on these materials.

As discussed in section 7.2.3, some PBA cathodes are capable of undergoing two-electron cycling. This has an obvious advantage: higher electrode capacity and resulting higher cell energy. However, there is no evidence that PBA electrodes cycled through two electrons of oxidation and reduction are stable for the tens of thousands of cycles observed during single electron transfer cycling. In the case of the popular manganese hexacyanoferrate cathode, full oxidation results in the formation of high spin Mn^{III} at the P sites. This valence and spin state is known to readily disproportionate and hydrolyze in other chemical systems. Even in an anhydrous electrolyte, there is a risk that Mn^{III} will gradually react with any water in the PBA structure. Similarly, full oxidation of PB to Berlin Green results in a material that is at least 100 times more water soluble. There is a risk that trace solubility of Berlin Green will limit the stability of cells containing a PB cathode during operation at full charge. Furthermore, to access a full two electrons worth of capacity, the PBA must be synthesized with the A sites fully occupied. In general, vacancy-free materials have non-cubic structures that undergo a reversible phase change to cubic when oxidized. To date, there is no data indicating that a PBA cathode may reversibly undergo such a phase change for tens of thousands of cycles. Further investigation of the stability of two-electron cycling in PBA cathodes is required to understand their viability for high energy density cells.

Finally, the environmental hazards of PBAs must be considered. These materials contain cyanide, which is acutely toxic as a free anion or gas. This makes PBAs different from other toxic battery materials such as lead or cadmium, for which the toxic species are innately nonvolatile. The cyanide ligands in PBAs are strongly bonded to the P and R site transition metals, making them much less hazardous than free cyanide. Hexacyanoferrate-based cathodes do not readily release cyanide and are non-toxic and non-volatile. For instance, PB itself is frequently used as an orally administered medicine to treat radiocaesium poisoning (Takahashi *et al.* 2019). Most hexacyanoferrates are stable against cyanide release in air and water up to 200–250°C. Hexacyanomanganate and hexacyanochromate have weaker R-CN bonds than hexacyanoferrate (see section 7.2.4). They release cyanide more readily, primarily by hydrolysis reactions with water to form transition metal oxides or hydroxides and free cyanide. For this reason, they are handled industrially under dry air (typically <10% relative humidity) or inert gas to avoid this decomposition. The dry rooms found in lithium-ion manufacturing plants typically operate at <1% relative humidity and are adequate for safe handling of hexacyanomanganate and hexacyanochromate materials at high volumes. Products based on PBA batteries are most commonly marketed and sold into industrial applications in which protective ventilation is already required for lead acid or lithium-ion batteries, each of which may emit toxic gases during high temperature failure. In these applications, PBA batteries may actually pose a lower risk of toxic gas release than other technologies because they are more thermally stable during abuse conditions such as overcharge or short circuit. Methods for disposal of PBA batteries at end-of-life remain an open question. As they contain no valuable rare earth or transition metals, they will likely be processed into slag via incineration. Such is the case for lithium-ion batteries after extraction of metals that have recycling value.

7.4. Conclusion and future outlook

PBAs have unique chemical and electrochemical properties that distinguish them from other classes of battery electrode materials. Principal among these are their high rate capability and long cycle life. These performance properties derive from the open framework PBA structure, which contains channels and interstices large enough to readily accommodate Na^+ and other cations without significant strain. Cathodes that operate via two-electron cycling achieve capacities over 150 mAh/g at potentials above 3 V versus sodium, placing them among the highest energy density sodium-ion cathodes. This makes them attractive targets for ongoing development of high energy density Na-ion batteries. The ability to substitute iron with manganese that is electrochemically active at low potential makes the PBA structure a rarity: both cathodes and anodes have the same structure and cycle by the same mechanism. The result is full cells that contain two PBA electrodes and have power and cycle life characteristics similar to those of individual PBA materials.

Modern investigation of PBAs for battery applications began in 2010. In less than a decade since then, the volume of research into these materials has increased by two orders of magnitude to hundreds of published reports per year. The first commercialization of products based on PBAs occurred in 2019. Considering these recent successes, the research and industrial communities appear poised for more discoveries and advances in the coming decade.

7.5. References

A123 Systems. (2011). Nanophosphate® high power lithium ion cell ANR26650m1-B. MD100113-01. A123 Systems.

Adamson, A.W., Welker, J., and Wright, W.B. (1951). Exchange studies with complex ions. II. The kinetics of the exchange of radiocyanide ion with potassium hexacyanomanganate (III) in aqueous solution. *J. Am. Chem. Soc.*, 73(10), 4786–4790.

Bagkar, N., Choudhury, S., Kim, K.-H., Chowdhury, P., Lee, S.-I., and Yakhmi, J.V. (2006). Crystalline thin films of transition metal hexacyanochromates grown under Langmuir monolayer. *Thin Solid Films*, 513, 325–330.

Brown, D.B., Shriver, D.F., and Schwartz, L.H. (1968). Solid-state reactions of iron(II) hexacyanochromate(III). *Inorg. Chem.*, 7 (1), 77–83.

Burns, J.C., Krause, L.J., Le, D.-B., Jensen, L.D., Smith, A.J., Xiong, D., and Dahn, J.R. (2011). Introducing symmetric Li-ion cells as a tool to study cell degradation mechanisms. *J. Electrochem. Soc.*, 158(12), A1417–A1422.

Buschmann W.E., Ensling, J., Gütlich, P., and Miller, J.S. (1999). Electron transfer, linkage isomerization, bulk magnetic order, and spin-glass behavior in the iron hexacyanomanganate Prussian blue analogue. *Chem. Eur. J.*, 5(10), 3019–3028.

Catala, L., Brinzei, D., Prado, Y., Gloter, A., Stéphan, O., Rogez, G., and Mallah, T. (2009). Core-multishell magnetic coordination nanoparticles: Toward multifunctionality on the nanoscale. *Angew. Chem.*, 121, 189–193.

Chae, M.S., Heo, J.W., Kwak, H.H., Lee, H., and Hong, S.-T. (2016). Organic electrolyte-based rechargeable zinc-ion batteries using potassium nickel hexacyanoferrate as a cathode material. *J. Power Sources*, 337, 204–211.

Deng, L., Yang, Z., Tan, L., Zeng, L., Zhu, Y., and Guo, L. (2018). Investigation of the Prussian blue analog $Co_3[Co(CN)_6]_2$ as an anode material for nonaqueous potassium-ion batteries. *Adv. Mater.*, 30, 1802510.

Dutta, P.K., Myung, Y., Venkiteswaran, R.K., Mehdi, L., Browning, N., Banerjee, P., and Mitra, S. (2019). Mechanism of Na-ion storage in BiOCl anode and the sodium-ion battery formation. *J. Phys. Chem. C*, 123(18), 11500–11507.

Erinmwingbovo, C., Palagonia, M.S., Brogioli, D., and La Mantia, F. (2017). Intercalation into a Prussian blue derivative from solutions containing two species of cations. *ChemPhysChem*, 18(8), 917–925.

Eftekhari, A. (2003). A high-voltage solid-state secondary cell based on chromium hexacyanometallates. *J. Power Sources*, 117, 249–254.

Ferrari, A., Morisi, E., and Tani, M.E. (1964). The crystal structure of hexacyanomanganates(III) of bivalent cations. *Acta Cryst.*, 17, 311.

Grabner, E.W. and Kalwellis-Mohn, S. (1987). Hexacyanoferrate layers as electrodes for secondary cells. *J. Appl. Electrochem.*, 17, 653–656.

He, G. and Nazar, L.F. (2017). Crystallite size control of Prussian white analogues for nonaqueous potassium-ion batteries. *ACS Energy Lett.*, 2, 1122–1127.

Hu, H., Liu, W., Zhu, M., Lin, Y., Zhang, J., Chen, T., and Liu, K. (2019). Yolk-shell Prussian blue nanoparticles with fast ion diffusion for sodium-ion battery. *Mater. Lett.*, 248, 206–209.

Herren, F., Fischer, P., Ludi, A., and Hälg, W. (1980). Neutron diffraction study of Prussian blue, $Fe_4[Fe(CN)_6]_3 \cdot xH_2O$. Location of water molecules and long-range magnetic order. *Inorg. Chem.*, 19, 956–959.

Hurlbutt, K., Wheeler, S., Capone, I., and Pasta, M. (2018). Prussian blue analogs as battery materials. *Joule*, 2, 1950–1960.

Itaya, K. and Shibayama, K. (1982). Prussian-blue-modified electrodes: An application for a stable electrochromic display device. *J. Appl. Phys. Chem.*, 53(1), 804–805.

Itaya, K., Shibayama, K., Toshima, S., Ataka, T., and Iwasa, K. (1983). Electrochemical cell. European Patent Application No. 86555A1.

Ji, Z., Han, B., Liang, H., Zhao, C., Gao, Q., Xia, K., and Wu, J. (2016). On the mechanism of the improved operation voltage of rhombohedral nickel hexacyanoferrate as cathodes for sodium-ion batteries. *ACS Appl. Mater. Interfaces*, 8, 33619–33625.

Kalwellis-Mohn, S. and Grabner, E.W. (1989). A secondary cell based on thin layers of zeolite-like nickel hexacyanometallates. *Electrochim. Acta*, 34(8), 1265–1269.

Kim, D.J., Jung, Y.H., Bharathi, K.K., Je, S.H., Kim, D.K., Coskunm, A., and Choi, J.W. (2014). An aqueous sodium ion hybrid battery incorporating an organic compound and a Prussian blue derivative. *Adv. Energy Mater.*, 4(12), 1400133.

Lee, H.-W., Wang, R.Y., Pasta, M., Lee, S.W., Liu, N., and Cui, Y. (2014). Manganese hexacyanomanganate open framework as a high-capacity positive electrode material for sodium-ion batteries. *Nat. Commun.*, 5, 5280.

Li, C.H., Peprah, M.K., Asakura, D., Meisel, M.W., Okubo, M., and Talham, D.R. (2015). Stepwise reduction of electrochemically lithiated core-shell heterostructures based on the Prussian blue analogue coordination polymers $K_{0.1}Cu[Fe(CN)_6]_{0.7}\cdot 3.5H_2O$ and $K_{0.1}Ni[Fe(CN)_6]_{0.7}\cdot 4.4H_2O$. *Chem. Mater.*, 27(5), 1524–1530.

Lu, Y., Lee, J.-J., Nishijima, M., and Kakimoto, S. (2017). Alkali and alkaline-earth ion batteries with hexacaynometallate cathode and non-metal anode. United States Patent No. 9,599,358.

McCargar, J.W. and Neff, V. (1988). Thermodynamics of mixed-valence intercalation reactions: The electrochemical reduction of Prussian blue. *J. Phys. Chem.*, 92, 3598–3604.

Messina, R. and Perichon, J. (1980). Mécanisme de la réduction électrochimique en milieu non aqueux de materiaux cathodiques utilisés dans les piles au lithium. *J. Appl. Electrochem.*, 10, 655–663.

Morimoto, Y., Urase, S., and Shibata, T., (2016). Enhanced battery performance in manganese hexacyanoferrate by partial substitution. *Electrochim. Acta*, 210, 963–969.

Nakamoto, K., Sakamoto, R., Kitajou, A., and Okada, S. (2017). Effect of concentrated electrolyte on aqueous sodium-ion battery with sodium manganese hexacyanoferrate cathode. *Electrochemistry*, 85(4), 179–185.

Nakamoto, K., Sakamoto, R., Sawada, Y., Ito, M., and Okada, S. (2018). Over 2 V aqueous sodium-ion battery with Prussian blue-type electrodes. *Small Methods*, 3(4), 1800220.

Neff, V. (1978). Electrochemical oxidation and reduction of thin films of Prussian blue. *J. Electrochem. Soc.*, 125, 886–887.

Neff, V. (1985). Some performance characteristics of a Prussian blue battery. *J. Electrochem. Soc.*, 132(6), 1382–1384.

Nie, P., Shen, L., Luo, H., Ding, B., Xu, G., Wang, J., and Zhang, X. (2014). Prussian blue analogues: A new class of anode materials for lithium ion batteries. *Chem. Mater. A*, 2, 5852–5857.

Pasta, M., Wessells, C.D., Huggins, R.A., and Cui, Y. (2012). A high-rate and long cycle life aqueous electrolyte battery for grid-scale energy storage. *Nat. Commun.*, 2, 1149.

Pasta, M., Wessells, C.D., Liu, N., Nelson, J., McDowell, M.T., Huggins, R.A., Toney, M.F., and Cui, Y. (2014). Full open-framework batteries for stationary energy storage. *Nat. Commun.*, 5, 3007.

Pasta, M., Wang, R.Y., Ruffo, R., Qiao, R., Lee, H.-W., Shyam, B., Guo, M., Wang, Y., Wray, L.A., Yang, W., Toney, M.F., and Cui, Y. (2016). Mangaese-cobalt hexacyanoferrate cathodes for sodium-ion batteries. *J. Mater. Chem. A*, 4, 4211–4223.

Robin, M.B. (1962). The color and electronic configurations of Prussian blue. *Inorg. Chem.*, 1(2), 337–342.

Rodríguez-Hernández, J., Reguera, E., Lima, E., Balmaseda, J., Martínez-García, R., and Yee-Madeira, H. (2007). An atypical coordination in hexacyanometallates: Structure and properties of hexagonal zinc phases. *J. Phys. Chem. Sol.*, 68, 1630–1642.

Rudola, A., Du, K., and Balaya, P. (2017). Monoclinic sodium iron hexacyanoferrate cathode and non-flammable glyme-based electrolyte for inexpensive sodium-ion batteries. *J. Electrochem. Soc.*, 164(6), A1098–A1109.

Scholz, A. and Dostal, A. (1995). The formal potentials of solid metal hexacyanometalates. *Angew. Chem. Int. Ed. Engl.*, 34(23/24), 2685–2687.

Shriver, D.F., Shriver, S.A., and Anderson, S.E. (1965). Ligand field strength of the nitrogen end of cyanide and structures of cubic cyanide polymers. *Inorg. Chem.*, 4(5), 725–730.

Shrivastava, A. and Smith, K.C. (2018). Electron conduction in nanoparticle agglomerates limits apparent Na+ diffusion in Prussian blue analogue porous electrodes. *J. Electrochem. Soc.*, 165, A1777–A1787.

Smith, A.J., Dahn, H.M., Burns, J.C., and Dahn, J.R. (2012). Long-term low-rate cycling of $LiCoO_2$/graphite Li-ion cells at 55° C. *J. Electrochem. Soc.*, 159(6), A705–A710.

Song, J., Wang, L., Lu, Y., Liu, J., Guo, B., Xiao, P., Lee, J.-J., Yang, X.-Q., Henkelman, G., and Goodneough, J.B. (2015). Removal of interstitial H_2O in hexacyanometallates for a superior cathode for a sodium-ion battery. *J. Am. Chem. Soc.*, 137(7), 2658–2664.

Sottman, J., Bernal, F.L.M., Yusenko, K.V., Merrmann, M., Emerich, H., Wragg, D.S., and Margadonna, S. (2016). In operando syncrotron XRD/XAS investigation of sodium insertion into the Prussian blue analogue cathode material $Na_{1.32}Mn[Fe(CN)_6]_{0.83}\cdot zH_2O$. *Electrochim. Acta*, 200, 305–313.

Su, D., McDonagh, A., Qiao, S.-Z., and Wang, G. (2017). High-capacity aqueous potassium-ion batteries for large-scale energy storage. *Adv. Mater.*, 29(1), 1604007.

Takahashi, A., Tanaka, H., and Kawamoto, T. (2019). Prussian blue nanoparticles and nanocomposites for Cs decontamination. *Prussian Blue-Type Nanoparticles and Nanocompositles*, 217–242.

Treadwell, W.D., Gübeli, O., and Huber, D. (1941). Zur Kenntnis des Doppelsalzes Kaliumtricyanomanganat (I) – Kaliumhexacyanomanganat (II). *Helv. Chim. Acta*, 24(1), 152–157.

Trocoli, R. and La Mantia, F. (2015). An aqueous zinc-ion battery based on copper hexacyanoferrate. *ChemSusChem*, 8, 481–485.

Tung, T.-S., Chen, L.-C., and Ho, K.-C. (2003) An indium hexacyanoferrate-tungsten oxide electrochromic battery with a hybrid K^+/H^+-conducting polymer electrolyte. *Solid State Ionics*, 165, 257–267.

Wang, B., Han, Y., Wang, X., Bahlawane, N., Pan, H., Yan, M., and Jiang, Y. (2018). Prussian blue analogs for rechargeable batteries. *iScience*, 3, 110–133.

Wang, H., Zhu, Q., Li, H., Xie, C., and Zeng, D. (2018). Tuning particle size of Prussian blue by a dual anion source method. *Cryst. Growth Design*, 10(18), 5780–5789.

Wang, R.Y., Shyam, B., Stone, K.H., Nelson Weker, J., Pasta, M., Lee, H.-W., Toney, M.F., and Cui, Y. (2015a). Reversible multivalent (monovalent, divalent, trivalent) ion insertion in open framework materials. *Adv. Energy Mater.*, 5(12), 1401869.

Wang, R.Y., Wessells, C.D., Huggins, R.A., and Cui, Y. (2013). Highly reversible open framework nanoscale electrodes for divalent ion batteries. *Nano Lett.*, 13, 5748–5752.

Wang, L., Song, J., Qiao, R., Wray, L.A., Hossain, M.A., Chuang, Y.-D., Yang, W., Lu, Y., Evans, D., Lee, J.-J., Vail, S., Zhao, X., Nishijima, M., Kakimoto, S., and Goodenough, J.B. (2015b). Rhombohedral Prussian white as cathode for rechargeable sodium-ion batteries. *J. Am. Chem. Soc.*, 137, 2548–2554.

Wardecki, D., Ojwang, D.O., Grins, J., and Svensson, G. (2017). Neutron diffraction and EXAFS studies of $K_{2x/3}Cu[Fe(CN)_6]_{2/3}\cdot nH_2O$. *Cryst. Growth Des.*, 17, 1285–1292.

Wessells, C.D., Huggins, R.A., and Cui, Y. (2011). Copper hexacyanoferrate battery electrodes with long cycle life and high power. *Nat. Commun.*, 2, 550.

Wessells, C.D., McDowell, M.T., Peddada, S.V., Pasta, M., Huggins, R.A., and Cui, Y. (2012b). Tunable reaction potentials in open framework nanoparticle battery electrodes for grid-scale energy storage. *ACS Nano*, 6(2), 1688–1694.

Wessells, C.D., Peddada, S.V., McDowell, M.T., Huggins, R.A., and Cui, Y. (2012a). The effect of insertion species on nanostructured open framework hexacyanoferrate battery electrodes. *J. Electrochem. Soc.*, 149(2), A98–A103.

Wheeler, S., Capone, I., Day, S., Tang, C., and Pasta, M. (2019). Low-potential Prussian blue analogues for sodium-ion batteries: Manganese hexacyanochromate. *Chem. Mater.*, 31, 2619–2626.

Wu, J., Song, J., Dai, K., Zhuo, Z., Wray, L.A., Liu, G., Shen, Z.-X., Zeng, R., Lu, Y., and Yang, W. (2017). Modification of transition-metal redox by interstitial water in hexacyanometalate electrodes for sodium-ion batteries. *J. Am. Chem. Soc.*, 139, 18358–18364.

Wu, X., Cao, M., Hu, C., and He, X. (2006). Sonochemical synthesis of Prussian blue nanocubes from a single-source precursor. *Cryst. Growth Design.*, 6(1), 26–28.

Wu, X., Cao, Y., Ai, X., Qian, J., and Yang, H. (2013). A low-cost and environmentally benign aqueous rechargeable sodium-ion battery based on $NaTi_2(PO_4)_3$-$Na_2NiFe(CN)_6$ intercalation chemistry. *Electrochem. Comm.*, 31, 145–148.

Wu, X., Lou, Y., Sun, M., Qian, J., Cao, Y., Ai, X., and Yang, H. (2015a). Low-defect Prussian blue nanocubes as high capacity and long life cathodes for aqueous Na-ion batteries. *Nano Energy*, 13, 117–123.

Wu, X., Sun, M., Guo, S., Qian, J., Liu, Y., Cao, Y., Ai, X., and Yang, H. (2015b). Vacancy-free Prussian blue nanocrystals with high capacity and superior cyclability for aqueous sodium-ion batteries. *ChemNanoMat*, 1(3), 188–193.

Wu, X., Wu, C., Wei, C., Hu, L., Qian, J., Cao, Y., Ai, X., Wang, J., and Yang, H. (2016). Highly crystallized $Na_2CoFe(CN)_6$ with suppressed lattice defects as superior cathode material for sodium-ion batteries. *ACS Appl. Mater. Interfaces*, 8 (8), 5393–5399.

Yang, Y., Liu, E., Yan, D., Ma, C., Wen, W., Liao, X.-Z., and Ma, Z.-F. (2016). Influence of structural imperfection on electrochemical behavior of Prussian blue cathode materials for sodium ion batteries. *J. Electrochem. Soc.*, 163(9), A2117–A2123.

You, Y., Wu, X.-L., Yin, Y.-X., and Guo, Y.-G. (2014). High-quality Prussian blue crystals as superior cathode materials for room-temperature sodium-ion batteries. *En. Environ. Sci.*, 7 (5), 1643–1647.

Zhang, L., Chen, L., Zhao, X., and Liu, Z. (2014). Towards high-voltage aqueous metal-ion batteries beyond 1.5 V: The zinc/zinc hexacyanoferrate system. *Adv. Energy Mater.*, 5(2), 1440930.

Zheng, J., Deng, W., Hu, Z., Zhuo, Z., Liu, F., Chen, H., Lin, Y., Yang, W., Amine, K., Li, R., Lu, J., and Pan, F. (2018). Asymmetric K/Li-ion battery based on intercalation selectivity. *ACS Energy Lett.*, 3, 65–71.

8

The Design, Performance and Commercialization of Faradion's Non-aqueous Na-ion Battery Technology

Ashish RUDOLA, Fazlil COOWAR, Richard HEAP and Jerry BARKER
Faradion Limited, Sheffield, UK

8.1. Introduction

Although the concept of sodium-ion (Na-ion) batteries has been around since the early 2000s (Barker *et al.* 2003), this energy storage technology has been largely overlooked due to the phenomenal success of lithium-ion (Li-ion) technology. Recent advances, however, have demonstrated that Na-ion technology is not only a viable alternative to Li-ion batteries but also competitive in terms of cost, performance and safety. Its similarity to Li-ion batteries makes it a drop-in replacement for this technology, where existing manufacturing equipment/facilities may be used for all aspects of cell production. Unlike Li-ion batteries, however, Na-ion cells are able to use aluminum for both current collectors, reducing both weight and cost. In addition, since aluminum is not subject to the dissolution issues experienced by copper, it is possible to discharge Na-ion cells to 0 V, making them significantly safer for both transportation and storage. The data presented in this article were submitted in June 2019, so the figures shown do not reflect Faradion's

Na-ion Batteries,
coordinated by Laure MONCONDUIT and Laurence CROGUENNEC. © ISTE Ltd 2020.

current level of development; as of February 2020, cell specific energies of >150 Wh/kg have been achieved in pouch cells, indicating that Faradion is on track to meet its cell specific energy target of 190 Wh/kg within 2020.

Faradion was established in 2011, with the aim of identifying new Na-based energy storage materials with lower cost, improved performance and increased intrinsic safety. Faradion's business focus is concentrated on high-energy applications, such as large-scale energy storage systems, storage for renewables, uninterruptable power supply (UPS) and low-cost electric vehicles where the cost reduction over Li-ion would be most apparent and thereby maximizing the cost benefit of this technology. However, while optimizing this technology for energy density and cycle life, encouraging rate capability data were also recorded, suggesting suitability for use in power applications, including power tools, household appliances and fast charge E-bus applications.

Since its founding, thousands of positive and negative active materials have been screened electrochemically, resulting in a wealth of new information and the generation of a significant IP portfolio. To date, Faradion has filed 29 patent applications, covering active materials (both positive and negative), their methods of manufacture and their use in commercial devices. Other patent applications have focused on novel electrolytes, alternate cell configurations and Na-ion specific designs for battery packs and battery management systems (BMS). More details on Faradion's patent portfolio will be described later in the chapter.

As part of its commercialization focus, Faradion has collaborated with its commercial partners to produce large-scale quantities of optimized active materials, electrolytes and electrodes for successful incorporation into a significant number of prototype sized cells. To date, more than 40 kWh of prototype production has been completed successfully. Moreover, this work has resulted in the production of several prototype demonstrators based on the use of the company's proprietary large format Na-ion pouch cells. For example, in 2015 Faradion produced a 400 Wh battery pack for an E-bike demonstrator. More recently, other demonstrator units have targeted stationary markets, energy back-up and some motive applications. Faradion anticipates that this demonstration phase of its proprietary Na-ion technology will result in significant commercial interest and will allow the licensing of its technology to both existing and emerging battery companies. This process will accelerate the adoption of non-aqueous Na-ion batteries into many new markets and applications.

It should be emphasized that a significant amount of the information presented in this article relates to historic materials and cell data. On the basis of improvements and optimization of its active materials and cell building operations, Faradion has made some very impressive recent progress in the performance of cells and batteries.

8.2. Experimental

8.2.1. *Active materials*

The Faradion cell chemistry utilizes either an in-house prepared, or a commercially sourced, hard carbon as a negative electrode. One of Faradion's proprietary cathode materials is then used as the positive electrode. The active cathode material may be described by the general formula $Na_aNi_{(1-x-y-z)}Mn_xMg_yTi_zO_2$, and is protected by a wide range of issued patents and patent applications (see, for example, Barker and Heap 2013a, 2013b; Barker and Heap 2015a, 2015b; Sayers *et al.* 2015a, 2015b). Depending on the synthesis conditions and the target composition, the cathode active material may be produced in one of these three versions: (1) in a pure O3 structural form; (2) in a pure P2 structural form; or most advantageously (3) in a mixed O3/P2 structural form. The O3 and P2 structural forms for Na-layered oxides were first described by Delmas *et al.* (1980).

The cathode materials are made at the multi-kilogram scale using industry standard processing and calcining equipment. Based on the synthesis conditions, these cathode products may then be used in both energy and power applications. Moreover, by careful control of the synthesis conditions, it is possible to use an identical precursor mixture to make different structural variants. Indeed, the precise O3:P2 content in a mixed-phase product can be reliably predicted using this synthesis control. This has obvious advantages in a manufacturing environment, where a common cathode precursor mixture may be used to target different application types.

Much of Faradion's current production utilizes a mixed-phase cathode material. This mixed-phase material, containing controlled amounts of an O3-type structural component and a P2-type structural component, is used in order to take advantage of their respective advantages; O3-type phases are able to provide high specific capacities, whereas P2-type phases are able to provide high voltages and high rate capability. In addition to its excellent performance, the material is designed to be low-cost, with the nickel content being minimized (<18 wt.%), and the manufacturing method being a single-step solid-state process, prior to powder processing, in order to maximize tap density.

Figure 8.1 shows the powder X-ray diffraction (CuK_α irradiation) of representative cathode active materials that possess (a) pure O3-type phase, (b) pure P2-type phase and (c) a mixed-phase O3/P2 material. From these data, we may conclude that the crystal structures of sodium-based layered oxides are highly versatile and can therefore be tailored in order to meet the requirements of any target application. More details on the structural properties of these cathode materials may be found in Faradion's published IP (Barker and Heap 2013a, 2013b; Sayers *et al.* 2015a).

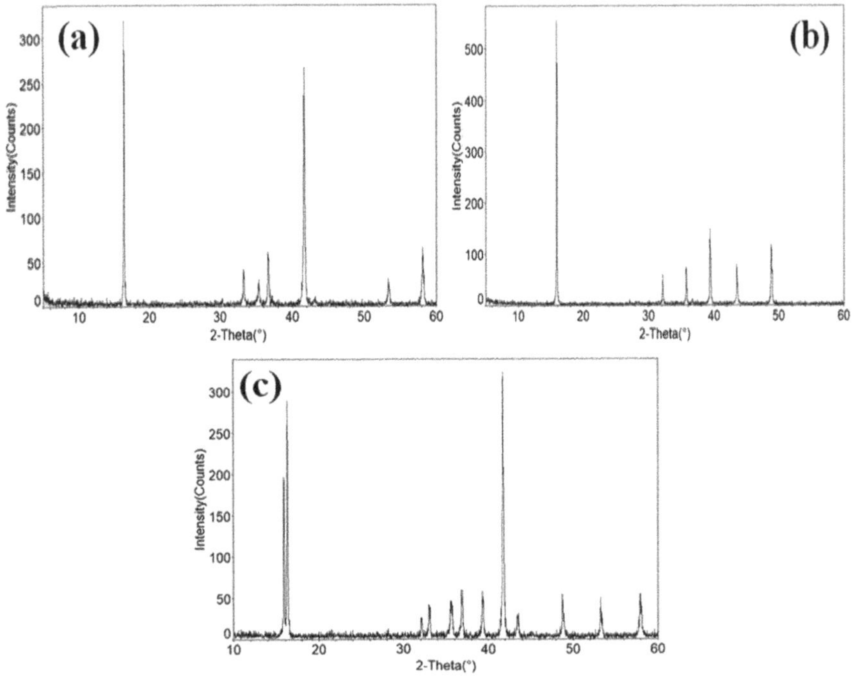

Figure 8.1. *X-ray diffraction pattern (CuK$_\alpha$ irradiation) of (a) a pure O3-type phase, (b) a pure P2-type phase and (c) a mixed-phase O3/P2 material*

Figure 8.2 shows the SEM images of a typical mixed-phase cathode material under different magnifications (figures courtesy of Haldor-Topsoe A/S, Denmark). Figure 8.2(a) indicates that the active material comprises an agglomerate of around 20 μm diameter. Closer view of this figure confirms that the agglomerate is composed of smaller primary particles of approximately 1 μm in diameter. Careful optimization of the synthesis methods and conditions during scale-up has meant that the tap density of the materials falls in the range of 1.5–2.0 g/cm^3. This relatively high value helps produce electrodes with efficient particle packing characteristics and therefore good volumetric energy density.

Inspection of the active material under higher SEM magnification (see Figures 8.2(b) and (c)) indicates that the primary particles form an intriguing stacking morphology. In Figure 8.2(b), we can detect that the smooth part (predominantly containing O3 structure material) neatly sandwiches the faceted part (predominantly containing P2 structure material) to produce a compact layered stacking arrangement. We believe it is highly likely that some intergrowths exist between the O3 and P2 sections. We are currently confirming this hypothesis by detailed TEM measurements of active materials containing different relative contents of the O3 and P2 phases. In Figure 8.2(c), we may also detect multistack arrangements containing alternate O3 and P2 structures. It is likely that the intergrowths are at least partially responsible for the outstanding electrochemical performance of these mixed-phase active materials. It should also be noted that simple physical mixtures of O3 and P2 components do not produce equivalent electrochemical performance.

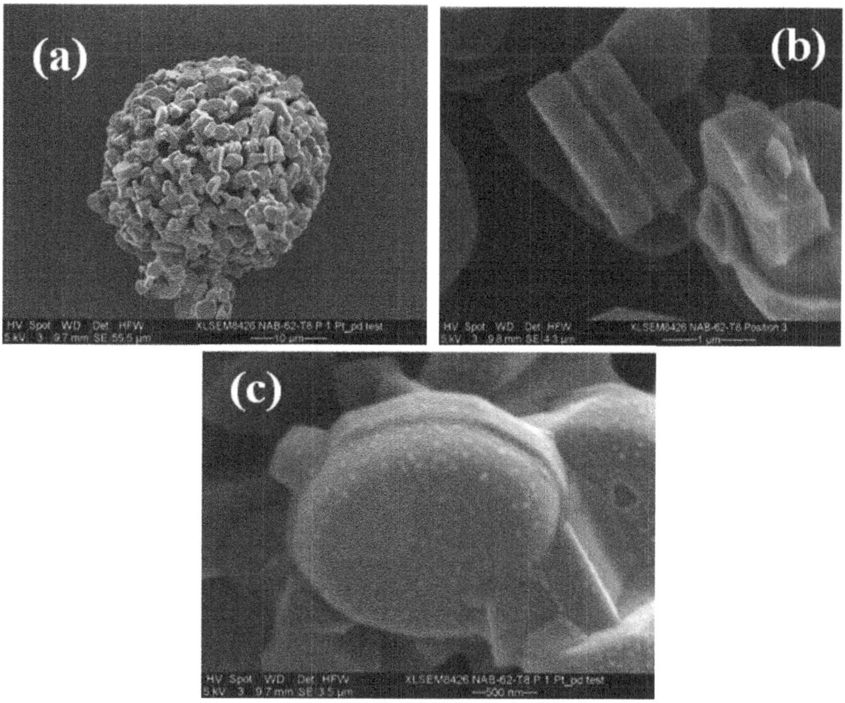

Figure 8.2. *SEM image of a typical mixed-phase cathode material showing (a) a ~20 µm agglomerate made up of ~1 µm primary particles and (b and c) morphology of the primary particles. The SEM micrograph images are courtesy of Haldor-Topsoe A/S, Denmark*

8.2.2. *Electrode fabrication*

The fabrication of cathode and anode electrodes and the production of full Na-ion cells have been carried out at several research laboratories and prototyping facilities. Only typical experimental conditions are described here. The target nominal capacities of the Na-ion pouch cells vary in the range of small (up to 10 mAh) to large (>10 Ah). Cylindrical and prismatic cell designs have also been made using similar electrode preparative methods.

8.2.2.1. *Cathode slurry and coating*

The fabrication of cathode slurry is carried out in two steps. The first step consists of dispersing the active material and carbon conductive additives in N-Methyl-2-pyrrolidone (NMP), under dynamic vacuum, in a high torque mixer model. This process enables the high viscosity mix to be processed with a high solid content. Shear forces between the container and mixer are controlled for optimizing the effect of mixing and dispersing.

The second step consists of high energy mixing of the slurry in a high shear mixer. This process is carried out under dynamic vacuum to break down the lumps and ensure that the paste is free of agglomerates. A polymer polyvinylidene fluoride (PVDF) solution containing a stabilizer is added to the paste and mixed for a further 2 h. The slurry is then transferred to the feeder tank attached to a reel-to-reel coater.

Using the comma-bar or slot-die coating technique, the cathode slurry is coated onto an aluminum substrate dried in a three-zone dryer with air-impingement nozzles set at three different temperatures. The resulting cathode is finally further dried at 120°C under dynamic vacuum overnight.

8.2.2.2. *Anode slurry and coating*

The fabrication of anode slurry is obtained by dispersing the active material and carbon additives in the polymer PVDF binder solution in a high torque mixer for about 5 h. The viscosity of the slurry is adjusted by adding appropriate amounts of NMP. The anode slurry is then degassed for 1 h. Similarly to the cathode coating, the anode slurry is also coated onto an aluminum substrate on a reel-to-reel coater.

8.2.2.3. *Calendering and slitting*

The anode and cathode electrodes are calendered to a specified targeted thickness/porosity depending on the particular target application. This process is undertaken using a heated reel-to-reel commercial calender. The electrodes are thereafter slit to a specific width. Electrodes of specific size are then stamped out for fabrication of cells of varying capacity and footprint.

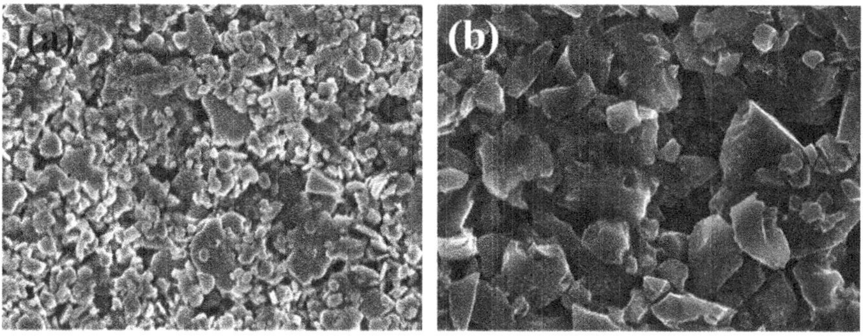

Figure 8.3. *SEM images of the composite Faradion: (a) cathode electrode and (b) anode electrode*

Figures 8.3(a) and (b) show, respectively, the SEM images of typical composite cathode and anode electrodes. As can be seen, the active particles of both cathode and anode are well connected by the carbon conductive additives and binder, leading to a uniform electrode conductive matrix.

8.2.3. *Pouch cell fabrication*

The fabrication of pouch cells of capacity >5 Ah is carried out by stacking multiple layers of cathode and anode separated by a separator (a typical commercial polyolefin material used widely in Li-ion batteries) using an automated stacking machine. The stacked cells are tab-welded and housed in a pre-formed aluminum pouch. Cells are filled under vacuum using an automated electrolyte filling machine. A simplified schematic of Faradion's pouch cell design is shown in Figure 8.4. Small-scale 10 mAh pouch cells are manually built, using single-sided electrodes.

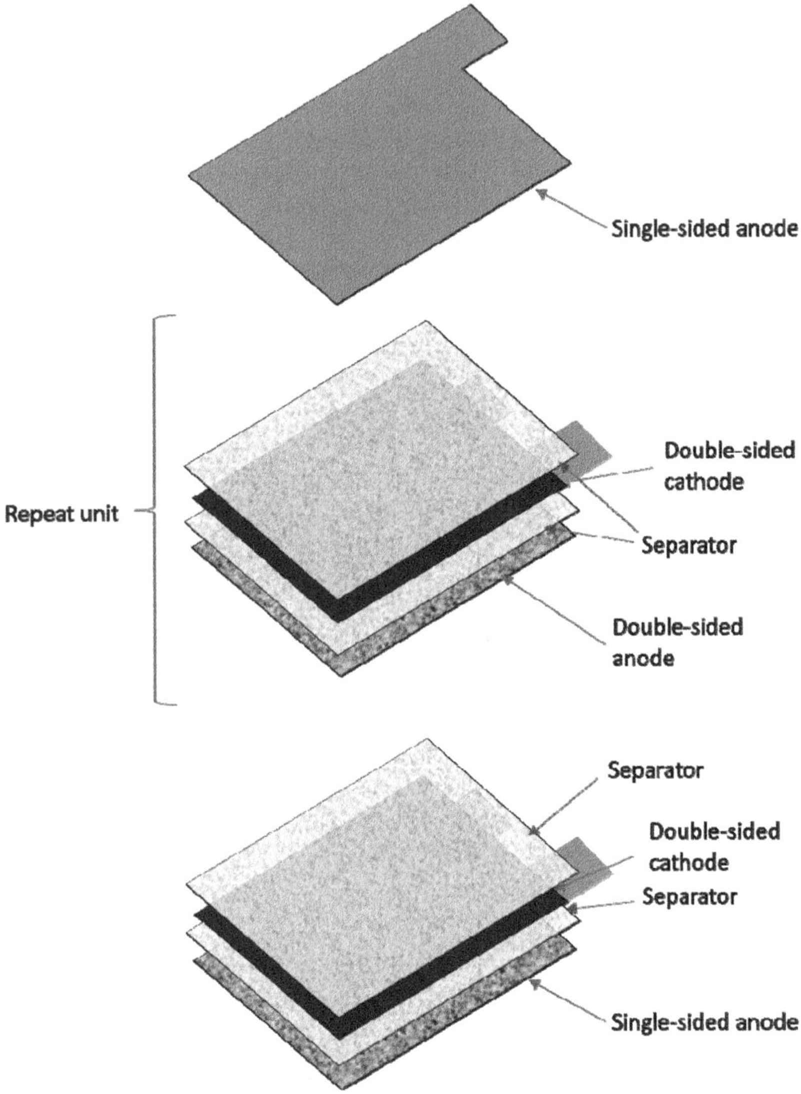

Figure 8.4. *Schematic representation of a stacked Faradion Na-ion pouch cell. For a color version of this figure, see www.iste.co.uk/monconduit/batteries.zip*

8.2.4. *Faradion electrolyte*

Faradion uses a proprietary electrolyte based on commercially available salts such as $NaPF_6$ in aliphatic and cyclic carbonate-ester solvents such as EC, DEC,

DMC, EMC, and PC. The ratio of these solvents is adjusted to suit different applications and optimized to meet ionic conductivity requirements. Further, electrolyte additives are used to boost specific Na-ion performance metrics, for example, increasing coulombic efficiency, extending cycle lifetime, decreasing parasitic heat flow and controlling impedance growth.

8.3. Cell performance

8.3.1. *Half-cell cycling*

As described previously, much of Faradion's current production utilizes one of its patented doped nickelates as cathode active material. Details about this substituted and structurally stabilized sodium nickel oxide, $Na_aNi_{(1-x-y-z)}Mn_xMg_yTi_zO_2$, are described in a number of Faradion patents (Barker and Heap 2013a, 2013b; Barker and Heap 2015a, 2015b; Sayers *et al.* 2015a, 2015b). The theoretical capacity of this material is 165 mAh/g based on the $Ni^{2+/4+}$ redox couple.

As shown in Figure 8.5(a), the material is capable of delivering in excess of 121 mAh/g when cycled between 4.1 and 2 V, 147 mAh/g when cycled between 4.25 and 2 V and near-theoretical 156 mAh/g between 4.35 and 2 V at ±C/5 rate in a sodium half-cell. Excellent performance at varying depths of discharge provide flexibility to tailor cell designs to suit specific applications, including both high-voltage and low-voltage systems.

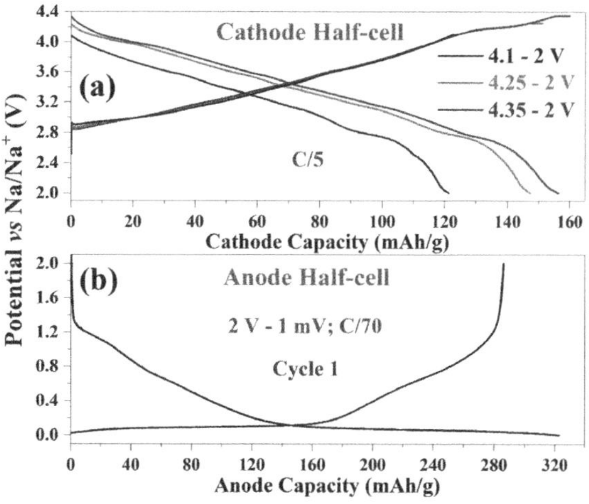

Figure 8.5. *Half-cell (vs. Na metal) performances of Faradion cathode and anode. For a color version of this figure, see www.iste.co.uk/monconduit/batteries.zip*

The first cycle of an anode half-cell is shown in Figure 8.5(b). The Faradion hard carbon anode can deliver a reversible specific capacity of 286 mAh/g with the first cycle coulombic efficiency close to 89% using Faradion electrolyte. This first cycle coulombic efficiency value for hard carbon is among the highest reported in the Na-ion literature (Hou *et al.* 2017). Please note that the hard carbon half-cell was cycled at very slow rates (C/70) to counteract the increased polarization of Na metal counter electrode as reported previously (Rudola *et al.* 2014, 2015). Owing to the flexibility of the cathode operating potentials, the cathode:anode active mass balance can be varied from as low as 1 to as high as 3 along with changing the voltage windows accordingly to suit different applications.

8.3.2. *Full Na-ion cell cycling: curves and stability*

Figure 8.6(a) shows a typical full cell cycling profile at 30°C when cycled at ±C/5 using voltage limits of 4.2–1.0 V using CC/CV (constant current/constant voltage) charging and CC discharging protocol. Under these conditions, a Faradion full Na-ion cell can deliver a cathode specific capacity of 128 mAh/g at an average cell voltage of 3.1 V, resulting in cathode specific energy density of 400 $Wh/kg_{cathode}$.

Figure 8.6(b) presents full cell cycling profiles if cycled between 4.3 and 1.0 V at ±C/7 also at 30°C. Using this expanded voltage window, the full cell can deliver 153 mAh/g at average voltage close to 3.2 V yielding almost 500 $Wh/kg_{cathode}$. As demonstrated later, such high specific energy density values are similar to commercial Li-ion cathodes, indicating that switching from Li-ion to Na-ion batteries would not necessarily mean sacrificing energy density.

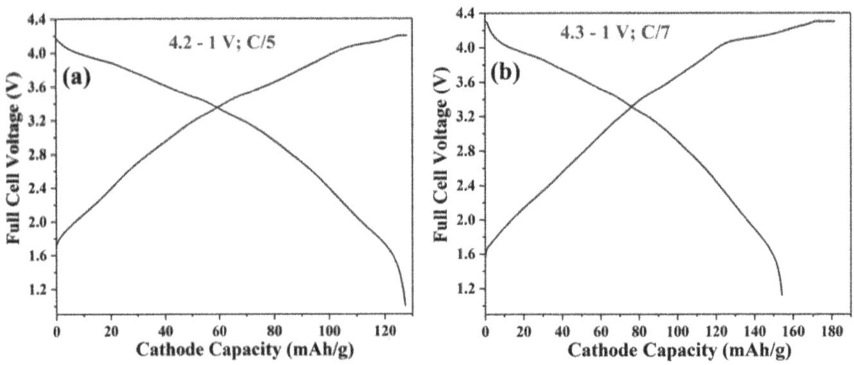

Figure 8.6. *Faradion full Na-ion cell voltage profile at 10 mAh scale. For a color version of this figure, see www.iste.co.uk/monconduit/batteries.zip*

Faradion full cells cycle with low capacity fade over thousands of cycles at 30°C. Latest optimization experiments have resulted in significant improvements to this cycling performance. As shown in Figure 8.7(a), after 750 cycles between voltage limits of 4.2 and 1.0 V at ±C/5, the cell retains 80% of its initial discharge capacities with stable coulombic efficiencies between 99.9 and 100.0%. Even at higher charge–discharge rates, the cycle life is excellent. Cycle life can be further enhanced if full cells are de-rated. For example (refer to Figure 8.7 (b)), if cycled between 4.0 and 1.0 V at ±C/5, Faradion full cells suffer only 7% capacity fade over 900 cycles (refer to Figure 8.7(c)). This translates to an extrapolated cycle life of over 3,500 cycles.

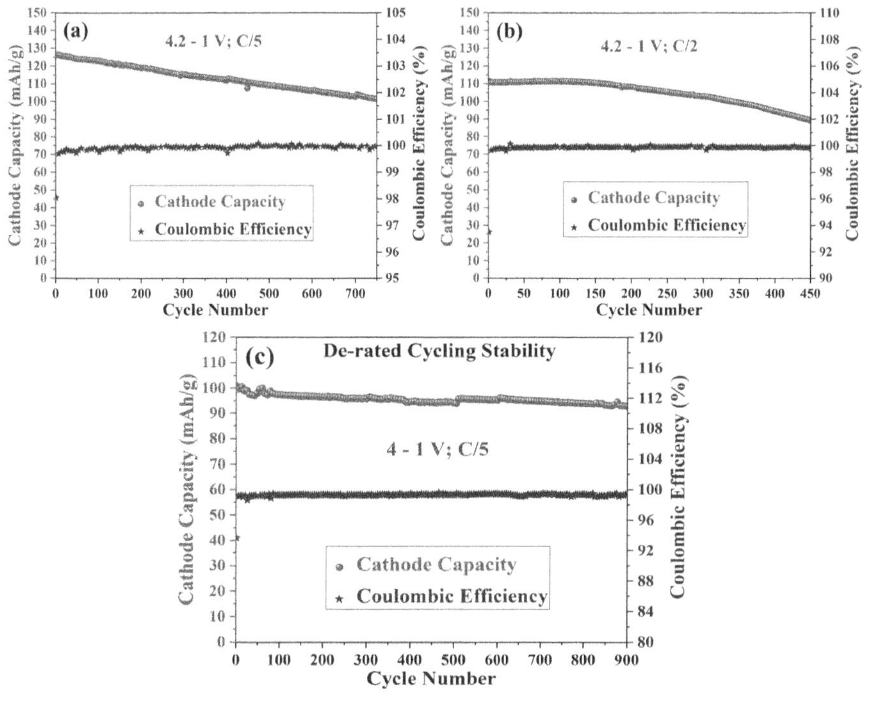

Figure 8.7. *Faradion full Na-ion cell cycling data at (a) C/5 at 4.2–1.0 V, (b) C/2 at 4.2–1.0 V and (c) C/5 at de-rated 4.0–1.0 V. For a color version of this figure, see www.iste.co.uk/monconduit/batteries.zip*

8.3.3. *Rate capability*

The Faradion cell design enables versatile performance at various discharge rates. Figure 8.8 shows the discharge voltage profiles at various rates from C/5 to 10 C (charging being performed at C/5 in all cases) using voltage limits of 4.2–1.0 V.

Even though the Faradion cell design is geared for energy applications, as opposed to power applications, Faradion full cells can still retain 91% of the rated capacities at 1 C discharge relative to the capacities obtained at C/5.

For many energy applications, this 1 C response is more than sufficient. For even faster rates, capacity retention of 83% at 3 C (20-min response) or 75% at 5 C (12-min response) can still be obtained. Figure 8.8 demonstrates that despite being designed for energy applications, Faradion energy cells can respond at a rapid 10 C rate (6-min response) albeit with a lower deliverable capacity retention. These results indicate that Faradion cells can be utilized for energy applications requiring responses across various time scales.

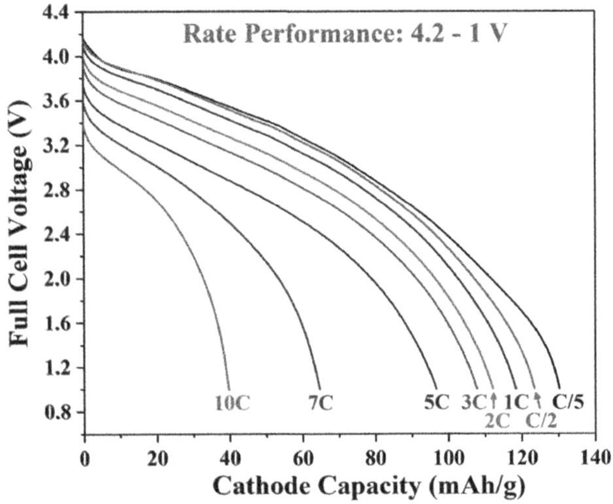

Figure 8.8. *Faradion full Na-ion cell rate capability for discharge at various rates (charging being conducted at C/5 for all discharge rates). For a color version of this figure, see www.iste.co.uk/monconduit/batteries.zip*

8.3.4. *Temperature studies*

For global application of Faradion Na-ion cells, it is imperative that the cell chemistry must perform in different climates and regions. As an important step toward addressing this, Faradion has undertaken extensive studies on how temperature affects cycling performance. As shown in Figure 8.9, Faradion Na-ion cells can cycle in an extremely stable fashion over 100 cycles at low temperatures such as –20°C as well as at quite high temperatures of 60°C. Lower cell capacities obtained at –20°C are expected for any battery system, as reaction kinetics become slower at such sub-zero temperatures. Despite this, it is encouraging to note that even at –20°C, 76% of the capacity delivered at 30°C can be obtained.

It may also be seen from Figure 8.9 that performance at 60°C is not compromised. At such high temperatures, reaction kinetics significantly increase (including those of undesirable side reactions such as electrolyte decomposition); the fact that the cycling performance is so stable even at 60°C points to the inherent stability of Faradion's Na-ion chemistry and cell design.

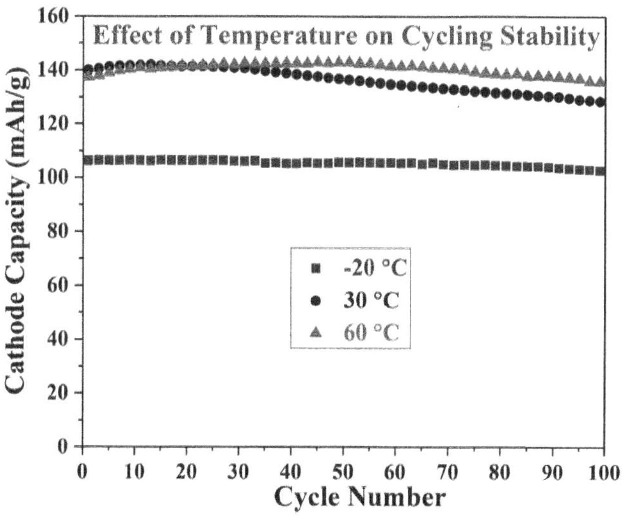

Figure 8.9. *Faradion full Na-ion cell temperature performance at 30, 60 and –20°C. For a color version of this figure, see www.iste.co.uk/monconduit/batteries.zip*

8.3.5. *Three-electrode cell studies*

A three-electrode cell can provide invaluable information about the individual behavior of the cathode, anode and also the full cell *in situ* without disturbing the full cell's cycling performance. Such knowledge can reveal whether cathode/anode potentials are too high/too low, thus informing how best to modify cell balance by changing the cathode:anode mass ratio. Especially for the hard carbon anode, too low potentials might incite plating, hence good cell design must incorporate a safety factor and ensure the hard carbon anode is not too close to 0 V versus Na/Na^+ at the fully charged state. Importantly, three-electrode cells circumvent the dangers of using Na metal as the counter and reference electrode (as occurs for half-cells) – the high polarization of Na metal can significantly affect evaluation of the working electrode in half-cells (Rudola *et al.* 2014). Hence, three-electrode electrochemical evaluation can provide insights half-cell evaluation cannot. Faradion has developed its own three-electrode pouch cell design that is routinely used to study and evaluate full cell performance.

As shown in Figure 8.10, the three-electrode cell utilizes two layers of separators with some overhang where a piece of Na metal can be situated via a third Al tab to form the reference electrode. This design has been found to be stable for several hundred charge–discharge cycles, with no adverse reactions being observed on the Na reference metal, even after 1 year of cell fabrication.

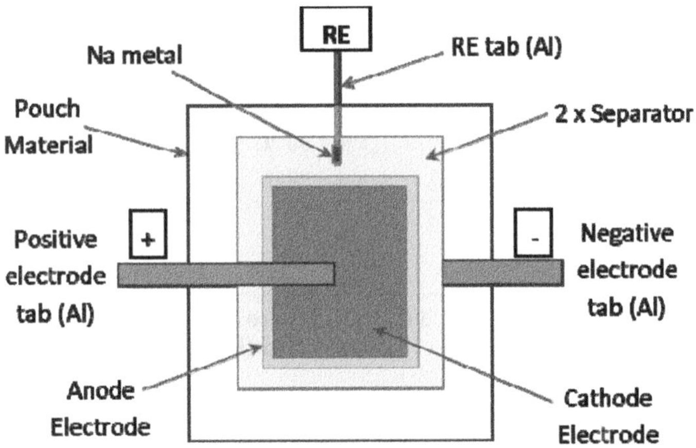

Figure 8.10. *Schematic of Faradion's three-electrode cell design*

Figure 8.11 presents the three-electrode cycling curves of a full cell cycled between voltage limits of 4.2–1.0 V at ±C/5 (similar conditions to that of the full cell shown in Figure 8.6(a)). Good reproducibility between two-electrode and three-electrode cells was obtained, along with results consistent with those of half-cells in terms of cathode (124 mAh/g) and anode (237 mAh/g) reversible capacities – the fact that the values in full cells are correspondingly lower than those from the half-cells can be explained after taking into account the first cycle coulombic inefficiencies of around 15% seen typically in Faradion full Na-ion cells.

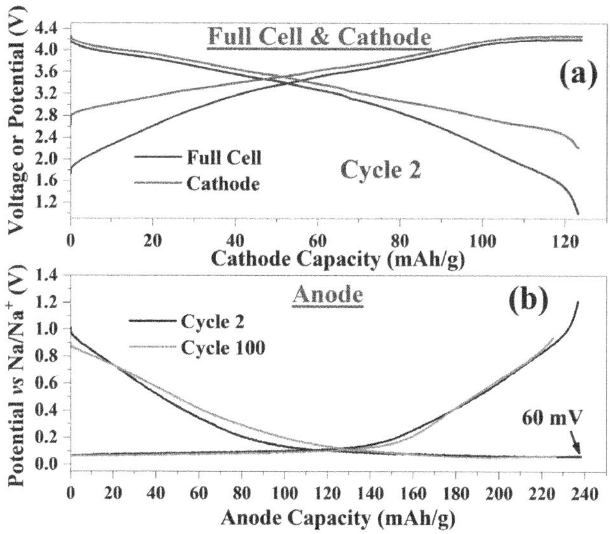

Figure 8.11. *Three-electrode cycling curves of a typical Faradion cell showing (a) cathode and full cell's cycling graphs for the second cycle and (b) anode cycling profiles at cycle 2 and 100. For a color version of this figure, see www.iste.co.uk/monconduit/batteries.zip*

As shown by Figure 8.11(b), the anode potentials at the fully charged state were around 60 mV versus Na. Such a favorably high anode potential at the fully charged state signifies that sodium plating will not occur in Faradion cells. Moreover, the anode potentials did not slip to lower values, even after 100 cycles, confirming the stable operation of Faradion cells. The three-electrode data also point to the cathode limited cell design – underutilization of the anode does negatively impact the full cell's energy density but is necessary from practical considerations, as it safeguards against sodium plating.

8.4. Safety and zero energy storage and transportation

An important safety advantage of Faradion's Na-ion technology over traditional Li-ion cells is the ability to store and transport cells at 0 V (i.e. under a condition of zero energy storage), as will be explained in further detail below. In addition, it is also important that the cells remain safe while in use, or if used incorrectly. In order to investigate this, Faradion has tested its cells in various ways. Faradion's cells were initially tested using accelerating rate calorimetry (ARC) by the industry experts

Thermal Hazard Technology (THT, Milton Keynes, UK). ARC is an effective adiabatic method of measuring the thermal stability of fully charged cells, and is directly related to the cell chemistry, allowing the investigation of a number of thermal degradation processes, such as separator melting, SEI breakdown, self-heating and thermal runaway. The tests simulate a worst-case adiabatic environment, which is similar to the conditions experienced by a cell in the middle of a large battery pack; in both cases, heat from the cell cannot escape to the surroundings. From a safety perspective, the most important outputs from this testing are the self-heating onset temperature (the temperature above which cells can undergo self-heating) and the maximum self-heating rate (the speed at which the energy is released).

Faradion pouch cells were provided to THT in a fully charged state, and these were tested alongside commercially available lithium cobalt oxide (LCO) and lithium iron phosphate (LFP) pouch cells. The results of the testing showed that the LCO cell began to undergo self-heating at 90°C, whereas the LFP cell showed a higher onset temperature of 100°C. The Faradion cells did not begin to react until 150°C, suggesting a much higher thermal stability. In addition to the temperature at which the cells become unsafe, the rate at which the energy is released is also important. After the onset of self-heating, the LCO cell reached a maximum rate of ~4,000°C/min, indicating an extremely fast release of energy. The LFP cell, however, was shown to be much safer, giving a maximum reaction rate of ~150°C/min. The Faradion cells showed an even lower rate of ~52°C/min, making them the safest of the three cell types tested.

Faradion has also had various cell types tested under abuse conditions. Note that 10 Ah and 2 Ah pouch cells were constructed within projects part-funded by Innovate UK, the UK's innovation agency, and these were tested by Warwick Manufacturing Group, following test protocols defined in *IEC 62660-2–Secondary lithium-ion cells for the propulsion of electric road vehicles–Part 2: Reliability and abuse testing*. Prior to abuse testing, the cells were formed, degassed and cycled 20 times. The tests performed are summarized in Table 8.1, the details and results of which are described further here. The external short test involved adjusting the state of charge (SoC) of the cell to 100%, before using a low-resistance connection to short-circuit the terminals for 10 min. This resulted in a minor expansion of the pouch cell, as the cell temperature rose to a maximum of 96°C. No violent reaction took place, and the subsequent contraction of the pouch suggests that the expansion was largely due to heating, rather than decomposition of cell components leading to gassing.

Test (fully charged cells)	Result	Max. temperature (°C)	Comments
External short circuit	No fire, no explosion	96	Minor pouch expansion
Overcharge	No fire, no explosion	24	Testing stopped at 2× rated capacity
Hot box (130°C)	No fire, no explosion	127	Some cell voltage drop (recoverable)
Flat crush	No fire, no explosion	27	Negligible V and T change
Edge-on crush	No fire, no explosion	20	Negligible V and T change
Nail penetration	No fire, no explosion	41	Internal short achieved

Table 8.1. *Safety/abuse testing of Faradion 10 Ah Na-ion pouch cells (IEC 62660-2 Testing Protocol)*

The overcharge test involved adjusting the cell to 100% SoC, followed by further charging at a rate of 1 C. No violent reaction was observed, with the cell reaching a maximum temperature of 24°C. The test was stopped when the cell capacity reached the pass criterion of twice its rated capacity.

The high temperature endurance test (commonly known as a hot box test) involved adjusting the SoC of the cell to 100%, before placing it in an oven and raising the temperature at 5 K/min to 130°C. This temperature was then maintained for a period of 30 min. During this time, a small voltage drop was observed temporarily, and no violent reaction occurred.

Crush tests were performed in two ways, applying force from a different direction in each case. During the flat crush test, the cell was adjusted to 100% SoC and laid on a flat surface, face-up. A crushing force was then applied, using a round crushing tool, until a force of 1,000 times the weight of the cell was achieved. During the edge-on crush test, the cell was adjusted to 100% SoC and the cell was placed on its side, between two plates. A crushing force was then applied to the edge of the cell, using a bar-shaped crushing tool, until the cell deformed by >15% of its original width. In both cases, no violent reaction occurred and there were negligible changes in temperature or voltage.

During the nail penetration test, the cell was adjusted to 100% SoC and a 3 mm diameter steel nail was driven through the cell. This was held in place until the cell temperature returned to ambient. No violent reaction was observed, and the maximum temperature reached was 41°C. In summary, the results obtained so far have been very positive, indicating the excellent safety characteristics of this technology.

An extremely important commercial and safety advantage of Faradion Na-ion battery technology is the ability to store and transport its cells at 0 V. This is a condition where the device has zero electrochemical energy storage. Under normal storage conditions, a physical short circuit is present between the positive and negative tabs. This renders the cell perfectly safe for transportation. Faradion has taken the opportunity to patent this concept (Barker and Wright 2017). This crucial and intrinsic safety aspect of the Na-ion battery chemistry is possible due to the use of aluminum current collector at both poles of the cell. By contrast, in order to minimize the possibility of Cu dissolution from the negative electrode current collectors, it is well established that graphite anode Li-ion cells must be stored and transported at a finite SoC, typically around 30–40%. This condition still represents a significant safety hazard, particularly for air transportation purposes. It is fair to assume that restrictions on air transportation of Li-ion batteries, particularly on passenger flights, will increase significantly in the near future.

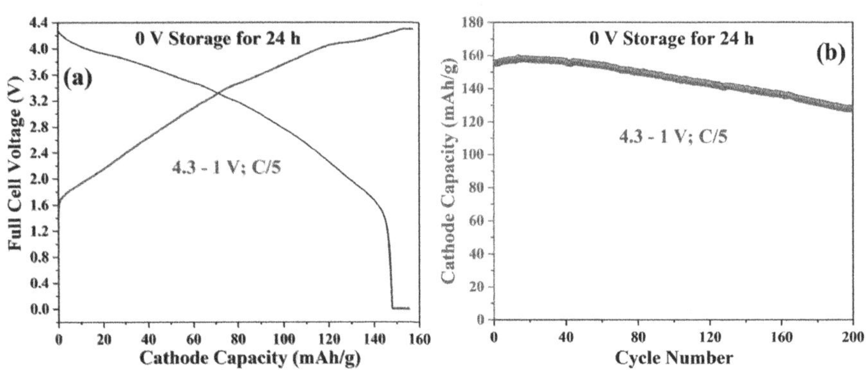

Figure 8.12. *Effect of 0 V short-circuit at the end of each discharge cycle for typical Faradion Na-ion cells. (a) Cell voltage profile and (b) cycling stability at ±C/5 cycled between 4.3 and 0 V at 30°C. After each discharge cycle, the cell was potentiostatically held at 0 V for 24 h. For a color version of this figure, see www.iste.co.uk/monconduit/batteries.zip*

In Figure 8.12(a), we show the voltage profile for a typical Faradion Na-ion cell that is held potentiostatically at a cell voltage of 0 V for 24 h at the end of every discharge process. This test protocol is used to represent the typical conditions that may be experienced by a Na-ion cell during a period of transportation. At the 0 V condition, the cell has zero electrochemical energy storage capacity and is effectively short-circuited. Note that the extended period of short circuit actually allows higher active material utilizations to be achieved. The cathode reversible specific capacity is now over 150 mAh/g, which approaches the figure achieved when the material is tested under extremely slow constant current charge/discharge rates.

Figure 8.12(b) represents the cycling behavior for this cell. From inspection we can conclude that the 0 V voltage hold has a negligible impact on the rate of capacity fade during cycling. Other tests (not shown here) have been conducted using extended 0 V hold periods of 48 h and 96 h with no detrimental impact on the cell performance. Indeed, the longest 0 V hold test period – that of over 6 months at room temperature – again had no negative impact on the subsequent cell performance. Indeed, somewhat surprisingly, there was negligible increase in the cell impedance over this extended storage period.

We believe that the 0 V capability is an important benefit for non-aqueous electrolyte Na-ion cells. Faradion now stores and ships all its prototype Na-ion cells with a physical short between the positive and negative terminals. We consider this to be the ultimate safety condition for these kinds of cells, and represents an important differentiating safety factor over current commercial graphite-based Li-ion cells.

8.5. Scale-up and prototyping

Faradion has successfully scaled its cell design across various scales from 10 mAh to 12 Ah. As an example of cell performance at large scale, Figure 8.13 displays ±C/5 cycling of a prototype 1 Ah cell at 30°C: extremely stable cycling was observed over 300 cycles at high coulombic efficiencies around 100%. The specific energy density for such a cell was 100 Wh/kg (based on total cell weight). It should be stressed that these are unoptimized cells and significantly higher specific energies have been realized at this scale already. Moreover, the specific energy of over 140 Wh/kg have been generated in 12 Ah pouch cells where there is a much-improved packing efficiency. Data on these 12 Ah cells are shown later in this chapter.

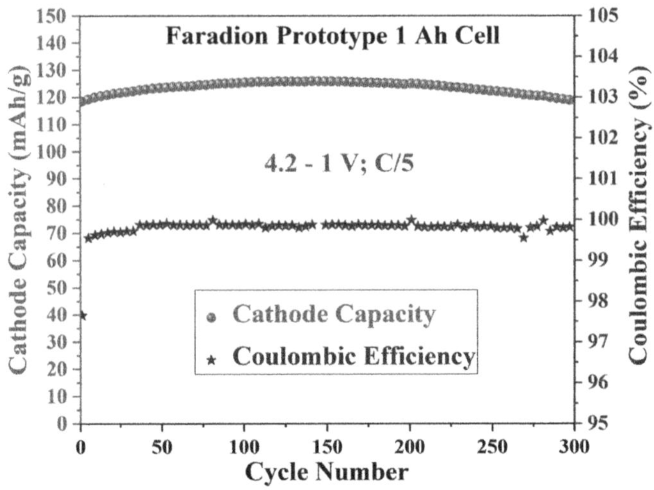

Figure 8.13. *Faradion 1 Ah Na-ion cell cycling performance. For a color version of this figure, see www.iste.co.uk/monconduit/batteries.zip*

8.6. Demonstrators: stacks and packs

Working with its commercial partners, Faradion has constructed well over 40 kWh of prototype Na-ion pouch cells for various demonstrator purposes. One of its first demonstrator packs was for an E-bike application and used multiple 3 Ah (nominal energy, 9 Wh) pouch cells. This culminated in the production of a 400 Wh battery pack (see Figure 8.14). This figure shows the completed battery pack, including the application housing, electronics and battery management system (BMS). The fully packaged battery system, including the control electronics, BMS and packaging, generated a specific energy performance of >80 Wh/kg and an estimated E-bike range in excess of 35 km. It should be stressed that the battery pack was oversized for demonstrational purposes where the emphasis was on successful operation, rather than space-saving considerations. The E-bike and battery pack are shown in operation in Figure 8.15. In a separate study, a 750 Wh E-scooter demonstration pack was also completed.

Figure 8.14. *Faradion's E-bike battery pack modules. For a color version of this figure, see www.iste.co.uk/monconduit/batteries.zip*

Figure 8.15. *Faradion's Na-ion powered E-bike in operation*

The current generation of Faradion prototype Na-ion pouch cells are generally in the range of 12–25 Ah capacity, which translates to about 35–75 Wh in energy terms. The specific energy of these cells is >140 Wh/kg, a performance that is comparable with some commercial graphite-based Li-ion batteries. An example of a 12 Ah Faradion pouch cell is shown in Figure 8.16(a). These cells utilize some of Faradion's latest cell fabrication techniques and also leverage the expertise of Faradion's commercial partners in large-scale cell manufacture.

Examples of performance obtained for 12 Ah pouch cell stacks are shown in Figure 8.16(b) and (c) at different charging and discharging rates. Figure 8.16(b) presents cycling performance at ±C/5 for three different voltage windows: 4.2–1.0 V and two different de-rated voltage windows of 4.1–1.0 V and 4.0–1.0 V. It can be seen that all three cells cycle in an extremely stable fashion over 250 cycles. Such excellent cycling stability was also seen for cells cycled between the aforementioned voltage windows but charged/discharged at a faster C/3 rate as displayed in Figure 8.16(c): cycling performance at ±C/3 was virtually indistinguishable from that at ±C/5 in terms of cycling stability as well as obtained discharge capacities.

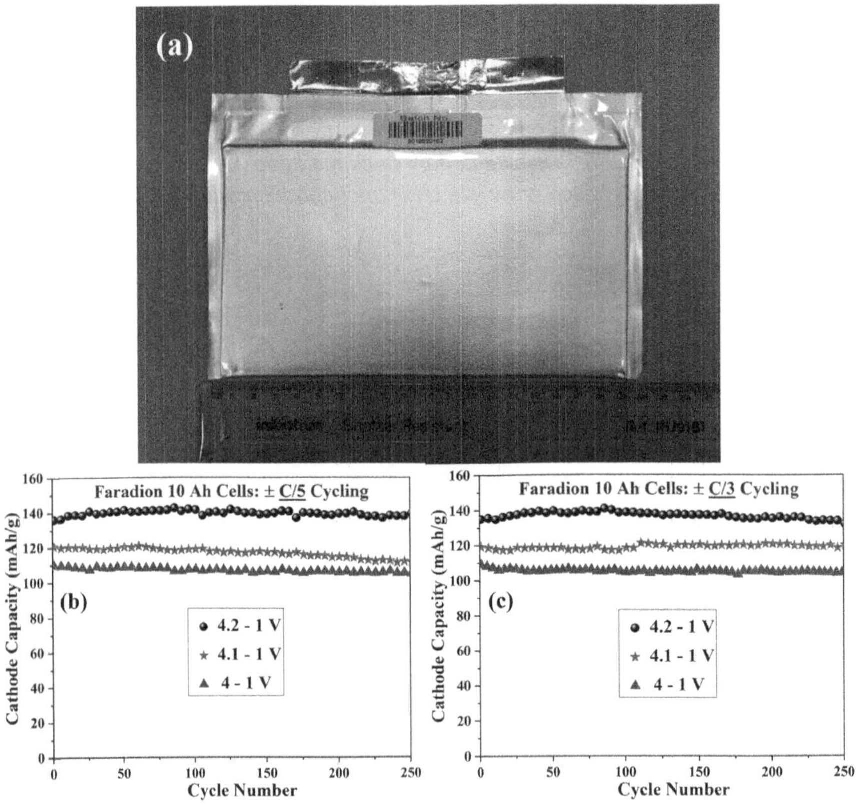

Figure 8.16. *Faradion 12 Ah Na-ion pouch cells. (a) Picture of such a cell containing Faradion's proprietary cell chemistry. The cycling performance of such 12 Ah cells at (b) ±C/5 and (c) ±C/3 at 4.2 and 1.0 V, 4.1 and 1.0 V and 4.0 and 1.0 V (data courtesy of Shangdong New Energy Technology Co. Ltd). For a color version of this figure, see www.iste.co.uk/monconduit/batteries.zip*

These results demonstrate the excellent scalability of Faradion's pouch cells. Future pouch cell production will target much higher cell capacities to maximize the packing efficiency of such cells and thus boost the specific energy performance. To meet future demand, Faradion is also planning to complement its pouch cell manufacturing capability with both prismatic and cylindrical format cells.

Faradion's technology roadmap indicates that within 2020, the specific energy of similarly sized pouch cells will reach in excess of 190 Wh/kg. This progress will be achieved by optimization of the manufacturing process and without a significant change in the cell chemistry. Example of this approach may include (1) using electrodes with higher areal capacity (>3 mAh/cm^2), (2) reducing the electrolyte loading in g/Ah and (3) minimizing the portion of "inactive" elements in the cell construction (i.e. the use of thinner separators, current collectors etc). The work underpinning this roadmap is currently underway in our laboratories in close collaboration with our commercialization partners. When complete we expect to be able to penetrate new markets that demand very high energy densities, such as full EV automotive applications. This will greatly extend our commercial offering.

Future demonstrator packs will include those for low-cost electric vehicles (LCEVs) – particularly relevant to some cost sensitive Asian markets – and UPS back-up systems. As mentioned before, to date, a combined prototype production of >40 kWh of Na-ion pouch cells has been produced by the company.

8.7. Business and IP strategy

Faradion's licensees are currently producing Na-ion pouch cells at a prototype level, with a specific energy of >140 Wh/kg. This is similar to the performance of commercial $LiFePO_4$-based Li-ion cells. Coupled with other favorable performance characteristics such as rate capability and cycle life, we believe that this level of specific energy is already at a level that is commercially viable for many potential applications such as stationary storage, telecoms, UPS and grid back-up.

The 2017 worldwide rechargeable battery market has been estimated to be >$75Bn (Pillot 2017). This market is currently dominated by Pb-acid and Li-ion battery technologies and includes large format, light industrial, military, Starting, Lighting and Ignition (SLI) and Electric Vehicle (xEV) applications (including Battery Electric Vehicles (BEV), Plug-in Hybrid Electric Vehicles (PHEV) and Hybrid Electric Vehicles (HEV)). Moving forward, Li-ion technology is expected to be become increasingly attractive compared to Pb-acid technology. Faradion believes that with the successful implementation of its technology and commercial roadmaps, its licensees will capture a significant portion of the world's rechargeable battery market. It should be possible to penetrate the markets currently dominated by both the Pb-acid and Li-ion technologies. Moreover, we believe that the performance and safety attributes of Faradion's unique Na-ion cell chemistry will mean it will be able to establish new markets and applications.

The initial target market for Faradion's technology will be for large-format energy storage applications where a combination of low materials cost, improved safety and transportation considerations are all at a premium. These applications will include those currently dominated by legacy Pb-acid battery technologies such as UPS, grid back-up, forklifts and telecoms. The current market value for this segment is estimated at >$26Bn per annum and rising. In addition, the wide operating temperature range (–30°C to +60°C) and good power capability characteristics mean that variants of the technology will may also be suitable for 12 V SLI applications, again in direct competition with the existing dominant Pb-acid battery solutions. Currently, the SLI market is amounting to >$25Bn per annum (Pillot 2017). Relatively minor modifications to the underlying cell chemistry means it will also be directed toward other potential applications, including military and automotive. Also, future military and medical applications may prove valuable, particularly leveraging the 0 V transportation and storage capability of Faradion's Na-ion chemistry – a distinct commercial and cost advantage over incumbent Li-ion technology.

We also anticipate significant demand for Faradion Na-ion cells in applications that may take advantage of the 0 V (zero-energy) storage and transportation capability (Barker and Wright 2017) of this technology. As mentioned previously, Faradion stores and ships all of its Na-ion cells and battery packs in a physically shorted state to ensure maximum safety benefit. This cannot be achieved using graphite-based Li-ion cells.

As we have previously noted, as a distinct advantage, Faradion's Na-ion technology may be fabricated on existing Li-ion manufacturing lines with minimum modification, meaning that potential cell manufactures won't incur a cost penalty to change to this cell chemistry. Faradion will offer its IP and associated know-how for the manufacture of its batteries in the various industry-standard formats: pouch, prismatic and cylindrical.

Faradion's patent portfolio covers many families of active materials (both cathode and anode), with different properties, so the cell chemistry can be tailored to the application, e.g. high-energy applications (such as grid storage), high-power applications (such as SLI batteries and power tools), high-energy and high-power applications (such as xEV batteries), or niche applications, such as in medical implants, in which a deep discharge must not be detrimental to the cell, or oil-drilling, where cells need to withstand high exposure and operational temperatures.

Faradion's current intellectual property portfolio covers active materials such as polyanions, sulfates, silicates, oxo-metallates, layered oxides and carbonaceous materials. On the cathode active material front, much recent work has been focused on mixed-phase nickelates, containing more than one layered oxide structure (e.g. O3-, P2-, P3-type). These are particularly attractive, due to their high energy, low cost, minimal use of nickel (<18 wt.% Ni in active cathode) and the fact that expensive materials such as cobalt or vanadium are not used. These typically comprise O3- and P2-type phases, which give rise to a good combination of high-energy and high-power characteristics, respectively.

Faradion has also developed an extensive IP portfolio covering all aspects of its proprietary Na-ion battery technology. On that basis, the IP portfolio coverage includes cathode and anode active materials (and associated low-cost preparative methods), electrolyte formulations, alternate cell configurations, designs for battery packs and BMS, the general use of Na-ion batteries and their safe storage and transportation. These patent applications are filed in many of the important territories worldwide, including Europe, the United States, Japan, China, India, Australia and South Korea. As the company moves forward with its developments, it will continue to file and refine its IP strategy to remain at the forefront of this energy storage sector. Table 8.2 lists some of Faradion's key issued U.S. patents.

#	US patent number	Title
1	US10263254	Tin-containing compounds
2	US10205168	Sodium transition metal silicates
3	US10196280	Compositions containing doped nickelate compounds
4	US10170212	Metal-containing compounds
5	US10115966	Metallate electrodes
6	US10050271	Metal-containing compounds
7	US9917307	Doped nickelate compounds
8	US9774035	Doped nickelate compounds
9	US9761863	Doped nickelate compounds
10	US9660263	Layered oxide materials for batteries
11	US9608269	Condensed polyanion electrode

Table 8.2. *A summary of some of Faradion's key issued U.S. patents*

8.8. Cost analysis

A major competitive advantage for Faradion's Na-ion technology will be its low materials costs. The company has undertaken a detailed cost analysis of its current cell chemistry using industry standard models (e.g. Argonne National Labs, BatPac Model). This analysis indicates that the materials cost per kWh at a manufacturing scale will be around 30% less than Li-ion technology using the $LiFePO_4$ active material. This cost estimate is based on the production of large format pouch cells of around 10–12 Ah cell capacity. This analysis makes Faradion's Na-ion technology cost competitive with even the most inexpensive Li-ion technologies, such as $LiFePO_4$.

The calculated cost breakdown for the various components for a Faradion 12 Ah pouch cell is as follows: cathode active material = 28%, anode active material = 26%, current collectors = 13%, electrolyte = 12%, separator = 3% and miscellaneous components to a fabricate pouch cell type battery = 18%. These numbers are very similar with the cost analysis conducted by Peters *et al.* (2016) for a cylindrical 18650 Na-ion battery, which is also based on a hypothetical Faradion cell chemistry. Further cost reductions, particularly focused on the use of inexpensive active materials, as well as improvements to the deliverable energy, allow the company to reliably predict that the materials cost per kWh will continue to be very competitively positioned against lithium cell chemistries. This will allow greater market penetration and encourage the replacement of legacy Pb-acid technology in even the most cost-sensitive applications.

8.9. Future developments

Faradion will continue to improve and optimize its Na-ion battery chemistry. For instance, the company expects its next generation active materials will comfortably outperform the current versions. Due to its early stage of development, Na-ion technology has often been considered a battery technology with a somewhat modest specific energy performance. Based on a 3.2 V average discharge voltage, and varying the reversible specific capacities for the active materials, the specific energy for a full Faradion Na-ion cell may be calculated. For instance, a calculation for a 16 Ah battery stack, based on a pouch cell construction designed for energy applications (i.e. using relatively thick electrodes with high areal capacity, >2.5 mAh/cm^2), was performed. Assuming a reversible cathode specific capacity of 140 mAh/g and an anode active material operating at 250 mAh/g, a well-balanced Na-ion cell should generate a specific energy of well over 160 Wh/kg.

The specific energy projections assume the use of double-sided electrodes and a stacked electrode-separator configuration within the pouch design – an electrode format that ensures an excellent packaging efficiency in volume terms. This Na-ion specific energy performance would be considered intermediate between commercial Li-ion cells based on the LFP and LCO cathode active materials. Fairly modest improvements in the active material performance will push the specific energy performance well beyond 170 Wh/kg and the Faradion technology roadmap indicates that specific energies in excess of 190 Wh/kg are achievable within 2020.

Faradion has made great advances in cathode optimization since its inception as evidenced by its 17 patents covering different cathode families. Apart from cathode active material optimization, excellent progress has been made in electrode engineering with ever increasing active material wt% in the electrode (such as 96 wt%), along with enhancements of the tap density of the cathode active material (tap density improvements of 33% have recently been achieved). On the anode side, Faradion has been able to devise its second-generation (Gen 2) anode, which is also hard carbon based but with higher specific capacity performance. For example, when the Gen 2 anode was cycled in sodium half-cells, at a rate approximately four times faster than that shown in Figure 8.17(a), rather than giving a lower capacity, it achieved an increase of ~ 50 mAh/g. Three-electrode full cells confirmed that reversible anode capacity increased from 234 mAh/g for the Gen 1 system to above 300 mAh/g using the Gen 2 anode in full cells (refer to Figure 8.17(b)). Furthermore, as shown in the inset, the anode potential at the fully charged state of the full cell was much higher for the Gen 2 anode as compared to that of the Gen 1 anode (around 80 vs. 60 mV). This indicates that the Gen 2 anode not only delivers higher capacity, but will also be safer as it would be further away from "plating potentials".

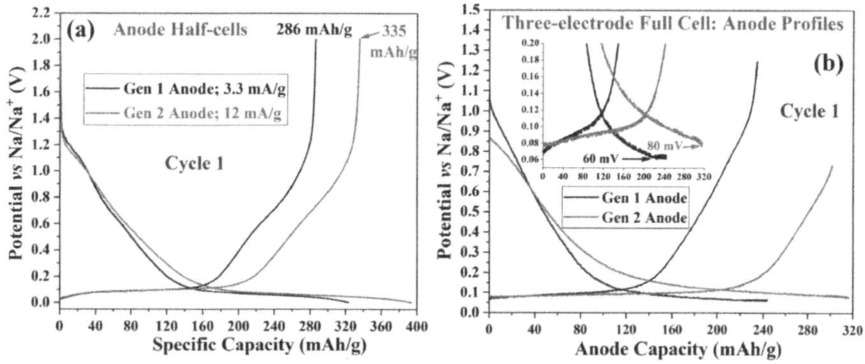

Figure 8.17. *Faradion Gen 2 anode performance in (a) half-cell and (b) three-electrode full cell. For a color version of this figure, see www.iste.co.uk/monconduit/batteries.zip*

The enhanced Gen 2 anode capacity translated to a much higher energy density of the full cell. Figure 8.18 presents the performance of 1 Ah prototype cells using Gen 1 and Gen 2 anodes. As exhibited, the cell using the Gen 2 anode demonstrated a 23% improvement over the specific energy delivered by the Gen 1 system, as well as stable cycling. Attractively, the round-trip energy efficiency (RTEE) of the Gen 2 system was also higher than that of the Gen 1 system (95% vs. 92%). Such high RTEE values are significantly higher than those of Pb-acid cells and even compare favorably against those of Li-ion batteries, indicating that Faradion Na-ion cells would make for economically efficient energy storage options for a range of applications.

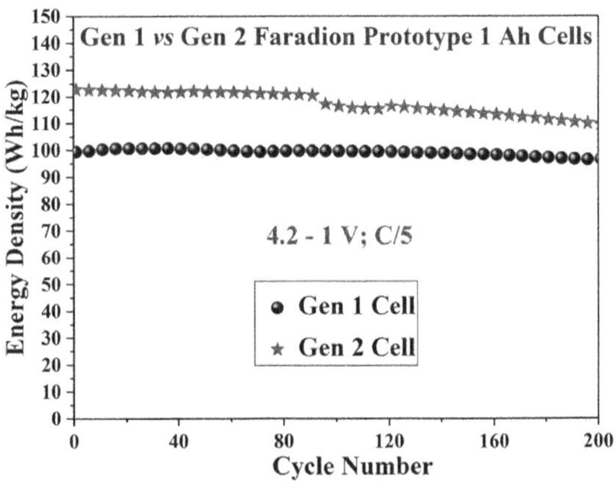

Figure 8.18. *Comparison of cell energy densities obtained between Faradion's unoptimized Gen 1 and Gen 2 sodium-ion 1 Ah prototype cells. For a color version of this figure, see www.iste.co.uk/monconduit/batteries.zip*

As more developments are made in the field of Na-ion technology, it is becoming increasingly clear that this is a viable alternative to Li-ion technology. As such, Na-ion R&D is growing at a rapid rate, with many organizations now working on Na-ion cathode development. In order to show a direct comparison of Faradion's Na-ion cathode technology with other cathode materials, Table 8.3 and Figure 8.19 show the cathode specific energies achieved for a selection of positive active materials, tested in full cells (vs. graphite for Li-ion and vs. hard carbon for Na-ion). This graph was created in order to show a direct comparison of the realistic energy deliverable for different cathode materials. Data were selected based on commercial relevance, e.g. half-cell data were not used, nor were data for pre-sodiated negative electrodes, as these do not give a realistic representation of the reversible energy achievable in a commercial environment.

Cathode active material	Cathode specific capacity (mAh/g)	Average discharge voltage in full cell (V)	Cathode specific energy (Wh/kg)	Reference
Faradion O3/P2	153	3.2	490	Faradion data
$NaVPO_4F$	82	3.7	303	Barker *et al.* (2003)
$Na_{0.9}[Cu_{0.22}Fe_{0.30}Mn_{0.48}]O_2$	100	3.2	320	Li *et al.* (2016)
$NaNi_{0.5}Mn_{0.3}Ti_{0.2}O_2$	106	3	318	Wang *et al.* (2016)
$NaNi_{0.5}Ti_{0.5}O_2$	82	3	246	Wang *et al.* (2015)
$Na[Ni_{0.6}Co_{0.2}Mn_{0.2}]O_2$	120	2.5	300	Hwang *et al.* (2017)
$Na_{7/9}Cu_{2/9}Fe_{1/9}Mn_{2/3}O_2$	89	3.6	320	Li *et al.* (2015)
$NaFe_{0.4}Mn_{0.3}Ni_{0.3}O_2$	120	3	360	Kuze *et al.* (2013)
$LiFePO_4$	150	3.2	480	Faradion data

Table 8.3. *Comparison of cathode specific discharge capacities and average discharge voltages in full Na-ion cells, along with corresponding cathode specific energies*

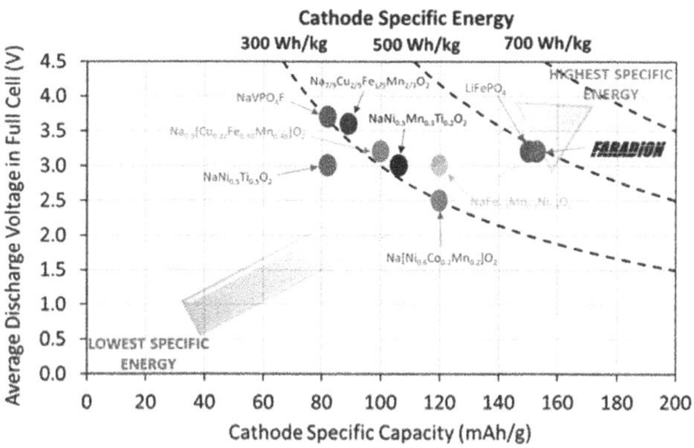

Figure 8.19. *Comparison of cathode-specific discharge capacities and average discharge voltages in full Na-ion cells. The corresponding cathode specific energies are shown by the dotted lines. For a color version of this figure, see www.iste.co.uk/monconduit/batteries.zip*

8.10. Conclusion

Faradion's unique non-aqueous electrolyte Na-ion battery technology is based on a layered nickelate cathode coupled to a hard carbon anode. This cell chemistry and its manufacturing and operation are protected by an extensive intellectual property portfolio. Advantageously, the Na-ion prototype cells are fabricated on existing Li-ion manufacturing lines meaning its adoption by the industry should represent a low-cost option. The outstanding performance of its prototype pouch cells means that the company expects the initial target markets to be energy-rich applications (for example, ESS, UPS, and stationary), although minor modifications to the cell chemistry and cell design will also allow power applications to be addressed in the near future.

Working closely with its licensees, Faradion has successfully scaled up its proprietary Na-ion battery technology such that current 12 Ah prototype pouch cells are delivering a specific energy performance of >140 Wh/kg. This is the highest specific energy for any known non-aqueous Na-ion battery technology. In addition, the company's technology roadmap suggests that a specific energy of >190 Wh/kg will be achieved within 2020 from a fully optimized pouch cell design. This will represent outstanding specific energy performance, directly competitive with some industry standard Li-ion cell chemistries. This is a significant achievement from a battery technology that began serious development around 25 years later than the commercially successful Li-ion technology. It should be noted that due to its similarities, the emerging Na-ion battery industry has learnt significant lessons from the development of Li-ion industry and this has helped to shorten the developmental cycle.

From a commercial standpoint, we believe that when coupled with other important performance attributes such as excellent cycle life, good rate capability and wide operational temperature range, the Faradion Na-ion battery technology will soon be able to penetrate both new and existing commercial markets. It should be recognized that the Faradion Na-ion cell chemistry also demonstrates some other important attributes such as low materials cost, 0 V storage and transportation capability and outstanding intrinsic safety.

In the coming years, Faradion is committed to further improving and optimizing the performance of its world-leading Na-ion battery technology. Recent efforts into developing Faradion's technology have already shown a significant boost to cell performance, most notably an enhancement of the cell energy density and cycle life. In conclusion, we believe the future is very bright for Faradion's Na-ion battery technology.

8.11. Acknowledgments

The authors would like to thank Faradion Limited for kind permission to publish this chapter. We would also like to acknowledge the important contributions from the Faradion R&D team and from Haldor-Topsoe A/S (SEM micrographs) and Shangdong New Energy Technology Co. Ltd. (cycling data for some of the Na-ion prototype cells).

8.12. References

Barker, J. and Heap, R. (2013a). Doped nickelate compounds. Faradion Limited. US Patent 9,917,307.

Barker, J. and Heap, R. (2013b). Doped nickelate compounds. Faradion Limited, US Patent Application 20150243983.

Barker, J. and Heap, R. (2015a). Doped nickelate materials. Faradion Limited, US Patent Application 20170174527.

Barker, J. and Heap, R. (2015b). Doped nickelate compounds. Faradion Limited, US Patent Application 20160329564.

Barker, J., Saidi, M.Y., and Swoyer J.L. (2003). A sodium-ion cell based on the fluorophosphate compound $NaVPO_4F$. *Electrochem. Solid-State Lett.*, 6(1), A1–A4.

Barker, J. and Wright, C. (2017). Storage and/or transportation of sodium-ion cells, Faradion Limited. US Patent Application 20170237270.

Delmas, C., Fouassier, C., and Hagenmuller, P. (1980). Structural classification and properties of the layered oxides. *Physica B+C*, 99(1–4), 81–85.

Hou, H., Qiu, X., Wei, W., Zhang, Y., and Ji, X. (2017). Carbon anode materials for advanced sodium-ion batteries. *Advanced Energy Materials*, 7, 1602898.

Hwang, J.-Y., Myung, S.-T., Choi, J.U., and Yoon, C.S., Yashiro, H., and Sun, Y.-K. (2017). Resolving the degradation pathways of the O3-type layered oxide cathode surface through the nano-scale aluminum oxide coating for high-energy density sodium-ion batteries. *J. Mater. Chem. A*, 5, 23671–23680.

Kuze, S., Kageura, J.-I., Matsumoto, S., Nakayama, T., Makidera, M., Saka, M., Yamaguchi, T., Yamamoto, T., and Nakane, K. (2013). Development of a sodium ion secondary battery (translated from R&D Report, "SUMITOMO KAGAKU"). Sumitomo Chemical Co., Ltd., Available at: https://www.sumitomo-chem.co.jp/english/rd/report/files/docs/2013E_3.pdf

Li, Y., Hu, Y., Li, H., Chen, L., and Huang, X. (2016). A superior low-cost amorphous carbon anode made from pitch and lignin for sodium-ion batteries. *J. Mater. Chem. A*, 4, 96–104.

Li, Y., Yang, Z., Xu, S., Mu, L., Gu, L., Hu, Y.-S., Li, H., and Chen, L. (2015). Air-stable copper-based P2-$Na_{7/9}Cu_{2/9}Fe_{1/9}Mn_{2/3}O_2$ as a new positive electrode material for sodium-ion batteries. *Adv. Sci.*, 2, 1500031.

Peters, J., Buchholz, D., Passerini, S., and Weil, M. (2016). Life cycle analysis of sodium-ion batteries. *Energy Environ. Sci.*, 9, 1744–1751.

Pillot, C. (2017). The rechargeable battery market. *33rd International Battery Seminary and Exhibits*, FL. Avicenne Energy.

Ren, W., Yao, X., Niu, C., Zheng, Z., Zhao, K., An, Q., Wei, Q., Yan, M., Zhang, L., and Mai, L. (2016). Cathodic polarization suppressed sodium-ion full cell with a 3.3V high-voltage. *Nano Energy*, 216–223.

Rudola, A. (2015). Energy storage using sodium-ion batteries. Available at: https://scholarbank.nus.edu.sg/handle/10635/121763.

Rudola, A., Aurbach, D., and Balaya, P. (2014). A new phenomenon in sodium batteries: Voltage step due to solvent interaction. *Electrochemical Communications*, 46, 56–59.

Sayers, R., Barker, J., and Heap R. (2015a). Compositions containing doped nickelate compounds. Faradion Limited, US Patent 10,196,280.

Sayers, R., Barker, J., and Heap, R. (2015b). Compositions containing doped nickelate compounds. Faradion Limited. US Patent Application 20170190595.

Wang, H., Gu, M., Jiang, J., Lai, C., and Ai, X. (2016). An O3-type $NaNi_{0.5}Mn_{0.3}Ti_{0.2}O_2$ compound as new cathode material for room-temperature sodium-ion batteries. *J. Power Sources*, 327, 653–657.

Wang, H., Xiao, Y., Sun, C., Lai, C., and Ai X. (2015). A type of sodium-ion full-cell with a layered $NaNi_{0.5}Ti_{0.5}O_2$ cathode and a pre-sodiated hard carbon anode. *RSC Adv.*, 5, 106519–106522.

List of Authors

Jerry BARKER
Faradion Limited
Sheffield
UK

Adrian BEDA
Université de Haute-Alsace
CNRS
Institut de Science des Matériaux de Mulhouse (IS2M) UMR 7361
Mulhouse
and
University of Strasbourg
France

Fan CHEN
Laboratoire de Réactivité et de Chimie des Solides
CNRS-UMR 7314
University of Picardie Jules Verne
and
RS2E, Réseau Français sur le Stockage Electrochimique de l'Energie
FR CNRS 3459
Amiens
France

Fazlil COOWAR
Faradion Limited
Sheffield
UK

Laurence CROGUENNEC
CNRS
University of Bordeaux
Bordeaux INP
ICMCB UMR 5026
Pessac
and
RS2E, Réseau Français sur le Stockage Electrochimique de l'Energie
FR CNRS 3459
and
ALISTORE-ERI European Research Institute
FR CNRS 3104
Amiens
France

Carolina DEL MAR SAAVEDRA RIOS
Université Grenoble Alpes
CEA LITEN
DEHT - LM
Grenoble
France

Maria FORSYTH
Institute for Frontier Materials
Deakin University
Burwood
Australia

Vincent GABAUDAN
Institut Charles Gerhardt Montpellier
University of Montpellier
CNRS
and
RS2E, Réseau Français sur le Stockage Electrochimique de l'Energie
FR CNRS 3459
Amiens
France

Richard HEAP
Faradion Limited
Sheffield
UK

Patrik JOHANSSON
Dept. of Physics
Chalmers University of Technology
Gothenburg
Sweden
and
ALISTORE-European Research Institute CNRS FR 3104
Amiens
France

Shinichi KOMABA
Department of Applied Chemistry
Tokyo University of Science
Japan

Kei KUBOTA
Department of Applied Chemistry
Tokyo University of Science
Japan

Le Anh MA
Department of Chemistry
Ångström Laboratory
Uppsala University
Sweden

Faezeh MAKHLOOGHIAZAD
Institute for Frontier Materials
Deakin University
Burwood
Australia

Christian MASQUELIER
Laboratoire de Réactivité et de Chimie des Solides
CNRS-UMR 7314
University of Picardie Jules Verne
and
RS2E, Réseau Français sur le Stockage Electrochimique de l'Energie
FR CNRS 3459
and
ALISTORE-European Research Institute CNRS FR 3104
Amiens
France

Camélia MATEI GHIMBEU
Université de Haute-Alsace
CNRS
Institut de Science des Matériaux
de Mulhouse (IS2M) UMR 7361
Mulhouse
and
Réseau sur le Stockage
Electrochimique de l'Energie (RS2E)
FR CNRS 3459
Amiens
France

Ronnie MOGENSEN
Department of Chemistry
Ångström Laboratory
Uppsala University
Sweden

Laure MONCONDUIT
Institut Charles Gerhardt Montpellier
University of Montpellier
CNRS
and
RS2E, Réseau Français sur le
Stockage Electrochimique de
l'Energie
FR CNRS 3459
and
ALISTORE-ERI European Research
Institute
FR CNRS 3104
Amiens
France

Andrew J. NAYLOR
Department of Chemistry
Ångström Laboratory
Uppsala University
Sweden

Long H.B. NGUYEN
CNRS
University of Bordeaux
Bordeaux INP
ICMCB UMR 5026
Pessac
and
Laboratoire de Réactivité et de
Chimie des Solides
CNRS-UMR 7314
University of Picardie Jules Verne
and
RS2E, Réseau Français sur le
Stockage Electrochimique de
l'Energie
FR CNRS 3459
Amiens
France

Cristina POZO-GONZALO
Institute for Frontier Materials
Deakin University
Burwood
Australia

Ashish RUDOLA
Faradion Limited
Sheffield
UK

Loic SIMONIN
Université Grenoble Alpes
CEA LITEN
DEHT - LM
Grenoble
France

Moulay Tahar SOUGRATI
Institut Charles Gerhardt Montpellier
University of Montpellier
CNRS
and
RS2E, Réseau Français sur le Stockage Electrochimique de l'Energie
FR CNRS 3459
and
ALISTORE-ERI European Research Institute
FR CNRS 3104
Amiens
France

Lorenzo STIEVANO
Institut Charles Gerhardt Montpellier
University of Montpellier
CNRS
and
RS2E, Réseau Français sur le Stockage Electrochimique de l'Energie
FR CNRS 3459
and
ALISTORE-ERI European Research Institute
FR CNRS 3104
Amiens
France

Colin D. WESSELLS
Natron-Energy
Santa Clara
USA

Reza YOUNESI
Department of Chemistry
Ångström Laboratory
Uppsala University
Sweden

Index

M, N, P

S, T, V, X